Advances in Experimental Medicine and Biology

Volume 1071

More information about this series at http://www.springer.com/series/5584

Estelle B. Gauda • Maria Emilia Monteiro
Nanduri Prabhakar • Christopher Wyatt
Harold D. Schultz
Editors

Arterial Chemoreceptors

New Directions and Translational Perspectives

Springer

Editors
Estelle B. Gauda
Department of Pediatrics
The Hospital for Sick Children
University of Toronto
Toronto, ON, Canada

Nanduri Prabhakar
Department of Medicine
University of Chicago
Chicago, IL, USA

Harold D. Schultz
Cellular and Integrative Physiology
University of Nebraska
Omaha, NE, USA

Maria Emilia Monteiro
Faculty of Medical Sciences
Nova University of Lisbon
Lisbon, Portugal

Christopher Wyatt
Department of Neuroscience, Cell Biology
Wright State University
Dayton, OH, USA

ISSN 0065-2598 ISSN 2214-8019 (electronic)
Advances in Experimental Medicine and Biology
ISBN 978-3-030-08184-3 ISBN 978-3-319-91137-3 (eBook)
https://doi.org/10.1007/978-3-319-91137-3

Softcover re-print of the Hardcover 1st edition 2018

Printed on acid-free paper

This Springer imprint is published by the registered company Springer Nature Switzerland AG
The registered company address is: Gewerbestrasse 11, 6330 Cham, Switzerland

Preface

What a pleasure it was for Dr. Machiko Shirahata and I to host the XXth International Society for Arterial Chemoreceptors at Johns Hopkins University in Baltimore, Maryland, from July 23 to 27, 2017. The theme of the meeting was "Making Sense Out of Sensing Hypoxia, New Directions and Translational Perspectives Across the Lifespan." It was a wonderful reunion of friends and colleagues from 11 countries and 4 continents who came together to share their love for science and discovery connected by the carotid body. This book is the compilation of the science that was presented, a historical overview of ISAC meetings, by Dr. Robert Fitzgerald, and tributes to Dr. Machiko Shirahata, by Dr. James Sham, and Dr. Constancio Gonzalez by Dr. Nanduri Prabahakar, two giants in the field who had passed away since the last ISAC meeting in Leeds, in 2014. We are grateful to Springer for again helping us catalogue the events of the week, binding them together to join the other volumes on the shelf that have documented communications presented at the ISAC meetings since 1950, and thus the evolution of our understanding of how arterial chemoreceptors contribute to health and disease. In this volume the chapters cover chemoreception (peripheral and central) and mechanisms of O_2 sensing, embracing a variety of clinical-related issues including hypertension, obstructive sleep apnea, sudden infant death syndrome, obesity, heart failure, and the translational connection between laboratory science and clinical application. I encourage you to read the chapter "ISAC Historical Overview" by Dr. Robert Fitzgerald who chronicles the major contributions from each of the meetings since 1950.

There are many individuals who I have to thank for making the meeting a success. I had amazing talented and hardworking organizing and scientific committees. Rosie Silva, my administrative assistant at Hopkins who was the Senior Administrative Coordinator for the ISAC, effortlessly kept everything together (including me) and remembered every little detail to ensure the meeting ran smoothly and our guests felt welcomed. Without her attention to detail, I could not have moved to another country and started a new job 5 months prior to the meeting and pulled it off. I am also grateful to Sonia Dos Santos, the lead administrative assistant at SickKids. She orchestrated my seamless transition to a new country and job at SickKids, assisted in putting together the ISAC program, was able to come to the meeting to help Rosie, and meet all the wonderful people I had spoken about so fondly. I also knew she would be able to easily connect with the Portuguese group since she is also Portuguese. With Rosie and Sonia bookending the event not much could go wrong. Machiko and I had several friends who volunteered their ideas, energy, and time, and I cannot thank them more. Santokh (Sam) Singh, a very good friend, my personal trainer in Baltimore, former real-estate agent for high-end condos in Baltimore, led the negotiations for the hotels, organized the trip to Annapolis, and visited the venues. Setsuko Takase, a very close friend of Machiko's who many of the ISAC members met in Leeds, is a travel agent and assisted many Japanese in organizing their visit or move to Baltimore. With her flare for elegance and exceptional good taste, Setsko helped select the social venues. I especially appreciated all the help from George Kim, MD, ISAC Webmaster, who updated the website: carotidbody.org, and orchestrated all the electronic communications; he was invaluable. He also updated the ISAC members list with accurate contact information. I have known George throughout my career at Johns Hopkins and he too has been a close friend.

Although Dr. Machiko Shirahata passed away in April 2016, she made significant contributions to the organization of the meeting. Soon after the Leed's meeting, we met on multiple occasions over tea, wine, and sushi to talk about venues, potential funding sources for the meeting, and her legacy. Drs. Nanduri Prabahakar, Chris Wyatt, and Harold Schultz, members of the scientific committee and the selection committee for the Machiko Shirahata ISAC Trainee Travel Awards, were immensely helpful in creating the theme for the meeting "Making Sense Out of Sensing Hypoxia, New Directions and Translational Perspectives Across the Lifespan," selecting plenary speakers and organizing the sessions.

I would be remised if I did not take a moment to specially thank Dr. Nanduri Prabhakar for introducing me to the Carotid Body, allowing me to work in his laboratory when I was a Neonatology Fellow at Case Western University in 1989 when I only had a desire to understand and passion for learning and working hard, but no former training in research. He has mentored me throughout the years, always available for advice and encouragement (scientific, professional, and personal). He encouraged me to attend my first ISAC meeting in 1993, Dublin, Ireland, and submit an oral presentation (not a poster); little did I know then that I would become the first woman President of the ISAC Society in 2014. Thank you Nanduri for being such an amazing mentor, sponsor, and inspiration.

Dr. Machiko Shirahata was a dedicated scientist, wonderful mentor, and collaborator, had a love for life and travel, and was inspired by ISAC. Knowing that she had lung cancer, she reflected on how she might merge her passions by leaving a legacy. At the Leeds ISAC conference, she made a commitment to support young trainees and give them the opportunity to attend future ISAC conferences. To that end, through a generous donation from Machiko's estate made available to ISAC by her son, Akira Fitzgerald, 10 trainees received the Machiko Shirahata ISAC Trainee Travel Award to ISAC XX, and another 10 awards will be available for the 21st international conference in Lisbon, Portugal. Please refer to the chapter, Machiko tribute, written by Dr. James Sham for an overview of her life and contributions.

The Machiko Shirahata ISAC Trainee Travel Award compliments the Heymans-De Castro-Neil Awards which have been routinely given to two to three trainees for the best abstracts presented by a trainee since 1989 at the X ISAC meeting.

These meetings always need sponsors, and we thank the Division of Neonatology at The Hospital for Sick Children, the Division of Neonatology at Johns Hopkins Hospital, *Journal of Physiology*, and Dr. Robert Fitzgerald for their generous donations.

I invite those who attended the conference and those who were unable to join us in Baltimore to take a moment to peruse the chapters in this book. True to the theme of the conference, the chapters outline new cutting edge research involving the carotid body and how oxygen is sensed across the lifespan and the translational implications of the work. The full length plenary communications presented by Drs. Greg Semena, Jan Marino (Nino) Ramirez, Christopher Wyatt, Jayasri Nanduri, Rodrigo Iturriaga, Michael Joyner, Andrea Porzionato, Patrice Guyenet, and Cormac Taylor will be published in an upcoming issue of the *Journal of Physiology*.

At the business meeting, Emilia Monteiro, was elected the President of ISAC and will host XXI ISAC meeting in Lisbon, Portugal, in 2020.

The ISAC meeting from the beginning was created on the solid premise of sharing ideas, collaborations, and discovery to explain basic human physiology. The collaboration between Dr. Fernando De Castro from Spain who performed the exquisite anatomical-histology of the carotid body and Dr. Heymans who performed the definitive experiments showing its role in chemoreceptor for which he won the Nobel Prize in 1938 (incomplete thought). They too met at scientific meetings, shared ideas, and collaborated on experiments. So I end this preface where it all began using a quote from the Swedish Professor, G Liljestrand in 1940, when the Nobel Prize was presented to Heymans for his characterization of "a curious structure" which was so beautifully identified and anatomically detailed by De Castro.

"Since the end of the eighteenth century we know of the existence of a curious structure in the region of the sinus, the glomus caroticum or carotid body which, in man, extends over only a few millimetres. The glomus consists of a small mass of very fine intertwining vessels arising from the internal carotid and enclosing various different types of cells. It has been considered by some as being a sort of endocrine gland similar to the medulla of the suprarenal glands. De Castro, however, in 1927 demonstrated that the anatomy of the glomus could in no way be compared to that of the suprarenal medulla. De Castro suggested rather that the glomus was an organ whose function was to react to variations in the composition of the blood, in other words an internal gustatory organ with special chemo-receptors.

In 1931, Bouckaert, Dautrebande, and Heymans undertook to find out whether these supposed chemo-receptors were responsible for the respiratory reflexes produced by modifications in the composition of the blood. By localized destruction in the sinus area they had been able to stop reflexes initiated by pressure changes, but respiratory reflexes could still continue to occur in answer to changes in the composition of the blood. Other experiments showed that Heymans's concepts on the important role played by the glomus in the reflex control of respiration by the chemical composition of the blood were undoubtedly correct."

Prof. G. Liljestrand (Karolinska Institutet); presentation of the 1938 Nobel Prize in Physiology of Medicine to Prof. C. Hyemans (Ghent, Belgium; January 16, 1940)

ISAC President, 2014–2017 Estelle B. Gauda, MD
Bethesda, MD, USA

Contents

Contributors

Teresa Agapito Departamento de Bioquimica, Biologia Molecular y Fisiologia-IBGM, Universidad de Valladolid-CSIC, Valladolid, Spain

CIBER de Enfermedades Respiratorias (CIBERES), Instituto de Salud Carlos III, Madrid, Spain

Julio Alcayaga Laboratorio de Fisiología Celular, Universidad de Chile, Santiago, Chile

David C. Andrade Laboratory of Cardiorespiratory Control, Department of Physiology, Faculty of Biological Sciences, Pontificia Universidad Católica de Chile, Santiago, Chile

Alexis Arce-Álvarez Laboratory of Cardiorespiratory Control, Department of Physiology, Faculty of Biological Sciences, Pontificia Universidad Católica de Chile, Santiago, Chile

Aida Bairam Centre de Recherche de l'Institut Universitaire de Cardiologie et de Pneumologie de Québec, Université Laval, Quebec City, QC, Canada

Barbara L. Barr Department of Neuroscience, Cell Biology, & Physiology, Wright State University, Dayton, OH, USA

Jaques Belik The Hospital for Sick Children, Division of Neonatology, University of Toronto, Toronto, ON, Canada

Christina D. Bruce Department of Biology, Faculty of Science and Technology, Mount Royal University, Calgary, AB, Canada

Alexandra E. Chiew Department of Biology, Faculty of Science and Technology, Mount Royal University, Calgary, AB, Canada

Nuno R. Coelho CEDOC, Centro de Estudos Doenças Crónicas, Nova Medical School, Faculdade de Ciências Médicas, Universidade Nova de Lisboa, Lisbon, Portugal

Sílvia V. Conde CEDOC, Centro de Estudos Doenças Crónicas, Nova Medical School, Faculdade de Ciências Médicas, Universidade Nova de Lisboa, Lisbon, Portugal

Trevor A. Day Department of Biology, Faculty of Science and Technology, Mount Royal University, Calgary, AB, Canada

Elena Roces de Álvarez-Buylla Centro Universitario de Investigaciones Biomédicas, Universidad de Colima, Colima, Mexico

Miguel A. de la Fuente Departamento de Biologia Celular, Histologia y Farmacologia/IBGM, Universidad de Valladolid-CSIC, Valladolid, Spain

Rodrigo Del Rio Laboratory of Cardiorespiratory Control, Department of Physiology, Faculty of Biological Sciences, Pontificia Universidad Católica de Chile, Santiago, Chile

Centro de Envejecimiento y Regeneración (CARE), Pontificia Universidad Católica de Chile, Santiago, Chile

Centro de Excelencia en Biomedicina de Magallanes (CEBIMA), Universidad de Magallanes, Punta Arenas, Chile

Juliann M. Di Fiore Case Western Reserve University, Rainbow Babies & Children's Hospital, Cleveland, OH, USA

Clara G. Dias CEDOC, Centro de Estudos Doenças Crónicas, Nova Medical School, Faculdade de Ciências Médicas, Universidade Nova de Lisboa, Lisbon, Portugal

Hugo S. Díaz Laboratory of Cardiorespiratory Control, Department of Physiology, Faculty of Biological Sciences, Pontificia Universidad Católica de Chile, Santiago, Chile

Lucília N. Diogo CEDOC, Centro de Estudos Doenças Crónicas, Nova Medical School, Faculdade de Ciências Médicas, Universidade Nova de Lisboa, Lisbon, Portugal

Inmaculada Docio Departamento de Bioquimica, Biologia Molecular y Fisiologia/IBGM, Universidad de Valladolid-CSIC, Valladolid, Spain

CIBER de Enfermedades Respiratorias (CIBERES), Instituto de Salud Carlos III, Madrid, Spain

Anette Ebberyd Section for Anesthesiology and Intensive Care Medicine, Department of Physiology and Pharmacology, Karolinska Institute, Stockholm, Sweden

Lars I. Eriksson Section for Anesthesiology and Intensive Care Medicine, Department of Physiology and Pharmacology, Karolinska Institute, Stockholm, Sweden

Function Perioperative Medicine and Intensive Care, Karolinska University Hospital, Stockholm, Sweden

Malin Jonsson Fagerlund Section for Anesthesiology and Intensive Care Medicine, Department of Physiology and Pharmacology, Karolinska Institute, Stockholm, Sweden

Function Perioperative Medicine and Intensive Care, Karolinska University Hospital, Stockholm, Sweden

Tariq H. Fayyad Department of Neuroscience, Cell Biology, & Physiology, Wright State University, Dayton, OH, USA

Robert S. Fitzgerald Departments of Environmental Health & Engineering, of Physiology, of Medicine, The Johns Hopkins University Medical Institutions, Baltimore, MD, USA

Estelle B. Gauda Department of Pediatrics, The Hospital for Sick Children, University of Toronto, Toronto, ON, Canada

Angela Gomez-Niño Departamento de Biologia Celular, Histologia y Farmacologia/IBGM, Universidad de Valladolid-CSIC, Valladolid, Spain

CIBER de Enfermedades Respiratorias (CIBERES), Instituto de Salud Carlos III, Madrid, Spain

Elvira Gonzalez-Obeso Servicio de Anatomia Patologica, Hospital Clinico Universitario, Valladolid, Spain

CIBER de Enfermedades Respiratorias (CIBERES), Instituto de Salud Carlos III, Madrid, Spain

Patrícia Grácio CEDOC, Centro de Estudos Doenças Crónicas, Nova Medical School, Faculdade de Ciências Médicas, Universidade Nova de Lisboa, Lisbon, Portugal

Bryan T. Hackfort Department of Cellular and Integrative Physiology at the University of Nebraska Medical Center, Omaha, NE, USA

James O. Hogan Department of Physiology and Biophysics, Chicago Medical School, Rosalind Franklin University of Medicine and Science, North Chicago, IL, USA

Rodrigo Iturriaga Laboratorio de Neurobiología, Facultad de Ciencias Biológicas, Pontificia Universidad Católica de Chile, Santiago, Chile

Julijana Ivanovska The Hospital for Sick Children, Division of Neonatology, University of Toronto, Toronto, ON, Canada

Tara A. Janes Department of Pediatrics, Université Laval, Québec, QC, Canada

Institut Universitaire de Cardiologie et de Pneumologie de Québec, Québec, QC, Canada

Florine Jeton Laboratoire Hypoxie & Poumon, Universite Paris 13, Paris, France

M. João Correia CEDOC, Centro de Estudos Doenças Crónicas, Nova Medical School, Faculdade de Ciências Médicas, Universidade Nova de Lisboa, Lisbon, Portugal

Vincent Joseph Centre de Recherche de l'Institut Universitaire de Cardiologie et de Pneumologie de Québec, Université Laval, Quebec City, QC, Canada

Jessica Kåhlin Section for Anesthesiology and Intensive Care Medicine, Department of Physiology and Pharmacology, Karolinska Institute, Stockholm, Sweden

Function Perioperative Medicine and Intensive Care, Karolinska University Hospital, Stockholm, Sweden

Dawon Kang Department of Physiology and Institute of Health Sciences, College of Medicine, Gyeongsang National University, Jinju, South Korea

Na-Young Kang The Hospital for Sick Children, Division of Neonatology, University of Toronto, Toronto, ON, Canada

Donghee Kim Department of Physiology and Biophysics, Chicago Medical School, Rosalind Franklin University of Medicine and Science, North Chicago, IL, USA

Richard Kinkead Department of Pediatrics, Université Laval, Québec, QC, Canada

Institut Universitaire de Cardiologie et de Pneumologie de Québec, Québec, QC, Canada

Sofien Laouafa Centre de Recherche de l'Institut Universitaire de Cardiologie et de Pneumologie de Québec, Université Laval, Quebec City, QC, Canada

CNRS, UMR 5023, Université Claude Bernard Lyon 1, Villeurbanne, France

Jack K. Leacy Department of Biology, Faculty of Science and Technology, Mount Royal University, Calgary, AB, Canada

Department of Physiology, University College Cork, Cork, Ireland

Kian Leong Lee Cancer Science Institute of Singapore, National University of Singapore, Singapore, Singapore

Mónica Lemus Centro Universitario de Investigaciones Biomédicas, Universidad de Colima, Colima, Mexico

Robert Lewis A.T. Still University School of Osteopathic Medicine, Mesa, AZ, USA

Andrea M. Linares Department of Biology, Faculty of Science and Technology, Mount Royal University, Calgary, AB, Canada

Claudia M. Lucero Laboratory of Cardiorespiratory Control, Department of Physiology, Faculty of Biological Sciences, Pontificia Universidad Católica de Chile, Santiago, Chile

Peter M. MacFarlane Case Western Reserve University, Rainbow Babies & Children's Hospital, Cleveland, OH, USA

Carli R. Mann Department of Biology, Faculty of Science and Technology, Mount Royal University, Calgary, AB, Canada

François Marcouiller Centre de Recherche de l'Institut Universitaire de Cardiologie et de Pneumologie de Québec, Université Laval, Quebec City, QC, Canada

Noah J. Marcus Department of Physiology and Pharmacology, Des Moines University, Des Moines, IA, USA

Valery Melnikov Facultad de Medicina, Universidad de Colima, Colima, Mexico

Bernardete F. Melo CEDOC, Centro de Estudos Doenças Crónicas, Nova Medical School, Faculdade de Ciências Médicas, Universidade Nova de Lisboa, Lisbon, Portugal

Souren Mkrtchian Section for Anesthesiology and Intensive Care Medicine, Department of Physiology and Pharmacology, Karolinska Institute, Stockholm, Sweden

Cynthia Mojarro Centro Universitario de Investigaciones Biomédicas, Universidad de Colima, Colima, Mexico

Emília C. Monteiro CEDOC, Centro de Estudos Doenças Crónicas, Nova Medical School, Faculdade de Ciências Médicas, Universidade Nova de Lisboa, Lisbon, Portugal

Sergio Montero Centro Universitario de Investigaciones Biomédicas, Universidad de Colima, Colima, Mexico

Facultad de Medicina, Universidad de Colima, Colima, Mexico

Jayasri Nanduri Institute for Integrative Physiology and Center for Systems Biology of Oxygen Sensing, Biological Sciences Division, University of Chicago, Chicago, IL, USA

Colin A. Nurse Department of Biology, McMaster University, Hamilton, ON, Canada

Ken D. O'Halloran Department of Physiology, University College Cork, Cork, Ireland

Ana Obeso Departamento de Bioquimica, Biologia Molecular y Fisiologia-IBGM, Universidad de Valladolid-CSIC, Valladolid, Spain

CIBER de Enfermedades Respiratorias (CIBERES), Instituto de Salud Carlos III, Madrid, Spain

Elena Olea Departamento de Enfermeria, IBGM-CSIC, Universidad de Valladolid, Valladolid, Spain

CIBER de Enfermedades Respiratorias (CIBERES), Instituto de Salud Carlos III, Madrid, Spain

María Paz Oyarce Laboratorio de Neurobiología, Facultad de Ciencias Biológicas, Pontificia Universidad Católica de Chile, Santiago, Chile

Joel D. B. Peltonen Department of Biology, Faculty of Science and Technology, Mount Royal University, Calgary, AB, Canada

Ying-Jie Peng Institute for Integrative Physiology and Center for Systems Biology of Oxygen Sensing, Biological Sciences Division, University of Chicago, Chicago, IL, USA

Sofia A. Pereira CEDOC, Centro de Estudos Doenças Crónicas, Nova Medical School, Faculdade de Ciências Médicas, Universidade Nova de Lisboa, Lisbon, Portugal

Jamie R. Pfoh Department of Biology, Faculty of Science and Technology, Mount Royal University, Calgary, AB, Canada

Lorenz Poellinger Cancer Science Institute of Singapore, National University of Singapore, Singapore, Singapore

Department of Cell and Molecular Biology, Karolinska Institutet, Stockholm, Sweden

Nanduri R. Prabhakar Institute for Integrative Physiology and Center for Systems Biology of Oxygen Sensing, Biological Sciences Division, University of Chicago, Chicago, IL, USA

Jesus Prieto-Lloret Departamento de Bioquimica, Biologia Molecular y Fisiologia/IBGM, Universidad de Valladolid-CSIC, Valladolid, Spain

CIBER de Enfermedades Respiratorias (CIBERES), Instituto de Salud Carlos III, Madrid, Spain

Richard L. Pye Department of Neuroscience, Cell Biology, & Physiology, Wright State University, Dayton, OH, USA

Ryan J. Rakoczy Department of Neuroscience, Cell Biology, & Physiology, Wright State University, Dayton, OH, USA

Mario Ramírez-Flores Facultad de Medicina, Universidad de Colima, Colima, Mexico

Mauricio A. Retamal Centro de Fisiología Celular e Integrativa, Clínica Alemana Universidad Del Desarrollo, Santiago, Chile

Ricardo Rigual Departamento de Bioquimica, Biologia Molecular y Fisiologia-IBGM, Universidad de Valladolid-CSIC, Valladolid, Spain

CIBER de Enfermedades Respiratorias (CIBERES), Instituto de Salud Carlos III, Madrid, Spain

Asuncion Rocher Departamento de Bioquimica, Biologia Molecular y Fisiologia/IBGM, Universidad de Valladolid-CSIC, Valladolid, Spain

CIBER de Enfermedades Respiratorias (CIBERES), Instituto de Salud Carlos III, Madrid, Spain

Damien Roussel CNRS, UMR 5023, Université Claude Bernard Lyon 1, Villeurbanne, France

Joana F. Sacramento CEDOC, Centro de Estudos Doenças Crónicas, Nova Medical School, Faculdade de Ciências Médicas, Universidade Nova de Lisboa, Lisbon, Portugal

Shaima Salman Department of Biology, McMaster University, Hamilton, ON, Canada

Joseph M. Santin Division of Biological Sciences, University of Missouri, Columbia, MO, USA

Eliano Dos Santos CEDOC, Centro de Estudos Doenças Crónicas, Nova Medical School, Faculdade de Ciências Médicas, Universidade Nova de Lisboa, Lisbon, Portugal

Gary Saran Department of Biology, Faculty of Science and Technology, Mount Royal University, Calgary, AB, Canada

Harold D. Schultz Department of Cellular and Integrative Physiology at the University of Nebraska Medical Center, Omaha, NE, USA

Jacinta Serpa CEDOC, Centro de Estudos Doenças Crónicas, Nova Medical School, Faculdade de Ciências Médicas, Universidade Nova de Lisboa, Lisbon, Portugal

Instituto Português de Oncologia de Lisboa Francisco Gentil (IPOLFG), Lisbon, Portugal

James S. K. Sham Division of Pulmonary and Critical Care Medicine, Johns Hopkins University School of Medicine, Baltimore, MD, USA

Mingma T. Sherpa Kunde Hospital, Khunde, Solukhumbu, Nepal

Maria Simarro Departamento de Enfermeria/IBGM, Universidad de Valladolid/CSIC, CIBERES, Instituto de Salud Carlos III, Valladolid, Spain

Jorge Soliz Centre de Recherche de l'Institut Universitaire de Cardiologie et de Pneumologie de Québec, Université Laval, Quebec City, QC, Canada

Liran Tamir-Hostovsky The Hospital for Sick Children, Division of Neonatology, University of Toronto, Toronto, ON, Canada

Camilo Toledo Laboratory of Cardiorespiratory Control, Department of Physiology, Faculty of Biological Sciences, Pontificia Universidad Católica de Chile, Santiago, Chile

Nicolas Voituron Laboratoire Hypoxie & Poumon, Universite Paris 13, Paris, France

Jiaju Wang Department of Physiology and Biophysics, Chicago Medical School, Rosalind Franklin University of Medicine and Science, North Chicago, IL, USA

Christopher N. Wyatt Department of Neuroscience, Cell Biology, & Physiology, Wright State University, Dayton, OH, USA

Xiuli Zhang Institute for Integrative Physiology and Center for Systems Biology of Oxygen Sensing, Biological Sciences Division, University of Chicago, Chicago, IL, USA

Shaelynn M. Zouboules Department of Biology, Faculty of Science and Technology, Mount Royal University, Calgary, AB, Canada

The History of the International Society of Arterial Chemoreception (ISAC)

At the conclusion of the 20th meeting (12th meeting of ISAC formally so constituted) in Baltimore, MD, USA, I was asked to write a history of ISAC. Being the last active founding member of ISAC, I thought the archives of ISAC should have some record of our scientific labors/successes regarding arterial chemoreception. Because of the goodness of Camillo Di Giulio, Chris Wyatt, Ronan O'Regan, and Keith Buckler, I received a listing of chemoreceptor meetings that predated those under the formal auspices of ISAC. ISAC was founded in a St. Louis restaurant after a FASEB meeting in 1987 by Carlos Eyzaguirre, Sal Fidone, Sukhamay Lahiri, Donald McDonald, and myself. ISAC members would like to see those initial/original meetings predating ISAC included in this historical overview. And so they have been. We have a record of 20 such gatherings of which the final 12 are ISAC meetings.

With the data available, I have tried to be consistent in presenting the facts and the ambiance of each meeting at least briefly: the meeting's date, number, primary mover or ISAC president, title of volume containing presentations, editors of volume, location of publisher, publisher, date of publication, volume number in *Advances in Experimental Medicine and Biology*. There follows: location of meeting, number of presentations, and number of contributors to the presentations. And, finally, an effort was made to spell out some of the issues discussed and debated, or some other facets of the ambiance.

1950 Meeting I: *Göran Liljestrand*

Co-organizers of the first ISAC meeting were Ulf von Euler and Yngve Zotterman. Held in Stockholm, Sweden, in August 1950, the meeting did highlight three investigators who contributed to chemoreceptor studies for several years: Fernando De Castro, the primary originator of carotid body studies, Corneille Heymans who produced so much regarding the respiratory reflex effects of carotid/aortic area stimulation that he was awarded the Nobel Prize in Physiology or Medicine for 1938, and Eric Neil who focused on the circulation. De Castro's 1928 paper described the CB: "organe sensoriel special dedie a percevoir quelques modifications qualitative du sang, plutot que d'un appareil destine a recevoir les variations de la pression sanguine." He was the first to demonstrate with classical nerve studies the *sensory* nature of the fibers abutting on the carotid body's type I cells and was the first to suggest the organ "tasted" the qualitative composition of the blood perfusing it. Heymans forcefully demonstrated that the hyperpnea consequent to hypoxia exposure was generated by peripheral chemoreceptors and not centrogenic. At that time CO_2-generated increases in ventilation needed more study. And Neil presented the role of the carotid bodies in the circulation especially during hemorrhage. He also spoke on autoregulation of the organ's blood flow. His ideas are found in the *Handbook of Physiology, Section 2: The Cardiovascular System, Volume III: Peripheral Circulation and Organ Blood Flow, Part 1*, Chapter I, "Peripheral blood flow: Historical aspects," pp. 1–20.

1966 Meeting II: *Bob Torrance*

Presentations are found in *Arterial Chemoreceptors*. Edited by R.W. Torrance. Oxford and Edinburgh, Blackwell Scientific Publications, 1968. These are the Proceedings of the Wates Foundation Symposium held at Oxford in July 1966. Bob's "Prolegomena" surveyed virtually all CB research up to that point. It is 40 pages long with 133 references and covers every facet of CB morphology, chemistry, neurophysiology, and systemic reflex responses. The earliest references to works of physiology appear to have been published in 1919 and 1922. Most come from the late 1940s, 1950s, and 1960s. To summarize the contents of this "Prolegomena" would be impossible. But for a young investigator interested in history and an understanding of the great work of earlier years, this would be a very worthwhile exercise. Thirty-four papers were presented by 50 investigators, but 70 attendees participated.

1973 Meeting III: *Michael Purves*

Presentations are found in *The Peripheral Arterial Chemoreceptors*. Edited by M.J. Purves. London/New York, Cambridge University Press, 1975. Bristol saw the presentation of 28 papers by 48 investigators of whom 35 participated in the meeting. Meant to afford opportunities for discussion of significant controversial data, the meeting was marked with a debated discussion of the existence, mechanisms, and functional significance of efferent inhibition of carotid body neural output. Donald McDonald and Bob Mitchell very elegantly presented ultrastructure of the organ and revealed the reciprocal synapses existing between the afferent fibers abutting on the type I cells of the organ. These fibers are the major type in the carotid sinus nerve. Eyzaguirre presented a reexplored version of De Castro's beautiful drawings of the carotid body. Of considerable interest to the investigators of ventilation control was the observation of Ponte and Purves that chemoreceptors should be relatively unresponsive to the oscillations of CO_2/H^+, at least during normal breathing.

1974 Meeting IV: *Autar Paintal*

Presentations are found in *Morphology and Mechanisms of Chemoreceptors*. Edited by A.S. Paintal. New Delhi, India, Navchetan Press (P) Ltd. of the Vallabhbhai Patel Chest Institute, 1976. These are the Proceedings of an International Satellite Symposium held in Srinagar, Kashmir, India, in October 1974, which preceded the IUPS meeting in Delhi, India. Ninety-four investigators attended this very successful chemoreceptor conference. The presentations focused on chemoreceptor (carotid body and aortic bodies) mechanisms at the organ, glomus cell, and mitochondria levels as well as on neurophysiological recordings from the stimulated chemoreceptors.

1976 Meeting V: *Dietrich Lubbers*

Presentations are found in *Chemoreception in the Carotid Body*. Edited by H. Acker, S. Fidone, D. Pallot, C. Eyzaguirre, D.W. Lubbers, and R.W. Torrance. Berlin/Heidelberg, Springer-Verlag, 1977. This workshop was held at the Max-Planck-Institut für Systemphysiologie in Dortmund. Forty papers were presented by 89 investigators of whom 43 were participants in the workshop. Highlighted were efforts to determine tissue PO_2 in the carotid body, an issue complicated by factors such as intercapillary distances and differential blood flow. Further, there was discussion of what was the primary chemosensor, afferent nerve or type I cell. The focus was also on trying to determine mechanisms of chemoreception. Microelectrode studies explored the glomus tissue to understand the conversion of a chemical stimulation into nervous activity. The measurement of turnover in catecholamine stores of the carotid body was also presented.

1979 Meeting VI: *Carlos Belmonte*

Presentations are found in *Arterial Chemoreceptors*. Edited by C. Belmonte, D.J. Pallot, H. Acker, S. Fidone. Leicester, UK, Leicester University Press, 1981. Held in Valladolid, Spain, in September, this conference saw 51 papers presented by 118 contributors. C. Eyzaguirre and R.W. Torrance proposed the meeting as a 50th anniversary celebration of the pioneering work of Fernando De Castro y Rodríguez with respect to the morphology of the carotid body and the hypotheses explaining its operation and purpose. In addition to the many papers, one finds a noteworthy tribute by Antonio Gallego, a personal friend of De Castro. He writes of the contributions to the discovery and study of the vascular baroreceptors and blood chemoreceptors. This truly most remarkable chapter is well worth reading.

1982 Meeting VII: *David Pallot*

Presentations are found in *The Peripheral Arterial Chemoreceptors.* Edited by D. J. Pallot. London, UK, Croom Helm/Oxford University Press (New York), 1984. The conference, held at the University of Leicester in the UK in September, offered 45 papers from 112 investigators. One major issue discussed was the role of the organ's catecholamines in the detection of CO_2, O_2, and H^+. A review of Dinger's work, the reinnervating of the carotid body with a foreign nerve, and Monti-Bloch's carotid body transplantations pretty much eliminated the hypothesis that the primary afferent endings are the site of chemotransduction. Isolated cell and patch clamp techniques were introduced to advance the understanding of how ion concentration affected the organ's processes.

1985 Meeting VIII: *Alexander Ribeiro*

Presentations are found in *Chemoreceptors in Respiratory Control.* Edited by J.A. Ribeiro and D.J. Pallot. London, Croom Helm, 1987. In this symposium, held at the Gulbenkian Institute of Science in Oeiras, Portugal, in October, 55 participants presented 45 papers covering morphology of the carotid bodies in various conditions and challenged with various agents. Cat, rabbit, and rat were the models used, in both in vivo and in vitro preparations.

1988 Meeting IX (ISAC Meeting I): *Carlos Eyzaguirre*

Presentations are found in *Arterial Chemoreception.* Edited by C. Eyzaguirre, S.J. Fidone, R.S. Fitzgerald, S. Lahiri, D.M. McDonald. New York, Springer-Verlag, 1990. This was the inaugural meeting of ISAC, formally so constituted, held in Park City, Utah. One hundred and eight scientists from 18 countries presented 87 studies. The theme was "Inasmuch as oxygen is the substrate *sine qua non* for the survival of all higher organisms, it is quite understandable that considerable interest surrounds investigations into mechanisms responsible for detecting dwindling oxygen supplies in the organism. This interest has intensified as the newer techniques of cell, sub-cell, and molecular biology have become available." Cardiopulmonary reflex responses were discussed. And the newer techniques, electrophysiological, morphological, biochemical, immunocytochemical, and molecular biological, were highlighted. A major problem presented was why one can block the exogenous application of a stimulating agent, but the animal's responses to hypoxia or hypercapnia remain. Consensus suggestions seemed to focus on the question of where the receptors are located; exogenous availability does not necessarily mean endogenous availability. Further, there are various subtypes of receptors. Finally, other stimuli received notice: the organ increases its output in response to increases in temperature and to increases in tonicity. Carlos Eyzaguirre introduced the notion of granting an award to young investigators which would be called the Heymans-De Castro award.

1989 Meeting X: *Andrzej Trzebski*

Presentations are found in *Chemoreceptors and Chemoreceptor Reflexes*. Edited by H. Acker, A. Trzebski, R.G. O'Regan. New York, Plenum Press, 1990. This Warsaw symposium, satellite to that year's IUPS meeting in Helsinki, Finland, recorded 59 papers from 229 contributors. It featured studies of cell electrophysiology, single ion channel recordings, and immunocytochemical work along with reflex responses to carotid body stimulation. Professor Trzebski thought it to be a 30-year anniversary of Brian Lloyd's 1959 "Oxford Conference." Warsaw was the first to see the C. Heymans, De Castro, and E. Neil awards being made. The awardees were M. Czyzyk-Krzeska (Poland), W. Kummer (FRG), and C. Peers (UK). Professor Ronan O'Regan wanted the name of Eric Neil added; all agreed.

1991 Meeting XI: *Pier Giorgio Data*

Presentations are found in *Neurobiology and Cell Physiology of Chemoreception*. Edited by P.G. Data, H. Acker, S. Lahiri. New York, Plenum Press, 1993. This is also Volume 337 of *Advances in Experimental Medicine and Biology*. Held at the University of Chieti in Chieti, Italy, this conference had five sections containing 61 papers presented by 143 participants. The Heymans-De Castro-Neil Award went to Agnes Görlach from Dortmund, FRG. And the Fernando Data Foundation Award was established to support investigators in underdeveloped areas to pursue chemoreception studies. In addition, plenary lectures were held in honor of Carlos Eyzaguirre and Robert Forster, a tribute to their many seminal contributions to the science of chemoreception.

1993 Meeting XII: *Ronan O'Regan*

Presentations are found in *Arterial Chemoreceptors: Cell to System*. Edited by R.G. O'Regan, P. Nolan, D.S. McQueen, D.J. Paterson. New York, Plenum Press, 1994. This is also Volume 360 of *Advances in Experimental Medicine and Biology*. The volume is dedicated to Eric Neil and introduced with a lovely "appreciation" of Eric Neil by former student Ronan O'Regan. Held at University College in Dublin, Ireland, this conference saw 75 papers presented by 219 participants. Also noteworthy are the summary overviews of previous chemoreceptor meetings in Stockholm (1950), Oxford (1966), Bristol (1973), Kashmir (1974), Valladolid (1979), Leicester (1982), and Oeiras (1985).

1996 Meeting XIII: *Patricio Zapata*

Presentations are found in *Frontiers in Arterial Chemoreception.* Edited by P. Zapata, C. Eyzaguirre, R.W. Torrance. New York, Plenum Press, 1996. This is also Volume 410 of *Advances in Experimental Medicine and Biology*. The conference, held in Santiago, Chile, saw 71 presentations offered by 223 investigators of whom 74 participated. The introductory tribute to Fernando De Castro on the centennial of his birth by Dr. Juan D. Vial, the President of the Catholic University of Chile and a former student of De Castro, showed remarkable figures of De Castro's carotid body drawings. This was followed by a Bob Torrance "Prolegomena" which served as a preface to some of the papers discussing PO_2, PCO_2, and H^+. And the third introductory piece is the oration given by Brian B. Lloyd at Daniel J.C. Cunningham's funeral in February 1996. D.J.C. Cunningham did much on CO_2's effect on human respiration.

1999 Meeting XIV: *Sukhamay Lahiri*

Presentations are found in *Oxygen Sensing Molecule to Man.* Edited by S. Lahiri, N.R Prabhakar, R.E. Forster II. New York, Kluwer Academic/Plenum Publishers, 2000. This is also Volume 475 of *Advances in Experimental Medicine and Biology*. The conference, held in Philadelphia, saw 80 presentations authored by 277 investigators. A hallmark of the volume: three tributes to Robert W. Torrance who passed away in January 1999, just 6 months before the meeting. Three former students of Bob, Charles Michel, Mark Hanson, and Prem Kumar, paid tribute to their mentor. And Chilean colleagues of Bob, Patricio Zapata and Carolina Larrain, wrote a reminiscence article. This meeting seemed to expand somewhat on the ordinary content of these conferences: the impact of hypoxia on cells other than the chemoreceptors and several explorations of the involvement of genes. The Heymans-De Castro-Neil Awards went to Beth Ann Summers (Case Western Reserve University, USA) and Roger J. Thompson (McMaster University, Canada). The awards specific for this meeting called the The Julius Comroe, Robert Forster, and Chris Lambertsen Awards went to Ricardo Pardal (Neuroscience Institute, Spain) and Nicholas A. Ritucci (Wright State University, USA).

2002 Meeting XV: *Jean-Marc Pequignot*

Presentations are found in *Chemoreception: From Cellular Signaling to Functional Plasticity*. Edited by J.M. Pequignot, C. Gonzalez, C.A. Nurse, N.R. Prabhakar, Y. Dalmaz. New York, Kluwer Academic/Plenum Publishers, 2003. This is also Volume 536 of *Advances in Experimental Medicine and Biology*. This meeting, held in November in Lyon, France, recorded 78 papers presented by 322 investigators. The organizers intended to bring together those working on the interface of cellular and molecular biology with those looking at the pathways of chemoreception and reflex responses. So the bacterial mechanisms of oxygen responsiveness as well as chemoreceptor cells and CNS neurons are studied, as well as the responses of organs, systems, and the entire organism. Special attention was paid to the plasticity of the entire oxygen sensing machinery as well as effector machinery in response to the continuous or intermittent exposure to altered oxygen levels. Finally, Professor Henry Gautier paid a special tribute to Pierre Dejours who had done so much work on the role of the arterial chemoreceptors in the regulation of respiration. His legacy, in part, is the "Dejours test" in which a double breath of hyperoxia, nitrogen, or CO_2 is provided to assess the chemosensitivity of the human or animal. The Heymans-De Castro-Neil prizes were awarded to Sílvia Conde (Portugal), Maria Garcia-Fernandez (Spain), and Michael G. Jonz (Canada).

2005 Meeting XVI: *Hisatake Kondo*

Presentations are found in *The Arterial Chemoreceptors.* Edited by Y. Hayashida, C. Gonzalez, H. Kondo. New York, Springer, 2006. This is also Volume 580 of *Advances in Experimental Medicine and Biology*. This conference, held at the Miyagi Zao Royal Hotel in Sendai, Japan, welcomed many investigators from around the world to the first Arterial Chemoreceptor meeting. The scientific presentations were preceded by a tribute to Professor Autar Singh Paintal, a giant investigator in arterial chemoreceptors, by K. Ravi and V.K. Vijayan. The conference presents an updated review of the physiology of the carotid body chemoreceptors. It contains the results on topics at the frontiers of future developments in O_2 sensing in chemoreceptor cells. Additionally, this volume provides data from studies carried out in other O_2-sensing tissues including pulmonary vasculature and erythropoietin-producing cells. It is a prime source of information and a guideline for arterial chemoreception researchers.

2008 Meeting XVII: *Constancio Gonzalez*

Presentations are found in *Arterial Chemoreceptors*. Edited by C. Gonzalez, C.A. Nurse, C. Peers. Berlin/Heidelberg, Springer Science + Business Media B.V., 2009. This is also Volume 648 of *Advances in Experimental Medicine and Biology*. This second ISAC conference to be held in Valladolid, Spain, presented 50 papers by 131 contributors. Introductory units of the volume are a photographic and brief text (by Patricio Zapata) tribute to Carlos Eyzaguirre who passed away on February 2, 2009. He was a true giant in the field of carotid body studies. A second unit is an invited article by F. De Castro presenting his father's work on the discovery of the sensory nature of the carotid bodies, rich with details. There follow two historical chapters: one facing rearward showing the rise of carotid body studies over the last 50 years and the other facing forward introducing a new preparation for carotid body study. The volume presented studies of the structure and development of the carotid body, of the molecular biology and biophysical aspects of the ion channels found in the glomus cells, of the peripheral neurotransmitters and their receptors, of the central integration of carotid body neural output, and of the reflexes it generates. Widening and enriching the meeting were the presentations on central chemoreceptors, hypoxic pulmonary vasoconstriction, lung receptors, and the latest gasotransmitter, hydrogen sulfide. The Heymans-De Castro-Neil prizes were awarded to Nikol Piskuric (Canada), Mark L. Dallas (UK), and Olaia Colinas (Spain).

2011 Meeting XVIII: *Colin Nurse*

Presentations are found in *Arterial Chemoreception: From Molecules to Systems*. Edited by C.A. Nurse, C. Gonzalez, C. Peers, N.R. Prabhakar. New York/Heidelberg, Springer Science+Business Media B.V., 2012. This is also Volume 758 of *Advances in Experimental Medicine and Biology*. Held at McMaster University in Hamilton, Ontario, Canada, in July, the conference members pondered 52 presentations by 193 contributors. The scientific presentations were introduced by Nanduri Prabhakar's tribute to Sukhamay Lahiri, a unique giant of the carotid body world lost in 2009. We mourn his passing away on May 2, 2009. Topics ranged widely as usual, including air vs. water breathers, impact of hormones in response to hypoxia, and further discussion of H_2S effects, of core temperature response to hypoxia, of the effect of development and age, of carotid body responses in sleep apnea, and congestive heart failure. The Heymans-De Castro-Neil prizes were awarded to Ana Rita Nunes (Portugal), Erica Wehrwein (USA), and Noah Marcus (USA). Additionally, Shaima Salman (Canada) and Edward O'Connor (Ireland) received honorable mention.

2014 Meeting XIX: *Chris Peers*

Presentations are found in *Arterial Chemoreceptors in Physiology and Pathophysiology*. Edited by C. Peers, P. Kumar, C.N. Wyatt, E. Gauda, C.A. Nurse, N. Prabhakar. Heidelberg, New York, Springer Cham, 2015. This is also Volume 860 of *Advances in Experimental Medicine and Biology*. This conference, held in Leeds, UK, saw 43 papers presented by 206 contributors. Several pathophysiological conditions were explored such as obstructive sleep apnea, the hypertension that results from the intermittent hypoxia, congestive heart failure, diabetes, and Duchenne's muscular dystrophy. Some newer physiological aspects were also introduced such as the effect of progesterone, its receptors and their location on respiratory control, the role of TRP channels in chemoreception, glomus cell mitochondria's role in oxygen sensing, and the usual number of very imaginative pharmacological investigations. The Heymans-De Castro-Neil prizes were awarded to Maria Joao Ribeiro (Portugal), Ma Carmen Fernandez-Aguera Rodriguez (Spain), and Jessica Hohlin (Sweden).

2017 Meeting XX: *Estelle Gauda*

The short presentations from the meeting are found in this current volume. *Arterial Chemoreceptors, New Directions and Translational Perspectives. Edited by E. B. Gauda, M.E. Monteiro, N. Prabhakar, C. Wyatt, and H.D. Schultz*. The program lists 45 presentations delivered by 190 investigators of whom 59 were attendees at the conference, which had tributes to deceased members, plenary lectures, standard oral presentations, and poster board presentations. The Heymans-De Castro-Neil prizes were awarded to Joana Sacramento (Portugal), Ryan Rakoczy (USA), and Elena Olea Fraile (Spain). Of considerable importance was the concluding discussions regarding the future of ISAC. The small number of attendees seemed to warrant such there was much discussion about being a satellite to other larger meetings related to respiratory control and chemoreception. These discussion will continue but no definitive decision was made to change the format of the ISAC meeting for 2020. The 2020 Meeting, ISAC XXI, will be held in Portugal. Emilia Monteiro is the current President of ISAC and will lead the organization of the meeting.

Respectfully submitted,
Robert S. Fitzgerald, Ph.D.
Professor

Departments of Environmental Health and Engineering, of Physiology, of Medicine
The Johns Hopkins University Medical Institutions
Baltimore, MD, 21205 USA

O_2/CO_2: Biological Detection to Homeostatic Control

1

Robert S. Fitzgerald

Abstract

Oxygen (O_2) and Carbon Dioxide (CO_2) are the two gases to be detected and controlled. Of interest might be a query of the evolutionary origin of each. From the cooling of the Big Bang (~13.8 Billion Years Ago [BYA]) came a quark-gluon plasma from which protons and neutrons emerged, producing H, He, Li. As H and He collapsed into the first stars at ~13.3 BYA carbon and monatomic oxygen were generated. Some 3 billion years ago greater amounts of diatomic oxygen (O_2) were provided by earth's photosynthesizing bacteria until earth's atmosphere had sufficient amounts to sustain the life processes of multicellular animals, and finally higher vertebrates. Origin of CO_2 is somewhat unclear, though it probably came from the erupting early volcanoes. Photosynthesis produced sugars with O_2 a waste product. Animal life took sugars and O_2 needed for life. Clearly, animal detection and control of each was critical. Many chapters involving great heroes describe phases involved in detecting each, both in the CNS and in peripheral detectors. The carotid body (CB) has played a crucial role in the detection of each. What reflex responses the stimulated CB generates, and the mechanisms as to how it does so have been a fascinating story over the last 1.5 centuries, but principally over the last 50 years. Explorations to detect these gases have proceeded from the organismal/system/ organ levels down to the sub-cell and genetic levels.

Keywords

Oxygen · Carbon dioxide · Great oxygen event · Chemoreceptors

1.1 Introduction

For the past 30 years the International Society for Arterial Chemoreception (ISAC) has addressed the issues of detecting O_2 and CO_2 in order to control them better for the homeostatic benefit of the organism. And even as far back as 1950 groups studying this have met quite frequently. To explore detection and homeostatic control of these gases immediately provokes two questions: How is this to be done? Why is it to be done? In the vertebrate the cardiopulmonary system is the tool which captures air from the external environment and delivers it to the cells of the organism, specifically to the mitochondria of the cells where the electron transport chain uses it as a last step in the oxidation of glucose. Is that important? And

R. S. Fitzgerald (✉)
Departments of Environmental Health & Engineering, of Physiology, of Medicine, The Johns Hopkins University Medical Institutions, Baltimore, MD, USA
e-mail: rfitzger@jhsph.edu

E. B. Gauda et al. (eds.), *Arterial Chemoreceptors*, Advances in Experimental Medicine and Biology 1071, https://doi.org/10.1007/978-3-319-91137-3_1

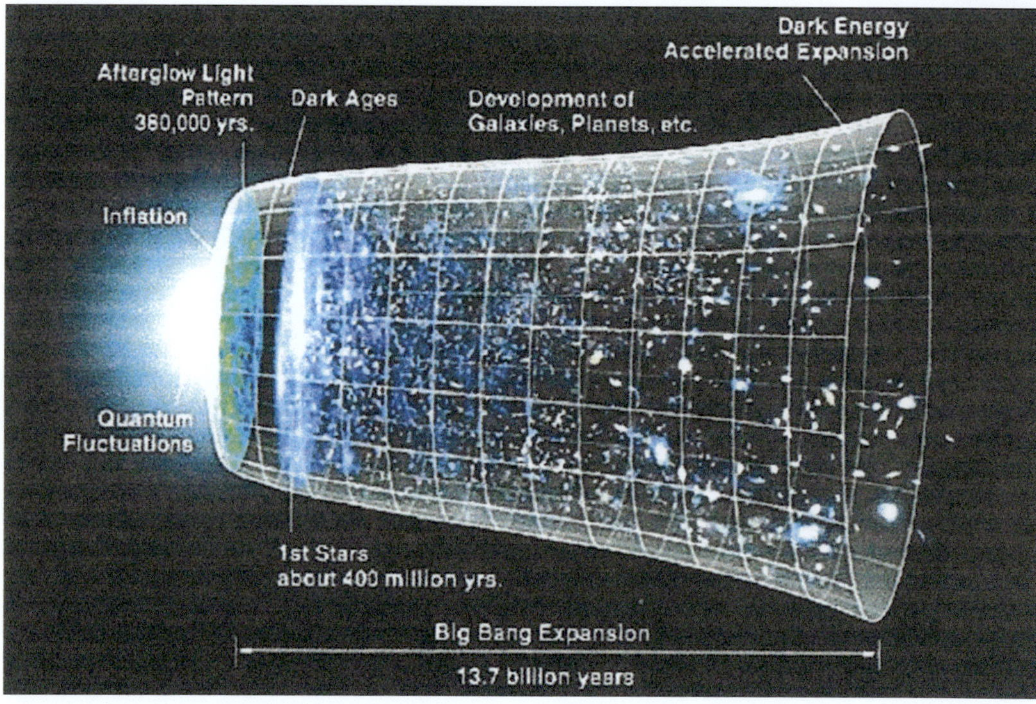

Fig. 1.1 Big bang. From hot quarks and an initial expansion the universe cooled, generating subatomic particles followed by simple atoms. Gravity brought these together to form the heavier atoms. Stars, galaxies were also formed. (From: Wikipedia, the Free Encyclopedia. Original version: NASA/WMAP Science Team, modified by Ryan Kaldan. With permission of JPL Image Use Policy)

if so, why? A homey experiment can give us a clue. How long can a subject hold her/his breath? Maybe 90 s would be an average duration.

So whereas one can go without food for a month, and without fluid for possiblly a week one needs air in seconds. The oxidation of glucose requires the air's oxygen (O_2) to generate ATP, the biological energy required for all the organism's living activities (e.g., heartbeat, diaphragm contracting, local motion, thinking). Mitochondria need O_2 in a hurry. Further, the metabolic resulting CO_2 and H_2O must be expelled. Hence it is clear the organism must be able to detect and control these two gases for life.

Since O_2 and CO_2 are so fundamental to life, one might initially query just what is the origin of these two gases on this planet. The standard model of the cosmos has space-time arising at the Big Bang ~13.8 billion years ago (BYA); monatomic oxygen (O) arose from the stars ~ 13.3 BYA. The Milky Way galaxy formed ~8.8 BYA. Earth finally appeared ~ 4.54 BYA. The only O on earth at that time was what had resulted from nuclear fusion, a very small amount (Fig. 1.1).

Cyanobacteria appeared ~2.5 BYA, and began to photosynthesize, the product of which was sugars. Their chloroplasts used sunshine, water, and CO_2. But from where did *this* CO_2 arise? It's really unclear, but many speculate it was a component of volcanic gas. The O generated by physicochemical fusion from smaller atoms and O_2 from primitive photosynthesis was chemically captured on earth by dissolved iron and organic matter in the sea. About 2.3 BYA The Great Oxidation Event occurred. At this point the oxygen sinks in the sea and on land had become saturated and O_2 began to accumulate in the atmosphere (Lane 2003) (Fig. 1.2).

The story of biological evolution from the primordial cell has been told many times. And many air-breathers have preceded *homo sapiens*.

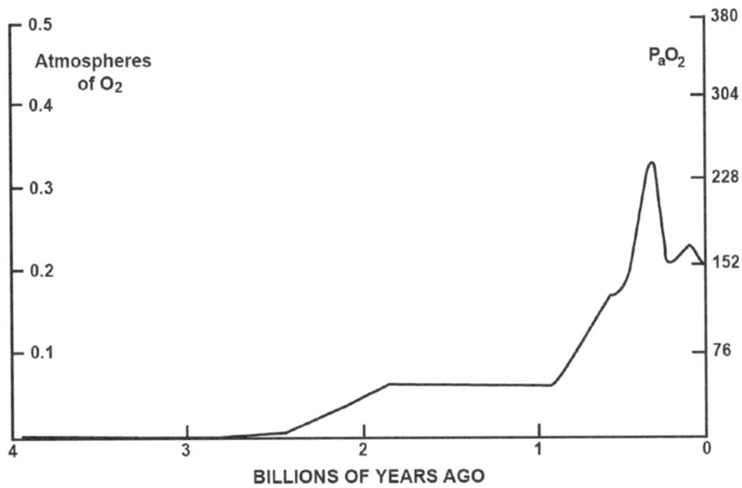

Fig. 1.2 Great oxygen event. From a virtually oxygen-free planet cyanobacteria began to produce O_2 as a waste product of photosynthesis about 2.5×10^9 years ago. But most was being absorbed into the oceans and seabed rock. With the saturation of these, about 1.85×10^9 years ago O_2 is becoming absorbed on land surfaces. About 850×10^6 years ago Earth's atmosphere starts to gain O_2. (Adapted from Wikipedia. Source: Holland, H [2006]. *Phil Trans Roy Soc: Biol Sci* 361: 903–915 with permission via Copyright Clearance Center)

However, this chapter will confine itself to how human subjects explored the issue of the biological detection of organismal O_2 and CO_2 in order to maintain homeostasis.

1.2 Oxygen/Carbon Dioxide and Respiration

The first method of biological detection was to observe general breathing movements, and later,the duration of breathing in an animal confined to a closed chamber. Parallel to this was the extinction of a flame or its effulgence in the same chamber.

But among the ancients of what was to become the European or western scientific tradition there was much ignorance concerning how the lungs and heart worked mechanically, and what each did for the organism. Plato thought breathing was involved in heat regulation, a balancing of the body's inner heat with the outside. Aristotle thought the arteries carried air, making the function of breathing unclear. Galen (131–201) also thought that since the heart drew air from the lungs and put it in the left ventricle breathing was meant to cool the blood. Most of the early, subsequent efforts to understand breathing were based on Galen's views. However, from the Arabic/Islamic medical/scientific traditions came the five volume *Canon of Medicine* in 1025 by the Persian philosopher Avicenna (980–1037) at age 21. The *Canon* brings together the successes of Hippocrates, Galen, as well as Egyptian, Persian, and Indian medical men plus his own research results and discoveries. The book was translated from Arabic into Latin and remained a text book or required background literature well into eighteenth century Europe. William Osler of McGill, Penn, and Hopkins fame described the *Canon* as "the most famous medical textbook ever written." Though a historical treasure, it is highly unlikely that detection of O_2 and CO_2 was ever considered. But Ibn-al Nafis (1213–1288), a second Arab physician and polymath deserves mention not for any work on the detection of O_2 and CO_2, but because his work on the pulmonary circulation, essential for rendering O_2 and CO_2 detectable, is the first. This is his description:

> *Blood from the right chamber of the heart must arrive at the left chamber, but there is no direct pathway between them. The thick septum of the*

heart is not perforated and does not have visible pores as some people thought or invisible pores as Galen thought. The blood from the right chamber must flow through the vena arteriosa (pulmonary artery) to the lungs, spread through its substances, be mingled there with air, pass through the arteria venosa (pulmonary vein) to reach the left chamber of the heart, and there form the vital spirit.

Ibn-al Nafis also commented on Avicenna's *Canon*. But we are unaware of any evidence supporting his thoughts on the detection of O_2 and CO_2. Although many important studies of the cardiopulmonary system's anatomy appeared in the first centuries of the second millenium (Leonardo de Vinci [1452–1519]), it would be safe to say none of them addressed the issue of the integrative control of oxygen & carbon dioxide homeostasis, or the organism's detection of these gases.

Robert Boyle (1627–1691) is thought to have performed the "fundamental experiment in the physiology of respiration". He recorded death in a variety of animals when they were enclosed in a chamber and the chamber evacuated. But if air was readmitted, they could be resuscitated. The chamber was designed and built by his colleague, Robert Hooke (1635–1703). An earlier version of this experiment was done by Leonardo da Vinci (1452–1519). On November 2, 1664 Hooke made holes in the lung of an anesthetized dog and passed air through the perforated lung. The dog remained alive, demonstrating it was air and not lung movement that was the essential contribution of the lung. Further, dark blood became red in the process. Richard Lower (1631–1691) had shown the same thing somewhat earlier. John Mayow (1643–1679) believed his observations led him to conclude that the "nitro-aerial particles with which spirit abounds" in air combined with the blood in the lung (like Ibn-al Nafis) in order to combine with combustible particles in the body. And though Mayow's best contributions are, perhaps, those he made regarding respiratory mechanics, there are historians who believe that Lavoisier (1743–1779) never cited the works of Mayow regarding respiratory chemistry of which he made significant use. Priestley's experiment on August 8, 1774 was to put an adult mouse in a chamber into which he had put the air he had gathered from buring *mercurius calcinatus* (HgO, mercuric oxide). He said if he had put the mouse into the chamber with common air, the mouse would have lasted only 15 min; but this mouse lasted a full hour. He did several more similar experiments. Priestley and Lavoisier were rivals. It seems probable that the former discovered what was to be called oxygen before the latter did. However, Lavoisier was a much more systematic investigator who organized both his own discoveries and those of Priestley into the complete oxygen story. Lavoisier is often called the Father of Modern Chemistry for his systematic thinking. He also was continually integrating oxygen with respiration. As a matter of fact Carl Scheele, a Swedish apothecary, discovered oxygen from his own studies before either Priestley or Lavoisier, but did not publish his discoveries until 1777. So by the end of the eighteenth century the need for oxygen (dephlogisticated air) in respiration was established quite firmly. Lavoisier is often criticized because he most often did not credit his predecessors. He did make an interesting discovery by recording oxygen consumption and CO_2 production over a 24 h period. O_2: 526 cc/min; CO_2: 204 cc/min, a RQ of 3.89, quite abnormal but possibly due to the measuring equipment at the time.

In addition to Lavoisier's awareness of CO_2 the Scotsman Joseph Black (1728–1799) made the close connection between CO_2 and the expirate in breathing, and found that CO_2 would not support either combustion or respiration. Finally, Daniel Rutherford (1749–1819) published in 1772 his report that once Black's "fixed air" (carbon dioxide) was removed from the expirate, a large amount of air remained which rapidly extinguished both life and flame. What remained was nitrogen. Henry Cavendish (1731–1810) also contributed to the discovery of nitrogen. By the end of the eighteenth century all respiratory-related gases had been detected and identified. The question remained as to *where* and *how* these gases were detected in the organism. But respiration (i.e., breathing) remained the method of detecting O_2 and CO_2 (Fig. 1.3).

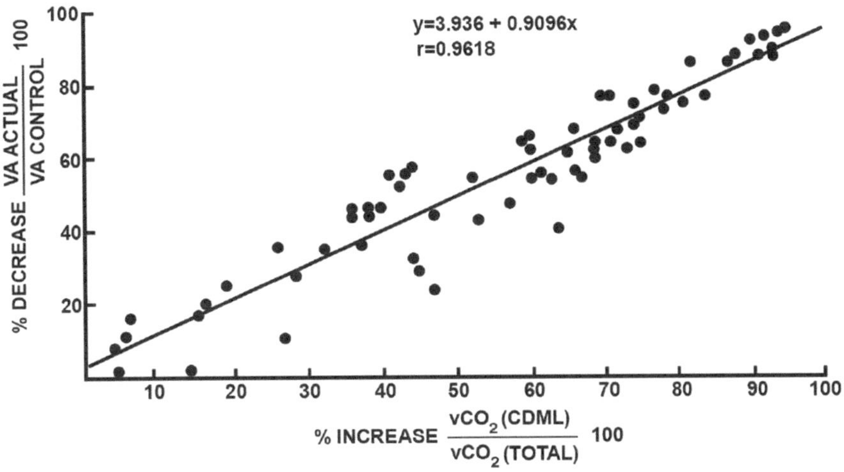

Fig. 1.3 Centrality of CO_2 on ventilation. In animal studies T. Kolobow used a device he called a "carbon dioxide membrane lung" (CDML) inserted into the venous circulation to scrub CO_2 from blood entering the right heart. Oxygen was supplied via a tracheostomy. The graph shows that of the total CO_2, the more of it scrubbed out by the CDML (abscissa), the more the alveolar ventilation decreased (ordinate) down to zero. (Kolobow et al. 1977; with permission Wolters Kluwer Health, Inc.)

1.3 CO_2 – Detecting Elements in the Respiratory System

What are the biological elements which will detect CO_2 and account for the cellular and integrative control of this gas to maintain homeostasis? In the absence of control, CO_2 could rise, as it does in obstructive sleep apnea, or in some cases of COPD. Respiratory acidosis develops. Any form of acute respiratory acidosis stimulates increases in both respiratory frequency and tidal volume. Further, it can impair the operation of both structural and functional proteins. Skeletal muscle (e.g., the diaphragm) decreases its force of contraction. Lowering CO_2 (respiratory alkalosis) results when ventilation exceeds metabolism. This can produce hypopnea, and eventually hypoxemia.

1.3.1 Detection in the CNS

In 1857 Kussmal and Tenner occluded the cerebral vasculature of an anesthetized dog. They observed the blood became dark; the dog hyperventilated, gasped, and eventually died (Kussmaul and Tenner 1857). This clearly suggested that CO_2 and hypoxia acted in the head. In 1885 Meischer-Rusch concluded from his experiments with human subjects (1885) that increases in CO2 was the ordinary stimulus to ventilation and not hypoxia since a small increase in CO_2 (1%) increased ventilation whereas a larger decrease in oxygen (3%) did not. The last decade of the nineteenth century saw the cross perfusion experiments in dogs by Leon Fredericq (1851–1935) in which he connected the proximal carotid arteries and jugular veins of Dog #1 to the distal counterparts in Dog #2; he then connected the proximals of Dog #2 to the distals of Dog #1. Occluding the trachea of Dog #1 increased the ventilation of Dog #2; this produced hypopnea and apnea in Dog 1. Further, hyperventilating Dog #1 made Dog #2 breathe less because he was receiving the depressed $PaCO_2$ from Dog #1. Using hyperoxia in other experiments with this preparation, Fredericq concluded to the landmark understanding for his day where and how CO_2 and O_2 influenced respiration (Fredericq 1890).

Working with human subjects, John Scott Haldane (1860–1936) once again demonstrated the power of CO_2 on ventilation. In human subjects a 3% rise in inspired CO_2 produced an increase in ventilation that could be matched only by a 7% reduction in inspired O_2. And from his work with colleague John Priestley in miners and divers, they concluded that CO_2 was acting on respiratory centers in the head in normal breathing (Haldane and Priestley 1905). To what were the respiratory centers responding? Was it CO_2 or H ion? In 1911 the first of Hans Winterstein's "reaction theories" was proposed based on injections of fixed acid into his dogs (Winterstein 1911). But an unrelated experiment showed that CO_2 could diffuse from an extracellular space to an intracellular space whereas H ion did not (Jacobs 1920). So the increase in *blood* [H ion] from increased $PaCO_2$ was not the stimulus to the respiratory centers. Winterstein's 1921 "reaction theory" proposed that it was [H ion] in the respiratory centers which increased ventilation upon exposure to increased inspired CO_2 (Winterstein 1921).

Now the question arose as to just *where* in the head were these respiratory centers which were to control CO_2 homeostasis? In the late 19th and very early twentieth centuries hypoxia was thought to act centrally, perhaps even at the same receptors, by virtue of the lactic acid hypoxia generated. In 1954 Leusen seems to have further defined this locus of the "centers" to some place in or near the ventricles with his ventriculocisternal perfusion of fluid containing increased [H ion]. This work (Leusen 1954) certainly influenced the subsequent studies of Hans Loeschcke, Robert Mitchell and their colleagues (Loeschcke et al. 1953; Mitchell et al. 1963) who yet further defined the locus to the floor of the fourth ventricle of the medulla in cats. From their studies the sensitive centers were located on the ventral surface, quite superficially (Fig. 1.4).

However, the simultaneous studies of Pappenheimer, Fencl, and colleagues perfusing the brains of unanesthetized goats with mock CSF placed the sensitive "centers" about two-thirds to three-fourths the distance along a bicarbonate concentration gradient from the CSF to the blood (Pappenheimer et al. 1965). In 1982 Dempsey and Forster reviewed with extraordinary comprehension the work of Loeschcke and his colleagues as well as that of Pappenheimer and his colleagues plus their own work (the Madison, Wisconsin Group) involving many species, conditions, and experiments. They con-

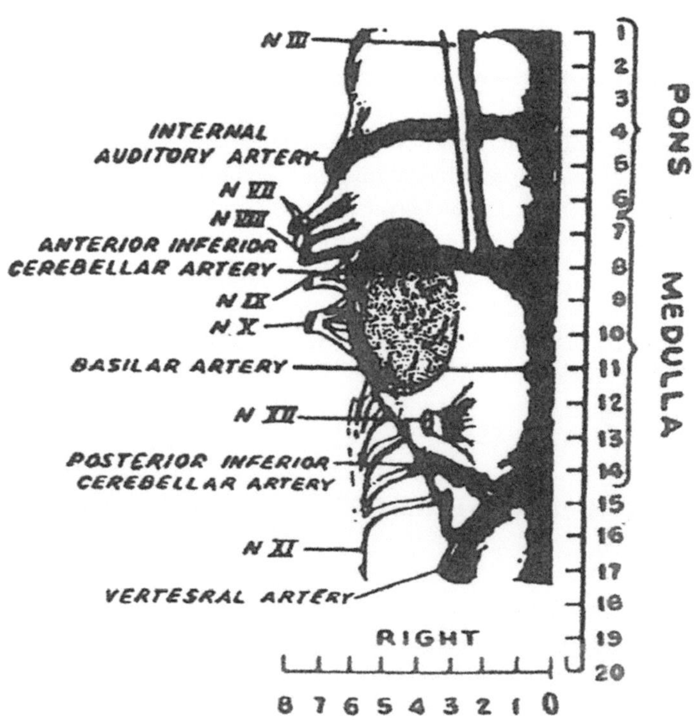

Fig. 1.4 CO_2-sensitive medullary areas. Ventral surface of the cat brain showing the areas to which Mitchell, Loeschcke, Severinghaus, and their colleagues applied CO_2/H^+ – soaked cotton pledgets to provoke an increase in ventilation, narrowing an important area for the biological detection of CO_2. (Mitchell et al. 1963; with permission Rightslink/John Wiley and Sons)

cluded that interstitial [H^+] was the stimulus for the medullary processes involved in the homeostatic control of ventilation in acid-base disturbances. And this area was also involved in eupneic breathing. But this [H^+]-sensitive structure was not on the surface but about 200 um below the ventrolateral surface of the medulla – closer to the interstitial fluid spaces (Dempsey and Forster 1982). At what is perhaps the molecular level for the biological detection of CO_2 Nattie has produced data showing that blocking an imidazole-histidine protein moiety in the ventrolateral medulla of in vivo preparation inhibits CO_2 sensitivity (Nattie 1995). But in more recent years efforts to identify CNS components of CO_2 detection met a problem. At a much more reduced level of organismal organization in the CNS our focus came to a specific location and on specific neurons and how these entities are to be challenged with the CO_2/H^+. Cells from the Phox2B retrotrapazoid nucleus, 5HT neurons of the medullary raphe, noradrenergic neurons of the locus ceruleus, and NK1R-ir cells of the raphe magnus…all are affected by CO_2/H^+, Excitation at a single site is modulated by subsequent hypocapnnic inhibition at other sites. And according to this long-term contributor to this controversy: central chemoreception is a distributed property; there is no unique chemoreceptor site nor is there a unique neuronal phenotype or mechanism" (Nattie and Li 2006). This information is interesting and helpful, but clearly indicates that more research is needed to clarify the picture of CO_2/H detection in the CNS.

1.3.2 Detection by the Carotid Body

Respiratory acidosis or metabolic acidosis according to one group is detected in an identical fashion. Both lower glomus cell pH (pH_i). This activates a Na^+/H^+ exchanger. H^+ is extruded bringing in Na^+. This activates and reverses the normal direction of the Na^+/Ca^{++} exchanger. Na^+ is extruded and Ca^{++} enters and attaches to the neurotransmitter-containing vesicles. The vesicles exocytose their contents into the synaptic cleft onto the appropriate post-synaptic receptors on the abutting afferent neuron (Gonzalex et al. 1995). A second view involves the H^+ – generated inhibition of background (TASK1/3) K^+ currents in the glomus cells. The resulting depolarization activates voltage-gated Ca^{++} channels. The entering Ca^{++} attaches to neurotransmitter-containing vesicles which dock on the inner surface of the glomus cell and exocytose their contents across the synapse to appropriate receptors on the abutting afferent neuron (Buckler 2015; Zhang and Nurse 2004).

1.4 O_2 – Detecting Elements in the Respiratory System

1.4.1 Detection in the CNS

Oxygen is sensed by several types of cells in the CNS. But it is not at all clear this influences homeostatic control via respiration. The detection of low oxygen by specialized receptors in the CNS to increase respiration does not seem to have significant data in support of that possibility, though hypoxia does affect cells in the CNS, as it does systemically. Some cells are excited; others, depressed. Again paradoxically, some depressed cells may cause excitation in some other part of the respiratory cell linkage chain. But generally hypoxia produces CNS neuronal depression (Bisgard and Neubauer 1995).

1.4.2 Detection by the Carotid Body

The principal detector of oxygen in vertebrates is the carotid body (CB) found bilaterally at the bifurcation of the common carotid artery into its internal and external branches (Fidone and Gonzalez 1986). The first known report of the carotid body (CB) structure appeared in the January 31, 1743 dissertation defense by Hartwig William Louis Taube in the lab of the great German physiologist, Albrecht von Haller. Occasional works in the CB area appeared in the nineteenth century; many focused on cardiorespiratory reflex responses. In 1900 Pagano (1900) and Siciliano (1900) also located the origin of reflex effects in the periphery. Heinrich Hering (1866–1948) in papers from 1923 to 1927 (Hering 1923) further established the role of the peripheral nerves in the reflex effects. So biological detection remained in the area of reflex effects. In the mid- to late 1920's two great centers pursued a deeper appreciation of the morphology and intercellular relations of the CB (Fig. 1.5) in the person of Fernando De Castro working in Madrid (DeCastro 1926, 1928). The second group of Corneille and J.F. Heymans and colleagues pursued the mechanisms of the cardiopulmonary reflexes by stimulating the carotid and aortic locations (Heymans J-F and Heymans C 1927; Heymans C et al. 1930). The story surrounding their interactions with each other and their colleagues has been told frequently in the past. Their work went on well into the 1930s and for which Corneille Heymans won the 1938 Nobel Prize in Physiology or Medecine. De Castro was such a strong influence on Heymans that many thought he should have shared the Prize. This thought does have credibility since it is clear from the reports of each that De Castro had correctly assigned the role of the CB before C. Heymans had done so. Each contributed to a deepening of our understanding of the CB's next door neighbor, the carotid sinus, principal detector and regulator of arterial blood pressure.

Later years has revealed several other reflex effects resulting from CB stimulation in addition to the control of O_2 and CO_2 homeostasis (Fitzgerald and Lahiri 1986). Recent studies have focused on the cellular and molecular mechanisms involved in the *detection* of hypoxia, hypercapnia, and acidosis (Fidone and Gonzalez 1986). Histological studies had confirmed that the carotid body's glomus cells were filled with neurotransmitter- containing vesicles.

Many studies confirmed the presence of ACh, ATP, DA, NE, 5HT, GABA, neurokinin A. Appropriate receptors were also located: Cholinergic muscarinic (M1, M2), ncotinic, dopamine (D2), GABA B, and purinergic (P2X2Y and others (Fig. 1.6)). Recent reports have uncovered what might be called modulators of carotid body neural output: NO, H_2S, adenosine. And other recent reports have identified TRPV1 channels and the olfactory receptor, Olf78, (Chang et al. 2015) as having roles in chemodetection. One commonly accepted cascade of events in detection of lowered levels of oxygen has been: hypoxia reduces K^+ traffic in channels including TASK K^+ channels (Buckler 2015;

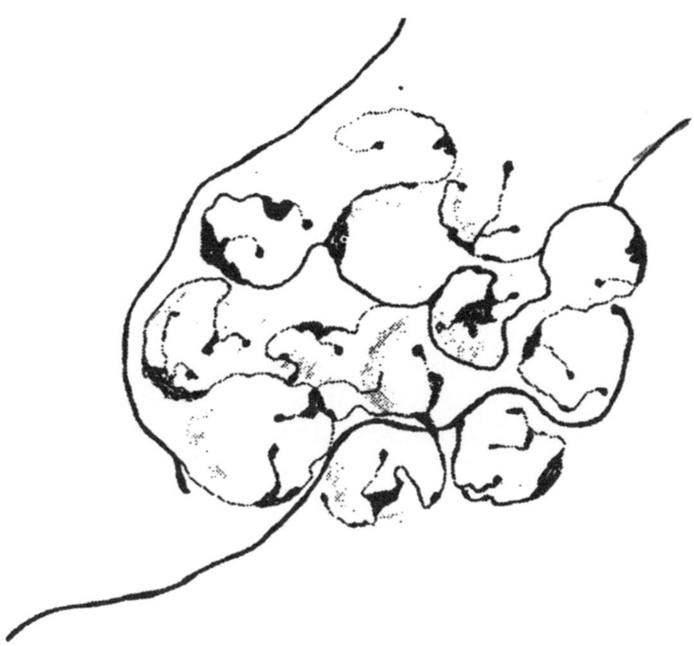

Fig. 1.5 Composite of several of De Casto's drawings showing the neurotransmitter-containing Type I cells with afferent neurons applied to them. Note the different type of endings…calyx, rete, simple fiber (with permission Oxford University Press/ American Physiological Society)

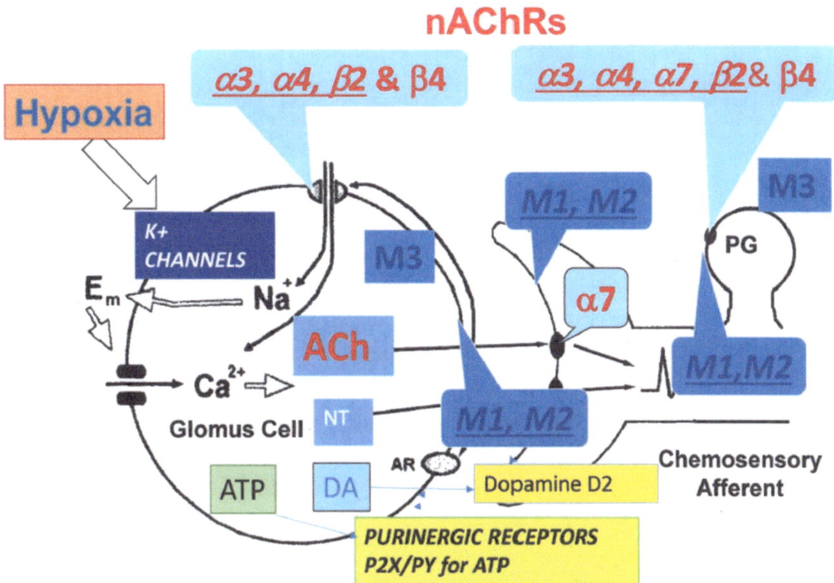

Fig. 1.6 Neurotransmitter receptors in the cat carotid body. A composite of studies, mostly provided by Shirahata, showing the location of various types of cholinergic receptors and highlighting the sub-units of the nicotinic receptors. These operate in the biological detection of O_2 by the carotid body (Shirahata et al. 1998; Hirasawa et al. 2003)

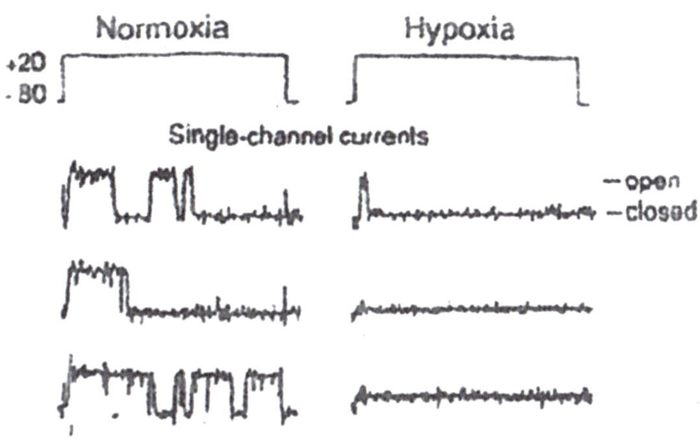

Fig. 1.7 K^+ channel behavior during hypoxia. Hypoxia inhibits current through this K^+ channel in the membrane of the Type I cell. In so doing the cell becomes depolarized which opens voltage-gated Ca^{2+} channels. Ca^{2+} rushes in, adhering to the transmitter-containing vesicles which exocytose their contents into the cleft between the cell and abutting afferent neuron. (Adapted from Lopez-Barneo 1994; with permission Rightslink/Elsevier)

Lopez-Barneo 1994), Fig. 1.7 elevating resting membrane potential; this depolarizing action opens voltage-gated calcium channels; the entering Ca^{++} binds to the vesicles; a protein, synaptophysin or physophilin docks the vesicles on the inner surface of glomus cells, and they exocytose their contents into a synaptic-like cleft between the glomus cell and the abutting afferent neuron. But in some species there is substantial evidence that H_2S, known to increase in an hypoxic environment, is thought to be the excitatory agent closing the maxi K^+_{BK} channel (Peng et al. 2010). However, other proposed mechanisms of hypoxic chemotranduction (e.g., “metabolic hypotheses”) have suggested that O_2 sensing by chemoreceptor cells is linked to mitochondrial structure and function (Buckler 2015). ATP is central to this explanation. As chemoreceptor neural output increases with the decreasing O_2, genesis of ATP must confront lowered O_2. A further problem is the link between a decrease in ATP and the Ca^{++}-triggered release of neurotransmitters has remained unclear. Finally, there are several recent reports of genetic differences in the detection of hypoxia as well as in carotid body structure and function (Weil 2003; Weil et al. 1998; Balbir et al. 2007, 2006; Yamaguchi et al. 2006). The above treatment is really focused on *acute* exposure to lowered PaO_2.

The ventilatory response to long term exposure to hypoxia has some phases, and the story is not yet clear. There is the hypoxic ventilatory run-off or depression which occurs 10–15 min after the CB-mediated increase in ventilation. But then there is the acclimatization to both short term and long term exposure to hypoxia. CNS neurons are in general turned off by hypoxia and therefore not responsible for the increased ventilation of acclimatization. This phenomenon does seem to have its locus in the CB, perhaps in a reduced sensitivity of the D2 dopamine receptor on the CB's afferent neuron to the NTS. The complexity surrounding this situation is fully reviewed recently (Bisgard and Neubauer 1995). Both areas of ventilatory changes need further study. A final point of "integrative control" which must be included is the finding by Philipson (Phillipson et al. 1981), and others that for rhythmic breathing generated in the CNS there is an absolute need for input from peripheral receptors. One study (Sullvan et al. 1978) in healthy unanesthetized dogs combined metabolic alkalosis with slow wave sleep, vagal blockade, and hyperoxia. Respiratory frequency was reduced to one breath/minute. Such data clearly demonstrate that afferent respiratory stimuli from the periphery are essential for sustaining adequate ventilation.

1.5 Summary

After a brief look at where this planet's O_2 and CO_2 originated inorganically and organically, the method of detecting the two gases biologically so that their homeostatic control would benefit the organism was found to be breathing. Early investigators of the lung, heart, and breathing were quite confused, and many thought the lung and heart were involved in the cooling of the inner body. Major advances in the understanding of the lung, heart, and breathing came with the seventeenth century Oxford group. They made it perfectly clear that air passing into the lungs was necessary for life; and this air did something beneficial to the blood. A next step identified CO_2 as a stimulus to breathing; not quite clear yet was the impact of hypoxia *per se*. Finally determined was the discovery that it was the $[H^+]$ in the CSF generated by CO_2 that was detected by structures in the medulla. Peripheral detection of hypoxia was by the CB, which also senses CO_2. More advanced studies of the CB have uncovered subcellular and even genetic modification of this detection. Finally, though it was clear that CO_2 seems to be the major stimulus to respiration acting centrally and peripherally, work has shown that the rhythmic breathing in the homeostatic condition is produced by CNS elements in the pons and medulla. But they must have input from peripheral receptors.

References

Balbir A, Okumura M, Schofield B, Coram J, Tankersley CG, Fitzgerald RS, O'Donnell CP, Shirahata M (2006) Genetic regulation of chemoreceptor development in DBA/2J and A/J strains of mice. Adv Exp Biol Med 580:99–104

Balbir A, Lee H, Okumura M, Biswal S, Fitzgerald RS, Shirahata M (2007) A search for genes that may confer divergent morphology and function in the carotid body between two strains of mice. Am J PhysiolLung Cell Mol Physiol 292:L704–L715

Bisgard GE, Neubauer JA (1995) Peripheral and central effects of hypoxia. In: Dempsey JA, Pack AI (eds) Regulation of breathing, 2nd edn, Revised and Expanded. Marcel Dekker Inc, New York, pp 617–668

Buckler KJ (2015) TASK channels in arterial chemoreceptors and their role in oxygen and acid sensing. Pflugers Arch – Eur J Physiol 467:1013–1025

Chang AJ, Ortega FE, Riegler J, Madison DV, Krasnow MA (2015) Oxygen regulation of breathing through an olfactory receptor activated by lactate. Nature 527:240–244

De Castro F (1926) Sur la structure et l'innervation de la glande intercarotidienne (Glomus caroticum) de l'homme et des mammiferes, et sur un noveau system d'innervation autonome du nerf glossopharyngien. Trav Lab Rech Biol Univ Madrid 24:365–432

De Castro F (1928) Sur la structure et l'innervation du sinus carotidien de l'homme et des mammiferes. Noveaux faits sur l'innervation et la fonction du glomus caroticum. Trav Lab Rech Biol Univ Madrid 25:331–380

Dempsey J, Forster H (1982) Mediation of ventilatory adaptations. Physiol Rev 62:262–346

Fidone SJ, Gonzalez C (1986) Initiation and control of chemoreceptor activity in the carotid body. In: Handbook of physiology section 3: The respiratory system volume II control of breathing. American Physiological Society, Bethesda, pp 247–312

Fitzgerald RS, Lahiri S (1986) Reflex responses to chemoreceptor stimulation. In: Handbook of physiology section 3: the respiratory system volume II control of breathing. American Physiological Society, Bethesda, pp 313–362

Fredericq L (1890) Sur la circulation cephalique croisse, ou echange de sang carotidien entre deux animaux. Arch Biol 10:127–130

Gonzalex C, Dinger B, Fidone S (1995) Mechanisms of carotid body chemoreception. In: Dempsey JA, Pack AI (eds) Regulation of Breathing, 2nd edn, Revised and Expanded. Marcel Dekker Inc, New York, pp 391–471

Haldane JS, Priestley JG (1905) The regulation of lung-ventilation. J Physiol Lond 32:225–255

Hering HE (1923) Der Karotisdruckversuch. Münch Med Wschr 42:1267–1290

Heymans J-F, Heymans C (1927) Sur les modificationes directes et sur la regulation reflexe de l'activite du centre respiratoire de la tete isolee du chien. Arch Int Pharmacodyn Ther 33:273–372

Heymans C, Bouckaert J-J, Dautrebande L (1930) Role reflexogene respiratoire des zones vaso-sensibles cardio-aortique et sino-carotidiennes. Ion Hydrogene,CO_2 sinus carotidiens et reflexes respiratoires. C R Soc Biol 105:881–884

Hirasawa S, Mendoza JA, Jacoby DB, Kobayashi C, Fitzgerald RS, Schofield B, Chandrasagaran S, Shirahata M (2003) Diverse cholinergic receptors in the cat carotid body unit. Adv Exp Med Biol 536:313–319

Jacobs MH (1920) The production of intracellular acidity by neutral and alkaline solutions containing carbon dioxide. Am J Phys 53:457–463

Kolobow T, Gattinoni L, Tomlinson TA, Pierce JE (1977) Control of breathing using an extracorporeal membrane lung. Anesthesiol 46:138–141

Kussmaul A, Tenner A (1857) Untersuchungen uber Ursprung und Wesen der fallsuchtartigen Zuckungen bei Verblutung sowie der Fallsucht uberhaupt. Utersuchungen zur Naturlehre Menschen Thiere 3:1–124

Lane N (2003) OXYGEN, the molecule that made the world. Oxford University Press Inc., New York

Leusen I (1954) Chemosensitivity of the respiratory center. Influence of CO_2 in the cerebral ventricles on respiration. Am J Phys 176:39–44

Loeschcke H, Koepchen H, Gertz K (1953) Uber den Einfluss von Wasserstoffionenkonzentration und CO_2-druck im Liquor cerebrospinalis auf die Atmung. Pfluger's Arch Gesamte Physiol 266:628–641

Lopez-Barneo J (1994) Oxygen-sensitive channels: how ubiquitous are they? Trends Neurosci 17:133–135

Meischer-Rusch F (1885) Bemerkungen zur Lehre von den Atmbewegungen. Arch Anat u Physiol, Leipzig 6:355–380

Mitchell R, Loeschcke H, Massion W, Severinghaus J (1963) Respiratory responses mediated through superficial chemosensitive areas on the medulla. J Appl Physiol 18:523–533

Nattie EE (1995) Central chemoreception. In: Dempsey JA, Pack AI (eds) Regulation of breathing. Marcel Dekker, New York, pp 473–510

Nattie EE, Li A (2006) Central chemoreception 2005: a brief review. Autonom Neurosc: Basic & Clin 126–127:332–338

Pagano G (1900) Sur la sensibilite du coeur et des vaisseaux sanguins. Arch Ital Biol 33:1–36

Pappenheimer J, Fencl V, Heisey S, Held D (1965) Role of cerebral fluids in control of respiration as studied in unanesthetized goats. Am J Phys 208:436–450

Peng YH, Nanduri J, Raghuraman G, Souvannakitti D, Gadalla MM, Kumar GK, Snyder SH, Prabhakar NR (2010) H_2S mediates O_2 sensing in the carotid body. PNAS 107:10719–10724

Phillipson EA, Duffin J, Cooper JD (1981) Critical dependence of respiratory rhythmicity on metabolic CO_2 load. J Appl Physiol Respirat Environ Exercise Physiol 50:45–54

Shirahata M, Ishizawa Y, Rudisill M, Schofield B, Fitzgerald RS (1998) Presence of nicotinic acetylcholine receptors in cat carotid body afferent system. Brain Res 814:213–217

Siciliano L (1900) Les effets de la compresssion des carotides sur la pression, sur le coeur et sur la respiration. Arch Ital Biol 33:338–344

Sullivan CE, Kozar LF, Murphy E, Phillipson EA (1978) Primary role of respiratory afferents in sustaining breathing rhythm. J Appl Physiol 45:11–17

Weil JV (2003) Variation in human ventilatory control – genetic influence on the hypoxic ventilatory response. Respir Physiol Neurobiol 135:239–246

Weil JV, Stevens T, Pickett CK, Tatsumi K, Dickinson MG, Jacoby CR, Robinson DM (1998) Strain-associated differences in hypoxic chemosensitivity of the carotid body in rats. Am J Physiol Lung Cell Mol Physiol 274:L767–L774

Winterstein H (1911) Die Regulierun der Atmung durch das Blut. Pfluger's Arch Gesamte Physiol 138:167–184
Winterstein H (1921) Die Reaktionstheorie der Atmungsregulation. Pfluger's Arch Gesamte Physiol 187:293–298
Yamaguchi S, Balbir A, Schofield B, Coram J, Tankersley CG, Fitzgerald RS, O'Donnell CP, Shirahata M (2006) Genetic influence on carotid body structure in DBA/2J and A/J strains of mice. Adv Exp Med Biol 580:105–109
Zhang M, Nurse CA (2004) CO_2/pH chemosensory signaling in co-cultures of rat carotid body receptors and petrosal neurons: role of ATP and ACh. J Neurophysiol 92:3433–3445

What Is the Point of the Peak? Assessing Steady-State Respiratory Chemoreflex Drive in High Altitude Field Studies

2

Christina D. Bruce, Gary Saran, Jamie R. Pfoh, Jack K. Leacy, Shaelynn M. Zouboules, Carli R. Mann, Joel D. B. Peltonen, Andrea M. Linares, Alexandra E. Chiew, Ken D. O'Halloran, Mingma T. Sherpa, and Trevor A. Day

Abstract

Measurements of central and peripheral respiratory chemoreflexes are important in the context of high altitude as indices of ventilatory acclimatization. However, respiratory chemoreflex tests have many caveats in the field, including considerations of safety, portability and consistency. This overview will (a) outline commonly utilized tests of the hypoxic ventilatory response (HVR) in humans, (b) outline the caveats associated with a variety of peak response HVR tests in the laboratory and in high altitude fieldwork contexts, and (c) advance a novel index of steady-state chemoreflex drive (SS-CD) that addresses the many limitations of other chemoreflex tests. The SS-CD takes into account the contribution of central and peripheral respiratory chemoreceptors, and eliminates the need for complex equipment and transient respiratory gas perturbation tests. To quantify the SS-CD, steady-state measurements of the pressure of end-tidal (P_{ET})CO_2 (Torr) and peripheral oxygen saturation (SpO_2; %) are used to quantify a stimulus index (SI; $P_{ET}CO_2/SpO_2$). The SS-CD is then calculated by indexing resting ventilation (L/min) against the SI. SS-CD data are subsequently reported from 13 participants during incremental ascent to high altitude (5160 m) in the Nepal Himalaya. The mean SS-CD magnitude increased approximately 96% over 10 days of incremental exposure to hypobaric hypoxia, suggesting that the SS-CD tracks ventilatory acclimatization. This novel SS-CD may have future utility in fieldwork studies

C. D. Bruce · G. Saran · J. R. Pfoh · S. M. Zouboules
C. R. Mann · J. D. B. Peltonen · A. M. Linares
A. E. Chiew · T. A. Day (✉)
Department of Biology, Faculty of Science and Technology, Mount Royal University, Calgary, AB, Canada
e-mail: tday@mtroyal.ca

J. K. Leacy
Department of Biology, Faculty of Science and Technology, Mount Royal University, Calgary, AB, Canada

Department of Physiology, University College Cork, Cork, Ireland

K. D. O'Halloran
Department of Physiology, University College Cork, Cork, Ireland

M. T. Sherpa
Kunde Hospital, Khunde, Solukhumbu, Nepal

E. B. Gauda et al. (eds.), *Arterial Chemoreceptors*, Advances in Experimental Medicine and Biology 1071, https://doi.org/10.1007/978-3-319-91137-3_2

assessing ventilatory acclimatization during incremental or prolonged stays at altitude, and may replace the use of complex and potentially confounded transient peak response tests of the HVR in humans.

Keywords
Hypoxic ventilatory response · Peripheral chemoreceptors · Central chemoreceptors · High altitude · Steady-state chemoreflex drive

2.1 Introduction

2.1.1 Respiratory Chemostimuli, Chemoreceptors and Chemoreflexes

Blood gases (O_2 and CO_2) are maintained at relatively constant values, in part through the coordination and interaction between central (brainstem) and peripheral (carotid body) chemoreceptors. Central chemoreceptors, located at various loci throughout the brainstem, detect the accumulation of $PCO_2/[H^+]$ in brain tissue (Guyenet and Bayliss 2015). Peripheral chemoreceptors are found within the carotid bodies, located bilaterally at the bifurcation of the common carotid arteries. The carotid bodies detect rapid changes in both O_2 and CO_2 in a synergistic fashion (Fitzgerald and Parks 1971; Lahiri and DeLaney 1975; López-Barneo et al. 2016).

In reduced preparations, intact animals and humans, chemoreflexes resulting from chemoreceptor stimulation can be measured in a variety of ways, depending upon the model system, context and experimental question (Duffin 2011; Teppema and Dahan 2010; Powell 2006, 2012). For example, in humans, the central chemoreflex can be elicited and measured by administering increases in CO_2 in a background of hyperoxia (>300 Torr PO_2), in a step-wise (steady-state) or incremental (rebreathing) fashion. The resulting linear response in ventilation quantifies the responsiveness. The relative hyperoxia silences the peripheral chemoreceptors to the increases in CO_2, isolating the central chemoreceptors. The peripheral chemoreflex can be elicited through tests exploiting both temporal domain and stimulus specificity (Pedersen et al. 1999; Smith et al. 2006). Because the carotid bodies detect rapid changes in blood gases, transient changes in O_2 and CO_2 are likely representative of the peripheral chemoreflex. In addition, because the central chemoreceptors are thought to limit their responsiveness to $PCO_2/[H^+]$, responses to changes in O_2, particularly in isocapnic conditions, likely represent the peripheral chemoreflex.

2.1.2 Hypoxic Ventilatory Response (HVR) Tests

Exposure to acute normobaric or prolonged hypobaric hypoxia presents a profound and sometimes life-threatening stressor. Many organ systems interact to coordinate and mount an acute response, including the sympathetic nervous system (SNS), cardiovascular and respiratory systems. In response to carotid body stimulation, a SNS response is elicited, increasing cardiac output and peripheral resistance, with a subsequent increase in mean arterial pressure, the driving force for maintaining blood flow to tissues. The respiratory system mounts a robust hypoxic ventilatory response (HVR), which can be measured and quantified in many ways (e.g., Teppema and Dahan 2010). A number of these tests with relevant caveats are outlined below.

2.1.2.1 Steady-State Isocapnic and Poikilocapnic Hypoxia Tests

Using chambers, large breathing bags with premixed gas concentrations or more sophisticated dynamic end-tidal forcing (DEF) systems, participants can be exposed to single or multiple steady-state reductions in the fraction of inspired $(F_I)O_2$. In the case of acute normobaric hypoxia in the isocapnic state, a large peak response is elicited within minutes, followed by hypoxic ventilatory decline (HVD; e.g., Powell et al. 1998; Steinback and Poulin 2007), possibly due to increases in brainstem blood flow and CO_2 washout from central chemoreceptors

(Berkenbosch et al. 1995; Hoiland et al. 2015) or other central or peripheral mechanisms (e.g., Honda 1995; Long et al. 1993; Poulin and Robbins 1998; Robbins 1995). In the poikilocapnic state, where CO_2 is allowed to drift downward as a result of the ventilatory response, the peak response is much smaller. In this poikilocapnic context, steady-state ventilation is similar to baseline conditions after approximately 10 min, likely due in part to the CO_2-O_2 stimulus interaction at the carotid body, and a new steady-state is achieved between hypoxia, hypocapnia and ventilation (e.g., Steinback and Poulin 2007; Pfoh et al. 2017). In either case, even though the hypoxic exposure is steady-state in nature, the response is quantified by visually identifying the peak in ventilation (e.g., Hoiland et al. 2015; Pfoh et al. 2016; Steinback and Poulin 2007), making these tests essentially peak response tests. In the context of high altitude ascent, these tests require sophisticated equipment that lacks portability (e.g., DEF system, gas tanks, gas analyzers, calibration gases). In the case of isocapnic hypoxia tests, it is an open question as to what level is appropriate to clamp the CO_2 at. Because of carotid body acclimatization, and the associated increase in ventilation, participants become incrementally more hypocapnic with ascent and acclimatization, departing from their sea level values during baseline measurements at altitude. In addition, because of the renal compensations (outlined below), there is a reduction in buffering capacity, and a given change in CO_2 will represent a larger relative chemoreceptor stimulus (e.g., Eger et al. 1968; Fan et al. 2010). Thus, participants simply cannot tolerate levels of CO_2 that would be eucapnic at sea level, prior to ascent. Because of these changes in CO_2 and bicarbonate, it is difficult to isolate the hypoxic responsiveness alone given the significant changes in other relevant variables.

2.1.2.2 Transient Hypoxic or Hyperoxic Tests

Given the reported temporal domain (i.e., fast) and stimulus specificity (i.e., hypoxia) of the carotid chemoreceptors, transient tests of the peripheral chemoreflex have been developed and utilized (Chua and Coats 1995; Edelman et al. 1973; Pfoh et al. 2016). Briefly, participants are exposed to multiple breaths of 100% N_2 to elicit a transient HVR. We recently compared a transient 100% N_2 test (TT-N_2) to both steady-state isocapnic and poikilocapnic HVR tests (SS-ISO and SS-POI respectively), and showed that (a) all three tests had different magnitudes, (b) the TT-HVR and SS-ISO test were moderately correlated with each other, within-individual and (c) the TT-N_2 had very little confounding sympathetic, cardiovascular and cerebrovascular responses, which are well-described with the SS-ISO (e.g., Steinback and Poulin 2008; Steinback et al. 2009). However, even though the TT-N_2 and SS-ISO tests are moderately correlated, the TT-N_2 was smaller in magnitude than that of SS-ISO tests and had a lot of within-individual variability between trials (Pfoh et al. 2016). From a fieldwork perspective, the TT-N_2 has similar limitations as the SS-ISO, given that a gas tank (100% N_2) and a calibrated gas analyzer is required, which limit portability. In addition, all hypoxic tests may be dangerous and uncomfortable for participants who are already hypoxic during baseline conditions at altitude, depending on the testing altitude.

In an attempt to overcome the portability and safety caveats of steady-state and transient hypoxic tests, we also recently characterized a transient hyperoxic withdrawal test (TT-O_2) in the context of steady-state hypoxia in the laboratory, and compared it to the TT-N_2 and a non-peak response SS-POI test (following 10-min of steady-state hypoxia). The TT-O_2, originally developed by Dejours (1962), requires only a 100% O_2 tank (readily available in cities prior to ascending to altitude), a spirometer and a pulse oximeter. Unlike the Dejours method, which only quantified a change in ventilation following transient 100% O_2 exposure, we indexed the ventilatory change against the change in peripheral oxygen saturation (SpO_2), providing normative values for this test in unacclimatized hypoxic conditions (Pfoh et al. 2017).

Unfortunately, these various steady-state and transient peak response tests described above have different magnitudes (Pfoh et al. 2016,

2017), are not well-correlated with each other (Pfoh et al. 2016, 2017), nor are their magnitudes related with oxygenation while in steady-state hypoxia (Pfoh et al. 2017), within-individuals. In addition, Milledge et al. (1988) found no relationship between the HVR prior to ascent with the severity of acute mountain sickness (AMS) symptoms upon rapid ascent to 5200 m. Thus, the validity and utility of peak response tests should be questioned with respect to assessing ventilatory acclimatization, oxygenation and avoiding AMS during exposure to chronic hypobaric hypoxia.

2.1.3 The Problem: Caveats for Tests of the HVR in High Altitude Fieldwork Contexts

Testing the HVR at altitude is often performed to assess ventilatory acclimatization (e.g., Forster et al. 1971; Sato et al. 1992; Smith et al. 1986). As a result of prolonged exposure to hypoxia, (e.g., high altitude ascent), ventilatory acclimatization ensues, increasing ventilation for a given combination of O_2 and CO_2. Similar to exposure to acute hypoxic conditions, during ascent to altitude, a new steady-state is achieved between hypoxia, hypocapnia and ventilation. Despite the competing nature of hypoxia (stimulatory) and hypocapnia (inhibitory), steady-state ventilation is eventually higher at rest than sea level due to carotid body plasticity (Smith et al. 1986).

Unfortunately, commonly applied peak response tests of the HVR utilized in many laboratory and field studies at altitude suffer from a number of important caveats, confounds and limitations, summarized below:

1. HVR tests typically require sophisticated equipment, lacking portability in many fieldwork contexts. These include feedback control systems, laptop computers, gas tanks (e.g., 100% O_2, 100% CO_2, 100% N_2, premixed gases), gas analyzers, and calibration gas tanks. Although possible in some high altitude regions, these are not trivial considerations when intending to ascend to high altitude laboratories, lodges or base camps.
2. It may be dangerous and uncomfortable to make participants more hypoxic when already hypoxic at rest while at altitude when carrying out HVR tests.
3. HVR tests are contaminated by chronic changes in CO_2 and arterial/CSF bicarbonate associated with ventilatory and renal acclimatization. Prolonged hypoxia-induced hypocapnia and respiratory alkalosis induces a compensatory metabolic acidosis through the renal excretion of bicarbonate (Dempsey et al. 2014; Krapf et al. 1991), potentially making central chemoreceptors more sensitive to changes in CO_2 due to the reduction in buffering capacity for a given change in CO_2 (Dempsey et al. 2014; Sato et al. 1992; Fan et al. 2010). These steady-state changes in resting CO_2 and acid-base homeostasis likely affect the magnitude of the peripheral chemoreflex independent from changes in O_2. These changes also likely affect the resting central chemoreceptor activation at altitude. Thus, it is difficult to make meaningful comparisons to unacclimatized sea level values.
4. The assumption of stimulus specificity (i.e., hypoxia) of the peripheral chemoreceptors may be incorrect. Recent studies have demonstrated the existence of central (i.e., brainstem) chemosensors that are both sensitive to hypoxia and elicit a hypoxic ventilatory response (Angelova et al. 2015; Curran et al. 2000; Gourine and Funk 2017).
5. Previous studies have demonstrated that the magnitude of different tests of the HVR are not well-correlated with each other, nor are they correlated with oxygenation (SpO_2 or $P_{ET}O_2$) while being exposed to steady-state hypoxia, within-individuals (Pfoh et al. 2016, 2017). If the HVR (broadly defined) is an intrinsic mechanism that protects oxygenation when hypoxic, then any valid test of the HVR should capture these relationships. Our previous study demonstrated that peak response tests are not (a) related to each other in magnitude and (b) are not related to oxygenation.

Thus, our observations should give investigators pause when considering the utility and validity of HVR tests in assessing ventilatory acclimatization when participants are exposed to prolonged hypoxia.

6. The assumption that separate tests of the central and peripheral chemoreflexes are distinct and can isolate chemoreceptors is also likely incorrect. For example, steady-state isocapnic tests have well-described sympathetic, cardiovascular and cerebrovascular effects, which may confound the resulting ventilatory response. Steady-state poikilocapnic tests, where CO_2 is reduced as a result of the HVR, causes a withdrawal of central chemoreceptor stimulation. In addition, prolonged hypoxic stimulation and hypocapnia causes renal compensation, likely affecting the responsiveness of central chemoreceptors through a reduction of $PCO_2/[H^+]$ buffering capacity. Lastly, there is a potential interaction between central and peripheral chemoreceptors, whereby the activation or withdrawal of stimulation of one chemoreceptor compartment may affect the sensitivity of the other compartment's resulting chemoreflex to a transient perturbation. On balance, experimental evidence suggests that in intact animals and humans, the interaction between chemoreceptors is likely simple addition (e.g., Forster and Smith 2010; Smith et al. 2010; Wilson and Teppema 2016). However, the interaction has not been tested in chronic hypoxia, and the interaction may be state-dependent and different than previously demonstrated when tested below eupneic levels of CO_2 (i.e., hypocapnia; e.g., Day and Wilson 2009; Wilson and Day 2013). Thus, HVR tests are likely contaminated by a host of integrative responses that confound the isolation of peripheral chemoreceptors in humans.
7. Lastly, peak HVR tests, whether steady-state or transient, do not represent the steady-state ventilatory strategy that an individual utilizes against a given combination of O_2, CO_2 and acid-base chemostimuli while at rest at altitude. Thus, it is difficult to interpret what a peak ventilatory response to a transient stimulus represents in this context.

Given the many caveats, confounds and limitations outlined above, ultimately these considerations raise doubts about the feasibility and ability of specific transient peak response tests to isolate and quantify specific peripheral chemoreflex responsiveness in human participants in the context of high altitude ascent and acclimatization.

2.1.4 The Solution? A Novel Index of Steady-State Respiratory Chemoreflex Drive

In response to the many challenges outlined above, and faced with a need for a feasible and portable metric to assess respiratory chemoreflexes in high altitude fieldwork studies, we recently developed a novel index of steady-state chemoreflex drive (SS-CD). The SS-CD takes into account the prevailing CO_2 and O_2 chemostimuli and the activation state of both central and peripheral chemoreceptors through their cumulative effect on resting ventilation (Pfoh et al. 2017). To make the SS-CD measurement and calculation, only three portable devices are required (see Fig. 2.1):

1. A spirometer for a measurement of steady-state minute ventilation (V_I; L/min)
2. A capnograph to measure the pressure of end-tidal (P_{ET})CO_2 (Torr)
3. A pulse oximeter to measure peripheral oxygen saturation (SpO_2; %)

First, steady-state representative measurements of $P_{ET}CO_2$ and SpO_2 are used to quantify a stimulus index:

$$SI = P_{ET}CO_2 / SpO_2$$

The SS-CD is then calculated by indexing resting ventilation (V_I; L/min) against the SI:

$$SS\text{-}CD = V_I / SI$$

In the context of respiratory chemoreflex activation state, the SI is applied to assess the steady-state prevailing $P_{ET}CO_2$ and SpO_2. The SS-CD

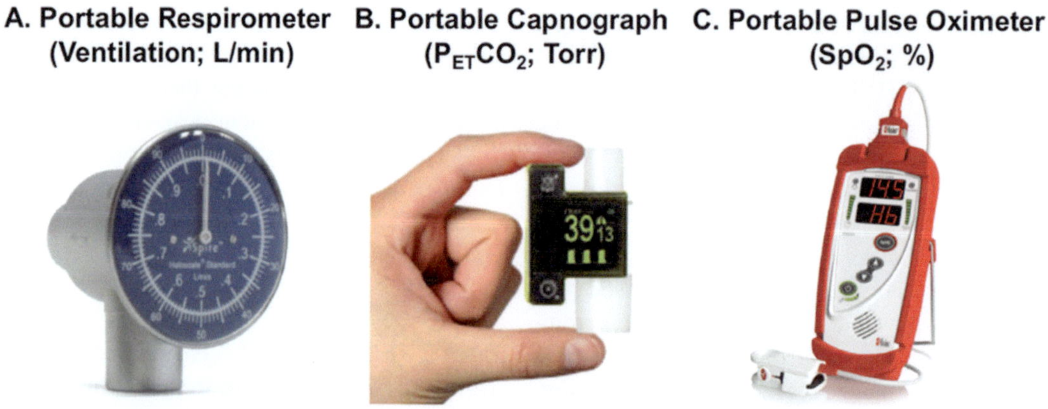

Fig. 2.1 Three portable devices to measure and calculate steady-state respiratory chemoreflex drive in humans in fieldwork contexts. (**a**) Portable respirometer to measure minute ventilation (V_I, L/min; picture from http://bit.ly/2F8xxRJ). (**b**) Portable capnograph to measure the pressure of end-tidal (P_{ET})CO_2 (Torr; picture from http://bit.ly/2BsmxA2). (**c**). Portable pulse oximeter to measure peripheral arterial oxygen saturation (SpO2, %; picture from http://bit.ly/2BuOabJ). These three devices, or others like them, make up a portable system that can be used to quantify resting steady-state respiratory chemoreflex drive in the laboratory, clinic or fieldwork contexts. All devices are non-invasive, do not require power, can be used simultaneously and require only a short-duration spot-check (i.e., a few minutes)

then gives the investigator insight into the steady-state ventilation against the prevailing chemostimuli, eliminating the need for transient gas challenges or peak response tests. This test can then be applied during incremental ascent or at multiple time points during acclimatization to a single altitude.

The strength of the SI lies not only in its simplicity of measurement and calculation, but also in the fact that both chemostimuli have well-documented linear relationships with ventilation. For example, CO_2 is linearly related to ventilation, as demonstrated by both steady-state and rebreathing tests (Duffin 2011; Nielsen and Smith 1952). In addition, although the relationship between O_2 and ventilation is curvilinear (e.g., Loeschchke and Gertz 1958; Weil et al. 1970), indexing ventilation against SpO_2 linearizes the relationship (e.g., Rebuck and Campbell 1974). Thus, CO_2 is proportionally and linearly related to ventilation, and SpO_2 is inversely and linearly related to ventilation, justifying the calculation of SI as $P_{ET}CO/SpO_2$ (Pfoh et al. 2017). Also by way of validation, we used this SI methodology previously in other contexts, specifically when quantifying cerebrovascular responsiveness to breath holding, where both CO_2 and O_2 are changing simultaneously at the metabolic rate (Bruce et al. 2016).

We recently compared the SS-CD between breathing room air and breathing steady-state normobaric hypoxia (~20 min of ~13.5% F_IO_2; ~90 Torr P_IO_2 in Calgary). Despite resting ventilation and the stimulus index being statistically higher in hypoxia, due to the competing effects of hypoxia and hypocapnia, there were no statistical differences in SS-CD between room air (i.e., baseline) and after 20 min of steady-state hypoxia (Pfoh et al. 2017). This demonstration is consistent with the fact that in response to acute poikilocapnic hypoxia, ventilation is not appreciably higher following the peak response, and HVD leads to a new steady-state following approximately 10-min, where ventilation is similar to baseline values breathing room air (e.g., Steinback and Poulin 2007).

Following this previous study comparing SS-CD while breathing room air and acute normobaric hypoxia in a laboratory setting, we aimed to assess the SS-CD during incremental ascent to high altitude. Specifically, we hypothesized that the SS-CD would track ventilatory acclimatization during incremental ascent over 10 days to high altitude hypoxia.

2.2 Methods and Materials

2.2.1 Participant Recruitment and Ethics

Participants were recruited on a voluntary basis to undergo serial measurements from a large group of participants (31) on a research expedition during a trek to Everest base camp (5300 m) in the Nepal Himalaya. Inclusion criteria included adult participants over 18 years of age who planned to trek the entire journey and those who were willing to provide free, informed, verbal, written and ongoing consent. In this study, we recruited 13 participants for inclusion (23.0 ± 3.1 years; BMI 24.0 ± 2.7 kg/m^2; 5 males). This study abided by the Canadian Government Tri-Council policy on research ethics with human participants (TCPS2) and the Declaration of Helsinki, except registration in a database. Ethical approval was received in advance through the Mount Royal University Human Research Ethics Board (Protocol 100012) and was harmonized with the Nepal Health Research Council (Protocol 109-2017). All participants were recruited via verbal communication and provided written and informed consent prior to voluntary participation in the study. Although this study took place in the context of a large research expedition to altitude, the specific study design, research question and data collection were planned *a priori*.

2.2.2 Study Protocol and Ascent Profile

Participants had baseline measurements performed at a low altitude of 1045 m (Calgary) or 1400 m (Kathmandu, Nepal), and these data were combined. They were then flown to 2800 m (Lukla) before beginning the trek. During the trek, all measurements were made on rest days (i.e., no altitude gain) following one night at each altitude, which occurred upon reaching altitudes of 3440 m (Namche; Day 3), 3820 m (Debuche; Day 5), and 4370 m (Pheriche; Day 7) and 5160 m (Gorak Shep; Day 10). Prior to measurements at 5160 m (Gorak Shep), participants trekked to ~5300 m (Everest base camp) and back (approximately 5 h of walking), after which measurements were taken at rest in the afternoon. No participants were taking prophylactic acetazolamide (Diamox) at any time during the trek.

2.2.3 Measurements

All data were collected between 06:00 and 17:00 on non-trekking days. Participants were seated in lodge bedrooms or dining rooms and provided with white noise through head phones to minimize distraction. They were instrumented to measure resting ventilation (V_I; L/min), the pressure of end-tidal (P_{ET})CO_2 and peripheral oxygen saturation (SpO_2). Ventilatory data were collected using a 16-channel PowerLab system (Powerlab/16SP ML880; AD Instruments; ADI; Colorado Springs, CO, USA) and analyzed offline using commercially available software (ADI LabChart Pro software version 8). Participants were instrumented with a mouthpiece, personal bacteriological filter, and nose clip. Respiratory flow was measured through the use of a calibrated pneumotachometer (HR 800 L flow head and spirometer amplifier; ADI ML141). Instantaneous inspired ventilation (V_I, L/min) was calculated as the product of breath-by-breath inspired volume (calculated from the integral of the flow signal) and respiratory rate (calculated by 60/period of the flow signal). $P_{ET}CO_2$ was measured using a portable, calibrated capnograph (Masimo EMMA, Danderyd, Sweden) with a personal mouthpiece and nose clip and SpO_2 was measured with a portable finger pulse oximeter (Masimo SET® Rad-5, Danderyd, Sweden).

2.2.4 Data Analysis and Statistics

Resting ventilation at each altitude was analyzed from a one-minute representative mean bin near the end of a 10-min baseline period, whereas $P_{ET}CO_2$ and SpO_2 measures were obtained after

Table 2.1 Baseline variables at rest during incremental ascent to high altitude in the Nepal Himalaya over 10 days

Variable/Day	0	3	5	7	10
Altitude (m)	1400	3440	3820	4370	5160
Atmospheric pressure (mm Hg)	648	509	486	454	411
Pressure of inspired O_2 (mm Hg)	136	107	102	95	86
V_I (L/min)	12.1 ± 0.6	14.4 ± 1.0	13.9 ± 0.9	14.3 ± 1.0	16.4 ± 1.3*†
$P_{ET}CO_2$ (Torr)	32.3 ± 1.5	26.8 ± 0.8*	25.8 ± 0.8*	21.8 ± 0.8*†	18.4 ± 0.6*†
SpO_2 (%)	96.5 ± 0.3	91.6 ± 0.7*	90.8 ± 0.9*	87.5 ± 0.8*†	78.0 ± 2.0*†
SI ($P_{ET}CO_2/SpO_2$)	0.34 ± 0.02	0.30 ± 0.01*	0.29 ± 0.01*	0.25 ± 0.01*†	0.24 ± 0.01*
SS-CD (V_I/SI)	36.8 ± 2.7	48.6 ± 3.9	48.9 ± 3.8	57.7 ± 5.3*	70.4 ± 6.7*†

Baseline data obtained from a one-minute bin near the end of a 10-minute room-air baseline, following at least one night at each altitude. The first measurement was made in Calgary (1045 m) or Kathmandu (1400 m), and each successive day during ascent is listed, with representative altitude, atmospheric pressure, and approximate pressure of inspired O_2
V_I minute ventilation, $P_{ET}CO_2$ partial pressure of end-tidal carbon dioxide, SpO_2 oxygen saturation via peripheral pulse oximetry, *SI* stimulus index ($P_{ET}CO_2/SpO_2$; Torr/%), *SS-CD* steady-state chemoreflex drive (V_I/SI)
*denotes different than 1400 m, $P < 0.001$
†denotes different than previous altitude, $P < 0.001$
These data n = 13, except V_I, SI and SS-CD: 1045/1400 m, n = 13; 3440 m, n = 10; 3820, n = 11; 4370 m, n = 11; 5160 m, n = 10
Data reported as mean ± SEM

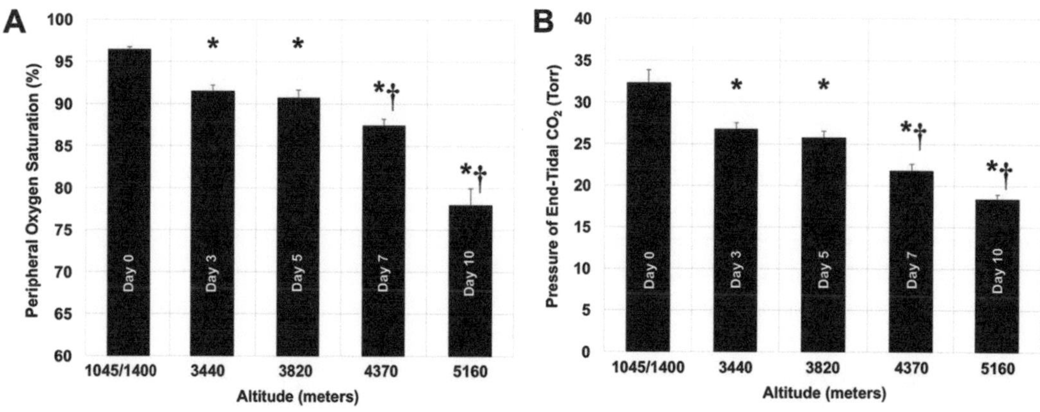

Fig. 2.2 Measures of peripheral oxygen saturation (SpO_2; %) and pressure of end-tidal CO_2 ($P_{ET}CO_2$; Torr) with ascent. Note the incremental decline in SpO_2 and $P_{ET}CO_2$ with incremental ascent, demonstrating hypoxia, and concomitant hypocapnia due to the hypoxic ventilatory response. The relative altitude and measurement day during ascent is labeled for each value. * denotes different than 1045/1400 m, $P < 0.001$. †denotes different than previous altitude, $P < 0.001$. These data, n = 13. Error bars represent mean ± SEM

steady-state was achieved. Data in Table 2.1 and Figs. 2.2 and 2.3 are presented as mean ± standard error of the mean (SEM). A one-factor repeated measures ANOVA was utilized to test for differences in $P_{ET}CO_2$, SpO_2, V_I, SI ($P_{ET}CO_2/SpO_2$) and SS-CD (V_I/SI) at each altitude. Statistical significance was assumed at $P < 0.05$ (SigmaPlot v14, Systat).

2.3 Results

Table 2.1 illustrates resting ventilation (V_I), the pressure of end-tidal (P_{ET})CO_2, peripheral oxygen saturation (SpO_2), stimulus index (SI) and measurement of steady-state chemoreflex drive (SS-CD) during incremental ascent to high altitude. All variables changed in predictable

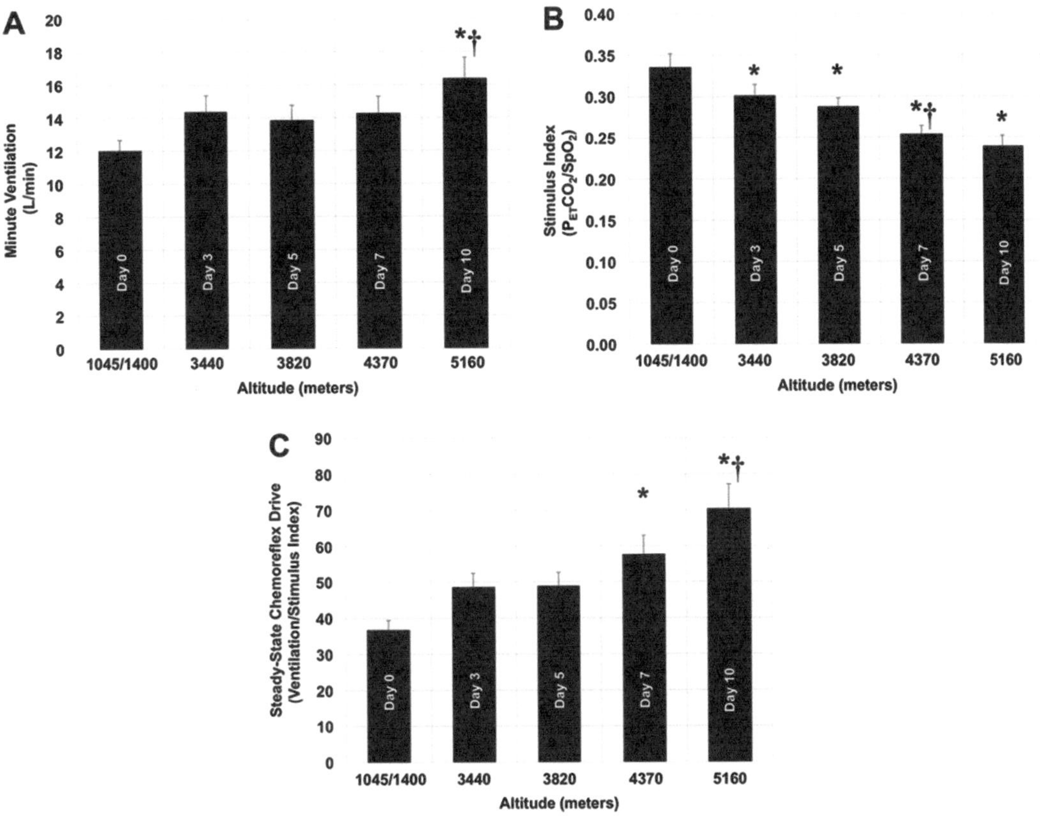

Fig. 2.3 **Steady-state chemoreflex drive during incremental ascent to high altitude**. These ventilatory (V_I) data in A. were derived from a one-min mean bin following 10-min of rest at each location, following at least one night at each respective altitude. The SI data in B. ($P_{ET}CO_2/SpO_2$) was calculated from data in Fig. 2.2. SS-CD in C. is calculated as V_I/SI at each location. The relative altitude and measurement day during ascent is labeled for each value. * denotes different than 1045/1400 m, $P < 0.001$. † denotes different than previous altitude. These data: 1045/1400 m, n = 13; 3440 m, n = 10; 3820, n = 11; 4370 m, n = 11; 5160 m, n = 10. Error bars represent mean ± SEM

ways with incremental ascent. Figure 2.2 illustrates the reductions in SpO_2 and $P_{ET}CO_2$ with ascent, illustrating hypoxia and concomitant hypocapnia due to the HVR. Following 10 days of incremental ascent, SpO_2 and $P_{ET}CO_2$ were 78.0 ± 2.0% and 18.4 ± 0.6 Torr at 5160 m, respectively, significantly lower than 1045/1400 m. Figure 2.3 illustrates resting V_I, SI and SS-CD during ascent to altitude. Following 10 days of incremental ascent, V_I, SI and SS-CD were 16.4 ± 1.3 L/min, 0.24 ± 0.01 and 70.4 ± 6.7 respectively, significantly different than 1045/1400 m. The mean SS-CD was approximately 96% larger (i.e. almost double) at 5160 m compared to 1045/1400 m, indicating appreciable ventilatory acclimatization in these participants.

2.4 Discussion

Following the development and validation of the SS-CD during exposure to normobaric hypoxia in a laboratory setting, we aimed to characterize the SS-CD during incremental and more prolonged exposure to high altitude. We found that the SS-CD index increased in magnitude with incremental ascent to altitude over 10 days to a maximum of 5160 m. On average, SS-CD increased 96% in this time frame, suggesting appreciable ventilatory acclimatization. This finding suggests that the SS-CD tracks the well-described ventilatory acclimatization during exposure to high altitude hypoxia.

The SS-CD captures the overall resting chemoreflex drive to prevailing chemostimui. Thus, this index includes contributions from both central and peripheral respiratory chemoreceptors and the prevailing O_2, CO_2 and acid-base status. Indeed, given the many considerations outlined above, the methodological and technical simplicity of the SS-CD illustrates its strength, and underlies its potential broad utility. Although the equipment we used here for measuring ventilation required a calibrated flow head, spirometer amplifier and a laptop with data acquisition hardware and software, we suggest that the same measures can be made using more portable respirometers (see Fig. 2.1), increasing the portability and utility in austere conditions.

There are many contexts where tests approximating the specific central and/or peripheral chemoreflex magnitude may be important, including mechanistic studies of cardiorespiratory control in animal models and humans. As with all experimental studies, the utility of any methodology lies in the experimental question. However, the novel SS-CD index outlined here may have future utility in assessing ventilatory acclimatization during incremental or prolonged stays at altitude, and may replace the use of complex and potentially confounded transient peak response tests of the HVR in some experimental or clinical contexts.

Acknowledgements This work was supported in part by a Natural Sciences and Engineering Research Council of Canada Discovery grant (TAD) and an MRU sabbatical (TAD). We would like to acknowledge and thank ADInstruments for their support of this project. We are grateful to our research participants for their time and effort.

References

Angelova PR, Kasymov V, Christie I, Sheikhbahaei S, Turovsky E, Marina N, Korsak A, Zwicker J, Teschemacher AG, Ackland GL, Funk GD, Kasparov S, Abramov AY, Gourine AV (2015) Functional oxygen sensitivity of astrocytes. J Neurosci 35(29):10460–10473

Berkenbosch A, Olievier CN, De Goede J (1995) Respiratory responses to hypoxia: peripheral and central effects. In: Semple SJG, Adams L, Whipp BJ (eds) Modeling and control of ventilation. Plenum Press, New York, pp 251–256

Bruce CD, Steinback CD, Chauhan UV, Pfoh JR, Abrosimova M, Vanden Berg ER, Skow RJ, Davenport MH, Day TA (2016) Quantifying cerebrovascular reactivity in anterior and posterior cerebral circulations during voluntary breath holding. Exp Physiol 101(12):1517–1527

Chua TP, Coats AJ (1995) The reproducibility and comparability of tests of the peripheral chemoreflex: comparing the transient hypoxic ventilatory drive test and the single-breath carbon dioxide response test in healthy subjects. Eur J Clin Investig 25:887–892

Curran AK, Rodman JR, Eastwood PR, Henderson KS, Dempsey JA, Smith CA (2000) Ventilatory responses to specific CNS hypoxia in sleeping dogs. J Appl Physiol (1985) 88(5):1840–1852

Day TA, Wilson RJA (2009) A negative interaction between central and peripheral respiratory chemoreceptors modulates peripheral chemoreflex magnitude. J Physiol 587(Pt 4):883–896

Dejours P (1962) Chemoreflexes in breathing. Physiol Rev 42:335–358

Dempsey JA, Powell FL, Bisgard GE, Blain GM, Poulin MJ, Smith CA (2014) Role of chemoreception in cardiorespiratory acclimatization to, and deacclimatization from, hypoxia. J Appl Physiol (1985) 116(7):858–866

Duffin J (2011) Measuring the respiratory chemoreflexes in humans. Respir Physiol Neurobiol 177(2):71–79

Edelman NH, Epstein PE, Lahiri S, Cherniack NS (1973) Ventilatory responses to transient hypoxia and hypercapnia in man. Respir Physiol 17:302–314

Eger EI 2nd, Kellogg RH, Mines AH, Lima-Ostos M, Morrill CG, Kent DW (1968) Influence of CO_2 on ventilatory acclimatization to altitude. J Appl Physiol 24(5):607–615

Fan JL, Burgess KR, Basnyat R, Thomas KN, Peebles KC, Lucas SJ, Lucas RA, Donnelly J, Cotter JD, Ainslie PN (2010) Influence of high altitude on cerebrovascular and ventilatory responsiveness to CO_2. J Physiol 588(Pt 3):539–549

Fitzgerald RS, Parks DC (1971) Effect of hypoxia on carotid chemoreceptor response to carbon dioxide in cats. Respir Physiol 12:218–229

Forster HV, Smith CA (2010) Contributions of central and peripheral chemoreceptors to the ventilatory response to CO_2/H^+. J Appl Physiol 108:989–994

Forster HV, Dempsey JA, Birnbaum ML, Reddan WG, Thoden J, Grover RF, Rankin J (1971) Effect of chronic exposure to hypoxia on ventilatory response to CO_2 and hypoxia. J Appl Physiol 31(4): 586–592

Gourine AV, Funk GD (2017) On the existence of a central respiratory oxygen sensor. J Appl Physiol (1985) 123(5):1344–1349

Guyenet PG, Bayliss DA (2015) Neural control of breathing and CO_2 homeostasis. Neuron 87(5):946–961

Hoiland RL, Ainslie PN, Wildfong KW, Smith KJ, Bain AR, Willie CK, Foster G, Monteleone B, Day TA (2015) Indomethacin-induced impairment of regional cerebrovascular reactivity: implications for respiratory control. J Physiol 593:1291–1306

Honda Y (1995) Ventilatory depression during mild hypoxia in adult humans. Jpn J Physiol 45:947–959

Krapf R, Beeler I, Hertner D, Hulter HN (1991) Chronic respiratory alkalosis. The effect of sustained hyperventilation on renal regulation of acid-base equilibrium. N Engl J Med 324(20):1394–1401

Lahiri S, DeLaney RG (1975) Stimulus interaction in the responses of carotid body chemoreceptor single afferent fibers. Respir Physiol 24:249–266

Loeschchke HH, Gertz KH (1958) Effect of oxygen pressure in inspired air on respiratory activity of the human, tested under the constant behavior of alveolar carbon dioxide pressure. Pflugers Arch Gesamte Physiol Menschen Tiere 267(5):460–477

Long WO, Giesbrecht GG, Anthonisen NR (1993) Ventilatory response to moderate hypoxia in awake chemodenervated cats. J Appl Physiol 74:805–810

López-Barneo J, González-Rodríguez P, Gao L, Fernández-Agüera MC, Pardal R, Ortega-Sáenz P (2016) Oxygen sensing by the carotid body: mechanisms and role in adaptation to hypoxia. Am J Physiol Cell Physiol 310(8):C629–C642

Milledge JS, Thomas PS, Beeley JM, English JS (1988) Hypoxic ventilatory response and acute mountain sickness. Eur Respir J 1(10):948–951

Nielsen M, Smith H (1952) Studies on the regulation of respiration in acute hypoxia; with a appendix on respiratory control during prolonged hypoxia. Acta Physiol Scand 24:293–313

Pedersen MEF, Fatemian M, Robbins PA (1999) Identification of fast and slow ventilatory responses to carbon dioxide under hypoxic and hyperoxic conditions in humans. J Physiol 521(1):273–287

Pfoh JR, Tymko MM, Abrosimova M, Boulet LM, Foster GE, Bain AR, Ainslie AN, Steinback CD, Bruce CD, Day TA (2016) Comparing and characterizing transient and steady-state tests of the peripheral chemoreflex in humans. Exp Physiol 101:432–447

Pfoh JR, Steinback CD, Vanden Berg ER, Bruce CD, Day TA (2017) Assessing chemoreflexes and oxygenation in the context of acute hypoxia: implications for field studies. Respir Physiol Neurobiol 246:67–75

Poulin MJ, Robbins PA (1998) Influence of cerebral blood flow on the ventilatory response to hypoxia in humans. Exp Physiol 83:95–106

Powell FL (2006) Lake Louise consensus methods for measuring the hypoxic ventilatory response. In: Roach RC et al (eds) Hypoxia and exercise. Springer, New York, pp 271–276

Powell FL (2012) Measuring the respiratory chemoreflexes in humans by J. Duffin. Respir Physiol Neurobiol 181(1):44–45

Powell FL, Milsom WK, Mitchell GS (1998) Time domains of the hypoxic ventilatory response. Respir Physiol 112:123–134

Rebuck AS, Campbell EJ (1974) A clinical method for assessing the ventilatory response to hypoxia. Am Rev Respir Dis 109(3):345–350

Robbins PA (1995) Hypoxic ventilatory decline: site of action (Commentary). J Appl Physiol 79:373–374

Sato M, Severinghaus JW, Powell FL, Xu FD, Spellman MJ Jr (1992) Augmented hypoxic ventilatory response in men at altitude. J Appl Physiol (1985) 73(1):101–107

Smith CA, Bisgard GE, Nielsen AM, Daristotle L, Kressin NA, Forster HV, Dempsey JA (1986) Carotid bodies are required for ventilatory acclimatization to chronic hypoxia. J Appl Physiol (1985) 60(3):1003–1010

Smith CA, Rodman JR, Chenuel BJ, Henderson KS, Dempsey JA (2006) Response time and sensitivity of the ventilatory response to CO_2 in unanesthetized intact dogs: central vs. peripheral chemoreceptors. J Appl Physiol 100:13–19

Smith CA, Forster HV, Blain GM, Dempsey JA (2010) An interdependent model of central/peripheral chemoreception: evidence and implications for ventilatory control. Respir Physiol Neurobiol 173:288–297

Steinback CD, Poulin MJ (2007) Ventilatory responses to isocapnic and poikilocapnic hypoxia in humans. Respir Physiol Neurobiol 155:104–113

Steinback CD, Poulin MJ (2008) Cardiovascular and cerebrovascular responses to acute isocapnic and poikilocapnic hypoxia in humans. J Appl Physiol 104:482–489

Steinback CD, Salzer D, Medeiros PJ, Kowalchuk J, Shoemaker K (2009) Hypercapnic vs. hypoxic control of cardiovascular, cardiovagal, and sympathetic function. Am J Physiol Regul Integr Comp Physiol 296:R402–R410

Teppema LJ, Dahan A (2010) The ventilatory response to hypoxia in mammals: mechanisms, measurement, and analysis. Physiol Rev 90(2):675–754

Weil JV, Byrne-Quinn E, Sodal IE, Friesen WO, Underhill B, Filley GF, Grover RF (1970) Hypoxic ventilatory drive in normal man. J Clin Invest 49:1061–1072

Wilson RJ, Day TA (2013) CrossTalk opposing view: peripheral and central chemoreceptors have hypoadditive effects on respiratory motor output. J Physiol 591(18):4355–4357

Wilson RJ, Teppema LJ (2016) Integration of central and peripheral respiratory chemoreflexes. Compr Physiol 6(2):1005–1041

Hypoxia Regulates MicroRNA Expression in the Human Carotid Body

3

Souren Mkrtchian, Kian Leong Lee, Jessica Kåhlin, Anette Ebberyd, Lorenz Poellinger, Malin Jonsson Fagerlund, and Lars I. Eriksson

Abstract

How hypoxia regulates gene expression in the human carotid body (CB) remains poorly understood. While limited information on transcriptional regulation in animal CBs is available, the impact of important post-transcriptional regulators, such as non-coding RNAs, and in particular miRNAs is not known. Here we show using *ex vivo* experiments that indeed a number of miRNAs are differentially regulated in surgically removed human CB slices when acute hypoxic conditions were applied. Analysis of the hypoxia-regulated miRNAs shows that they target biological pathways with upregulation of functions related to cell proliferation and immune response and downregulation of cell differentiation and cell death functions. Comparative analysis of the human CB miRNAome with the global miRNA expression patterns of a large number of different human tissues showed that the CB miRNAome had a unique profile which reflects its highly specialized functional status. Nevertheless, the human CB miRNAome is most closely related to the miRNA expression pattern of brain tissues indicating that they may have the most similar developmental origins.

Keywords

Carotid body · MicroRNA · Hypoxia

S. Mkrtchian (✉) · A. Ebberyd
Section for Anesthesiology and Intensive Care Medicine, Department of Physiology and Pharmacology, Karolinska Institute, Stockholm, Sweden
e-mail: souren.mkrtchian@ki.se

K. L. Lee
Cancer Science Institute of Singapore, National University of Singapore, Singapore, Singapore

J. Kåhlin · M. J. Fagerlund · L. I. Eriksson
Section for Anesthesiology and Intensive Care Medicine, Department of Physiology and Pharmacology, Karolinska Institute, Stockholm, Sweden

Function Perioperative Medicine and Intensive Care, Karolinska University Hospital, Stockholm, Sweden

L. Poellinger
Cancer Science Institute of Singapore, National University of Singapore, Singapore, Singapore

Department of Cell and Molecular Biology, Karolinska Institutet, Stockholm, Sweden

3.1 Introduction

The human carotid body (CB) oxygen sensing and signaling mechanisms are not well understood despite the existence of many and often conflicting hypotheses (Iturriaga and Alcayaga 2004; Lopez-Barneo et al. 2016). One of the contributing factors is the lack of human CB tissues

E. B. Gauda et al. (eds.), *Arterial Chemoreceptors*, Advances in Experimental Medicine and Biology 1071, https://doi.org/10.1007/978-3-319-91137-3_3

with notably one study (Ortega-Saenz et al. 2013) describing the morphology and chemosensory response of the organ to hypoxia. However, the contribution and targets of the transcriptional regulation driven by oxygen sensing in the human CB remain largely unexplored. A recently published map of the human CB transcriptome has demonstrated similarities but also significant differences in the expression of many important oxygen sensing CB genes as compared to animal CBs (Mkrtchian et al. 2012).

Among many unanswered questions in CB oxygen sensing, one of the most outstanding is the regulation of the expression of functionally important CB genes. In particular, very little is known about the post-transcriptional regulation by non-coding RNAs, especially microRNAs (miRNAs). These are short (20–22 nt) single-stranded RNA molecules that guide the RNA-induced silencing complex to complementary binding sites in the 3′-UTRs of messenger RNAs (mRNAs) thereby suppressing the translation of the corresponding protein (Wilczynska and Bushell 2015).

MicroRNA-mediated regulation is an important factor for cellular adaptation to a variety of environmental stresses including hypoxia. The hypoxia-inducible factor 1 and 2 alpha subunits HIF-1α and HIF-2α are the key transcription factors that are stabilized under hypoxic conditions and mediate the expression of many genes involved in metabolism, angiogenesis, cell cycle, growth and cell death in response to the lack of oxygen (Bristow and Hill 2008). It had also been demonstrated that hypoxia can regulate the expression of a distinct set of miRNAs that were termed hypoxamiRs (Kulshreshtha et al. 2007; Nallamshetty et al. 2013). HIF1-α was reported to activate a subset of these miRNAs and conversely, other hypoxamiRs were shown to regulate HIF-1 α expression (Loscalzo 2010; Bruning et al. 2011; Chen et al. 2016). Therefore, given the prominent role of the HIFs in the regulation of miRNAs, it was possible that miRNAs are also involved in the hypoxic response of the CB.

In the current study using a unique set of primary human CB specimens, we analyzed the CB miRNA expression under acute hypoxia relative to hyperoxia controls. Many of the hypoxia-regulated miRNAs were predicted to target mRNAs that lead to the upregulation of functions related to cell proliferation and immune response and downregulation of cell differentiation and cell death. Subsequently, we characterized the CB-specific pattern of miRNA expression in the control samples and identified the CB global miRNAome. Comparison of the CB global miRNA profiles with those from other human tissues revealed that while the CB was a highly distinct organ, it was most closely related to brain tissues.

3.2 Methods

3.2.1 Patient Carotid Body Collection

Experimental protocols were approved by the ethics committee on human research at the Karolinska Institutet in Stockholm, Sweden. Carotid bodies were collected from five male patients (age 42–80 years) with a BMI <32, scheduled for radical neck dissections not involving the CB. CBs were removed unilaterally under general anesthesia using sevoflurane and intravenous opioids with normoxic and normocarbic mechanical ventilation. Once surgically removed, the CBs were immediately placed in pre-oxygenated hyperoxic (95% O_2, 5% CO_2) ice-cold Krebs-Ringer Solution (KRS, 120 mM NaCl, 4 mM KCl, 2 mM $CaCl_2$, 1 mM $MgCl_2$, 21.4 mM $NaHCO_3$, 1.4 mM NaH_2PO_4, 10 mM glucose) and transported to the laboratory for further experiments within 15 min.

3.2.2 CB Sectioning and Treatments

After removal of connective tissues, CBs were sectioned along their longitudinal axis into 400 μm slices. Two to four CB slices from each of the 5 patients were used for hypoxia/hyperoxia treatments. Slices were distributed evenly into two 24-well plates containing 300 μl of pre-oxygenated KRS in each well. Pretreated hypoxic

KRS was added to one plate and moved into an H35 Hypoxia Station (Don Whitley Scientific) set at 10% O_2, 85% nitrogen and 5% CO_2 at 37 °C. The second control plate was fed with hyperoxic KRS and returned to the incubator with settings for hyperoxic conditions (95% O_2 and 5% CO_2 at 37 °C). Both plates were incubated for 1 h prior to harvesting for RNA extraction.

3.2.3 RNA Isolation and miRNA Analysis

Immediately after 1 h incubation, CB slices were immersed into 500 µl of Qiazol (Qiagen) and total RNA was extracted from the lysates using the miRNeasy Micro kit (Qiagen). RNA was reverse transcribed into cDNA using the miRCURY LNA Universal RT microRNA PCR, Polyadenylation and cDNA synthesis kit (Exiqon). Each microRNA was assayed by qPCR on the microRNA Ready-to-Use PCR Human panel I + II with the ExiLENT SYBR® Green master mix on a LightCycler® 480 Real-Time PCR System (Roche).

3.2.4 Data Analysis

Differential microRNA expression was identified on the basis of a fold change cutoff of >1.5 for the hypoxia treated carotid slices compared to the hyperoxia controls for each patient and with a t-test p-value <0.05. MicroRNA functional analysis was carried out using Ingenuity Pathway Analysis (IPA, Qiagen) and the mRNA targets of the microRNAs were identified using the microRNA Target Filter in IPA. Gene ontology and pathway analysis of these mRNAs for the determination of their cellular functions was subsequently carried out using IPA.

MicroRNA expression profiles of the human CBs (control hyperoxia samples) were compared with the miRNA profiles of various human tissues from two individuals (Human miRNA tissue atlas, (Ludwig et al. 2016). For direct comparison, we selected only those miRNAs from the tissue atlas that have corresponding matches with our identified CB miRNAs. This led to the isolation of 250 miRNAs that are overlapping between the human miRNA tissue atlas and the CB samples. The CB miRNA expression data were represented by the relative expression values $2^{\Delta Cq}$ where ΔCq = individual Cq – global mean Cq whereas the tissue atlas data were represented by the quantile normalized raw intensity values from the Agilent SurePrint Human miRNA array.

Furthermore, to harmonize the comparisons between these data sets, we used ranked expression values. Heatmap showing mean-centered sigma-normalized ranks for each sample using unsupervised hierarchical clustering and principal component analyses were generated using Qlucore Omics Explorer 3.2 (Qlucore).

3.3 Results

3.3.1 Hypoxia Regulates miRNA Expression in Primary Human CBs

CB slices were divided into two groups, one of which was kept under hypoxic, and another under control hyperoxic conditions for 1 h. The higher oxygen tension treatments at 95% O_2 were used to generate relatively well oxygenated normoxic controls as previous studies have shown that tissue pO_2 levels were less than half that of perfusates (Pepper et al. 1995). For the same reasons, incubations in 10% oxygen for hypoxic treatments are expected to generate oxygen tensions of less than 10% within the carotid slices that more closely reflect oxygenation levels in arterial blood during hypoxic challenge. The treated CBs were further processed for the isolation of RNA and analysis of miRNA expression was carried out using Exiqon qPCR miRNA panels that allow the simultaneous detection of 752 miRNAs. The number of miRNAs detected in each slice varied from 257 to 460 with 392 on average for each slice (Fig. 3.1), and 224 miRNAs were found to be expressed in all of the samples (data not shown).

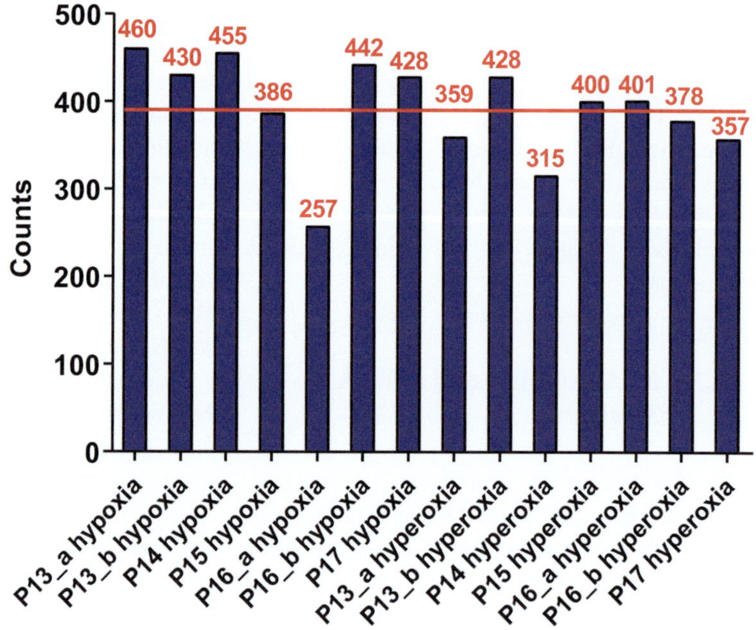

Fig. 3.1 Number of microRNAs detected in the human carotid body. Bar chart with numbers above each bar shows the number of miRNAs detected in each of 14 CB slices from five patients (P13–P17). Patients 13 and 16 are represented by two slices each, **a** and **b**. The red line shows the average count of 392 miRNAs per sample

Because of the possible contamination of CBs with blood, three miRNAs, miR-144, 142 and 451 were excluded from further analysis due to their known high expression in blood cells (Williams et al. 2013).

Thirty-nine miRNAs were identified to be differentially expressed in the hypoxia treated CB slices with a fold change of at least ±1.5 and t-test statistical significance of $p \leq 0.05$ (Fig. 3.2). Nevertheless, their physiological effects may be significant as each miRNA is capable of binding up to a few hundred target mRNAs bearing similar miRNA binding sites in their 3′-UTRs (Helwak et al. 2013). Indeed, using the miRNA Target Filter tool in IPA under high stringency conditions where only experimentally validated human targets are considered, this led to the identification of 550 possible mRNA targets (data not shown).

To address what were the biological functions regulated by these microRNA targeted mRNAs, the gene lists were subjected to gene ontology and pathway analysis using IPA. The IPA analysis of the upregulated mRNAs showed that the top most significantly upregulated functions are associated with cellular growth and proliferation, tissue development functions and cell cycle (Fig. 3.3).

Taken together, all of these analyses indicate that the cell proliferation function was predominantly driven by the mRNAs targeted by the hypoxia-induced microRNAs. In our previous study, we showed that human CB slices release pro-inflammatory cytokines under identical hypoxia conditions. In agreement with this, inflammatory response functions were also found to be among the significantly upregulated cellular processes (Fig. 3.3).

In further analysis of the functional significance of the mRNAs targeted by hypoxia-inducible miRNAs in the CB, we examined the upregulated miRNAs that would result in the downregulation of specific mRNA targets. Among various cellular processes inhibited by hypoxia via miRNA suppression, cell differentiation, and apoptosis were among the top significantly dysregulated functions as identified by IPA (Fig. 3.3). This suggested that the human CBs may be undergoing concerted hypoxic upregulation of cell proliferation at the expense of differentiation and the reduction of cell death.

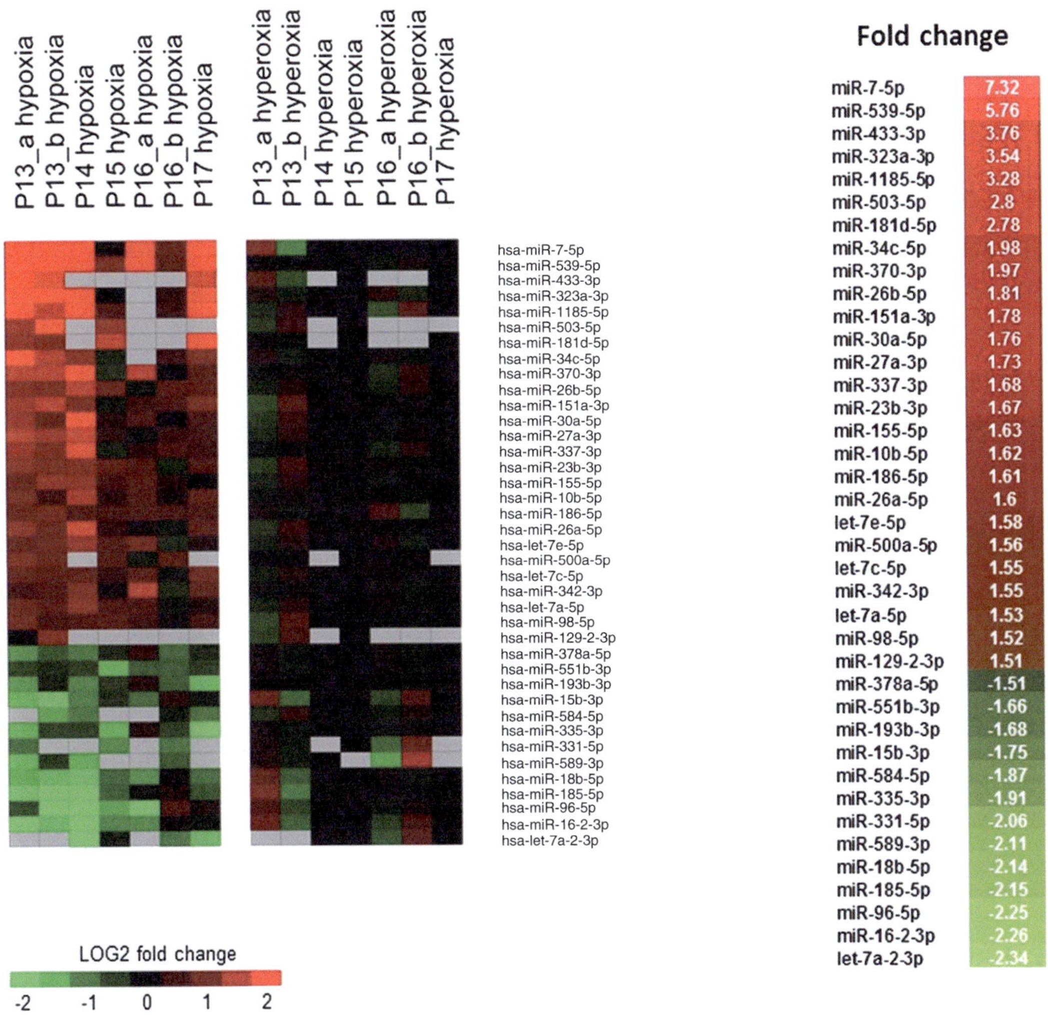

Fig. 3.2 Hypoxia regulates microRNA expression in human carotid bodies. Heatmap shows microRNAs that are differentially expressed by >1.5-fold and with t-test p value <0.05 in hypoxia treated CB slices compared to the average of the hyperoxia treated control slices (left panel). Color bar shows fold change on a Log2 scale in red for upregulation and green for downregulation. MicroRNAs with no data on their expression are indicated in gray. Average fold-change in hypoxia treated CB slices compared to the average of the hyperoxia treated control slices are shown in the right panel. The differentially expressed miRNAs are ranked by fold change in descending order. P13-P17, individual CB slices from five different patients

3.3.2 The miRNAome of the Human Carotid Body Is Highly Distinct from Other Human Tissues

The miRNA expression profile of human CBs remains poorly understood, yet this is important for the understanding of the underlying functions of the CB itself. To address what the CB expressed microRNAs may be involved with in terms of CB biology, we used the control CB slices incubated under hyperoxia that approximate for well oxygenated normoxic conditions and compared their miRNA profiles with those of different tissues. For this purpose, we used the human miRNA tissue atlas that consists of microarray-based miRNA expression from 61 tissue biopsies of two individuals (Ludwig et al. 2016).

We have examined the relation between the miRNA signature of CBs and diverse body tissues

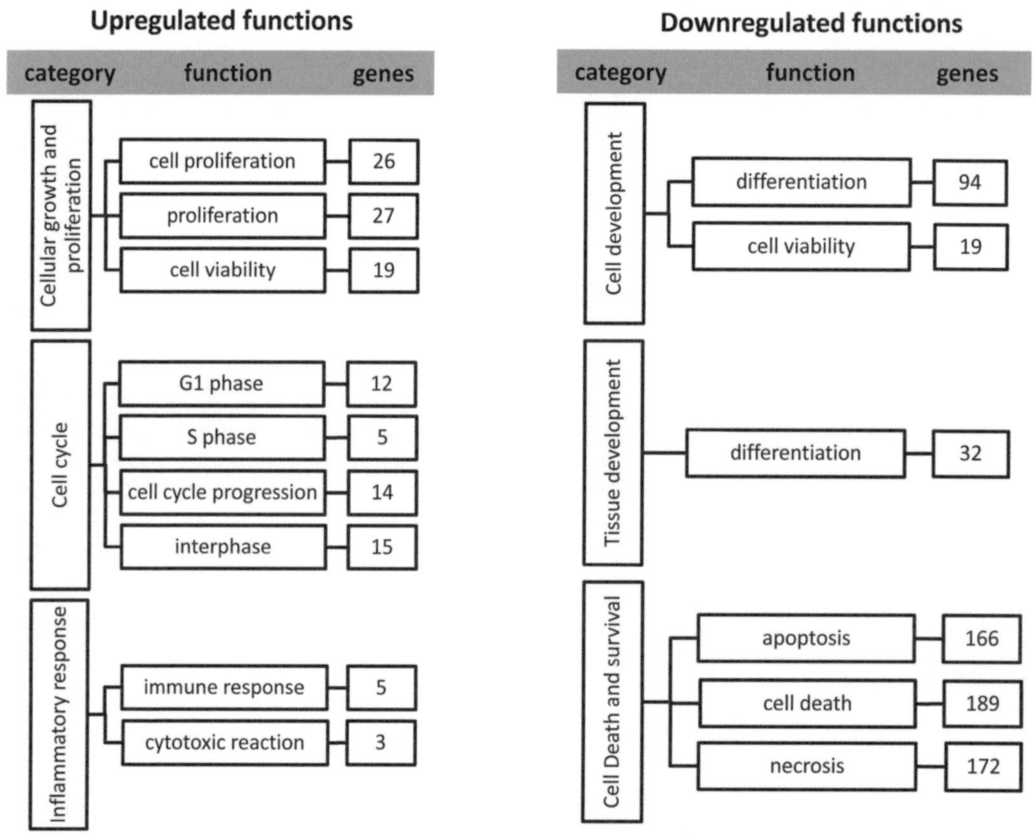

Fig. 3.3 Hypoxia regulated microRNAs target mRNAs that are involved in diverse cellular functions. Messenger RNA targets of the hypoxia regulated microRNAs were identified using the Ingenuity pathway (IPA) MicroRNA Target Filter and further subjected to the IPA core analysis for identification of gene ontology and biological pathways. Top ranking biological and cellular functions with Fisher's Exact Test $p < 0.05$ are shown in the dendrograms with the number of microRNA target genes predicted to be upregulated (left panel) and downregulated (right panel) in each functional category indicated

using principal component analysis (Fig. 3.4, right panel). We found all CB samples grouping closely together suggesting their miRNA profiles were consistent and similar to each other and there was comparatively little variation in the CB samples from different patients themselves. However, the grouping of CB samples was highly distinct with clear separation from other tissues. To further examine the relationship of the CBs with respect to the other body tissues, we next carried out hierarchical clustering analysis. The heatmap showing the clustering and miRNA expression patterns confirm the distinctiveness of the CBs from the other tissues (Fig. 3.4, left panel). Interestingly, despite the large differences between the CBs and other tissues, the CB microRNA signature appears to be most closely linked to the brain tissues supporting that they may have the most similar developmental origin (Hempleman and Warburton 2013).

3.4 Discussion

Cellular adaptation to hypoxia involves a complex interplay of transcriptional, post-transcriptional and post-translational events aimed at maintaining cellular homeostasis. In addition to the known regulatory mechanisms, another layer of regulation represented by non-coding RNAs including miRNAs, appears to play an important role in the hypoxic response. This

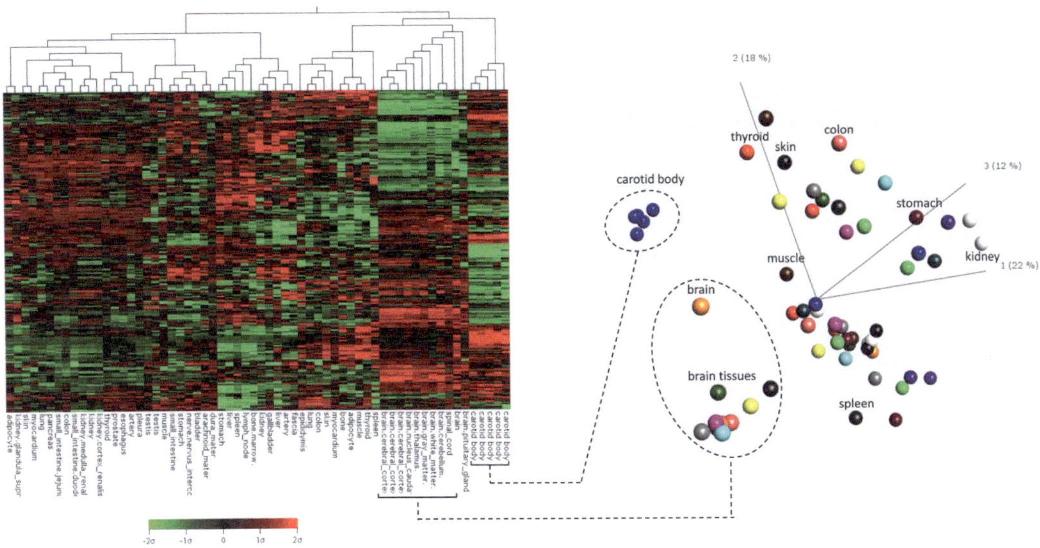

Fig. 3.4 The human carotid body miRNAome is most closely related to those of brain tissues. Heatmap (left panel) shows mean-centered, sigma-normalized ranks of miRNA expression data of individual patient CB samples under hyperoxia and human tissue expression data from the Human miRNA tissue atlas. Normalized gene expression data is shown in red for upregulation and green for downregulation over the mean. The right panel shows principle component analysis of the human carotid bodies and human tissue miRNA expression data (see Materials and Methods for details)

study is the first attempt to decipher the miRNA response under hypoxic conditions and also to map the miRNAome of primary human CBs.

Comparison of the differentially expressed CB miRNAs under hypoxic conditions with hypoxamiRs identified in other studies shows only small overlaps. The upregulation of miR-27a-3p, 26a, 7, 98, 155, let-7e and downregulation of let-7a-2-3p, miR-15b, 18a are often included in the miRNA signature of hypoxia (Kulshreshtha et al. 2007; Nallamshetty et al. 2013), and we found this was also the case for human CBs. However, other prominent constituents of the hypoxia miRNA signature such as miR-210 that is targeted by HIF-1α (Chan and Loscalzo 2010) was not affected in our CB model of hypoxia.

HIF-1α is one of the key downstream effectors of the oxygen response in cells that mediates their adaptation to hypoxic stress (Poellinger and Johnson 2004). Its expression is regulated by miRNAs amongst other factors. In particular, miR-18b is predicted to bind to the 3′-UTR of HIF-1α and thus inhibit *HIF1A* mRNA translation (Chen et al. 2016). Incidentally, under our experimental conditions miR-18b is downregulated by hypoxia, which may hypothetically lead to an increase in HIF-1α levels, thereby completing a positive feedback loop. It remains to be investigated whether regulation of HIF-1α level and activity does indeed occur in the human carotid body under hypoxia.

Along with the downregulation of miR-18b, hypoxia also induces the expression of another HIF-1α negative regulator miR-155. As HIF-1α itself is known to control the expression of miR-155, this mechanism may in principle represent a negative feedback loop to dampen and prevent runaway activation of hypoxia signaling (Bruning et al. 2011). Interestingly, miR-155 is also implicated in the activation of pro-inflammatory agents such as TNFα and interleukin 6 (IL-6) and in general the activation of innate immunity as well as positive regulation of antigen presentation (Rodriguez et al. 2007; Tili et al. 2007). Therefore, the upregulation of miR-155 may be one of the factors involved in the hypoxia-mediated release of pro-inflammatory cytokines including TNFα and IL-6 from the identical set of human CB slices characterized in our previous report (Kahlin et al. 2014).

It has been shown that enhanced HIF-1α signaling in the CB promotes hyperplasia (Bishop et al. 2013; Hodson et al. 2016). The analysis of biological processes potentially targeted by the differentially expressed miRNAs under hypoxia indicates upregulation of proliferative pathways with concomitant downregulation of differentiation and cell death related processes (Fig. 3.3). It is, therefore, possible that the phenomenon of hypoxic CB hyperplasia (Saldana et al. 1973; Kay and Laidler 1977) may be in part attributed to hypoxia-driven miRNA regulation.

This is the first study to demonstrate the global expression of miRNAs in the human carotid body where no data is presently available including for animal CBs. Similar to the human CB transcriptome, the CB miRNAome appears to have a unique profile as compared to other body tissues, which reflects its highly specialized functional status. Similar to the CB transcriptome (Mkrtchian et al. 2012), the human CB miRNAome shares to a certain extent the miRNA expression patterns of brain tissues. One of the highly expressed miRNAs in both the CB and the brain is miR-29b, which has been shown to be important for neuronal maturation (Kole et al. 2011). A group of brain-specific let-7 miRNAs and also miR-7 that are important for neuronal differentiation are also highly expressed in CBs (Nowak and Michlewski 2013; Zhu et al. 2016). Therefore, not only do these microRNAs demarcate the neural origin of the CBs, they may also direct the primary function of the CB in neural signaling.

Acknowledgements Supported by research grants from the Research Council for Medicine, Sweden, Stockholm County Council, Thorsten Söderberg Research Foundation, Gösta Fraenckels Foundation, Jeanssons Foundation, Tore Nilsons Fundation, Magnus Bergvalls Foundation, Capio Foundation, LPS Medical, Karolinska Institutet Funds and The Swedish Society for Medicine, all from Stockholm, Sweden.

References

Bishop T, Talbot NP, Turner PJ, Nicholls LG, Pascual A, Hodson EJ, Douglas G, Fielding JW, Smith TG, Demetriades M, Schofield CJ, Robbins PA, Pugh CW, Buckler KJ, Ratcliffe PJ (2013) Carotid body hyperplasia and enhanced ventilatory responses to hypoxia in mice with heterozygous deficiency of PHD2. J Physiol 591:3565–3577

Bristow RG, Hill RP (2008) Hypoxia and metabolism. Hypoxia, DNA repair and genetic instability. Nat Rev Cancer 8:180–192

Bruning U, Cerone L, Neufeld Z, Fitzpatrick SF, Cheong A, Scholz CC, Simpson DA, Leonard MO, Tambuwala MM, Cummins EP, Taylor CT (2011) MicroRNA-155 promotes resolution of hypoxia-inducible factor 1alpha activity during prolonged hypoxia. Mol Cell Biol 31:4087–4096

Chan SY, Loscalzo J (2010) MicroRNA-210: a unique and pleiotropic hypoxamir. Cell Cycle 9:1072–1083

Chen Y, Zhang Z, Luo C, Chen Z, Zhou J (2016) MicroRNA-18b inhibits the growth of malignant melanoma via inhibition of HIF-1alpha-mediated glycolysis. Oncol Rep 36:471–479

Helwak A, Kudla G, Dudnakova T, Tollervey D (2013) Mapping the human miRNA interactome by CLASH reveals frequent noncanonical binding. Cell 153:654–665

Hempleman SC, Warburton SJ (2013) Comparative embryology of the carotid body. Respir Physiol Neurobiol 185:3–8

Hodson EJ, Nicholls LG, Turner PJ, Llyr R, Fielding JW, Douglas G, Ratnayaka I, Robbins PA, Pugh CW, Buckler KJ, Ratcliffe PJ, Bishop T (2016) Regulation of ventilatory sensitivity and carotid body proliferation in hypoxia by the PHD2/HIF-2 pathway. J Physiol 594:1179–1195

Iturriaga R, Alcayaga J (2004) Neurotransmission in the carotid body: transmitters and modulators between glomus cells and petrosal ganglion nerve terminals. Brain Res Brain Res Rev 47:46–53

Kahlin J, Mkrtchian S, Ebberyd A, Hammarstedt-Nordenvall L, Nordlander B, Yoshitake T, Kehr J, Prabhakar N, Poellinger L, Fagerlund MJ, Eriksson LI (2014) The human carotid body releases acetylcholine, ATP and cytokines during hypoxia. Exp Physiol 99:1089–1098

Kay JM, Laidler P (1977) Hypoxia and the carotid body. J Clin Pathol Suppl (R Coll Pathol) 11:30–44

Kole AJ, Swahari V, Hammond SM, Deshmukh M (2011) miR-29b is activated during neuronal maturation and targets BH3-only genes to restrict apoptosis. Genes Dev 25:125–130

Kulshreshtha R, Ferracin M, Wojcik SE, Garzon R, Alder H, Agosto-Perez FJ, Davuluri R, Liu CG, Croce CM, Negrini M, Calin GA, Ivan M (2007) A microRNA signature of hypoxia. Mol Cell Biol 27:1859–1867

Lopez-Barneo J, Gonzalez-Rodriguez P, Gao L, Fernandez-Aguera MC, Pardal R, Ortega-Saenz P (2016) Oxygen sensing by the carotid body: mechanisms and role in adaptation to hypoxia. Am J Physiol Cell Physiol 310:C629–C642

Loscalzo J (2010) The cellular response to hypoxia: tuning the system with microRNAs. J Clin Invest 120:3815–3817

Ludwig N, Leidinger P, Becker K, Backes C, Fehlmann T, Pallasch C, Rheinheimer S, Meder B, Stahler C, Meese E, Keller A (2016) Distribution of miRNA expression across human tissues. Nucleic Acids Res 44:3865–3877

Mkrtchian S, Kahlin J, Ebberyd A, Gonzalez C, Sanchez D, Balbir A, Kostuk EW, Shirahata M, Fagerlund MJ, Eriksson LI (2012) The human carotid body transcriptome with focus on oxygen sensing and inflammation--a comparative analysis. J Physiol 590:3807–3819

Nallamshetty S, Chan SY, Loscalzo J (2013) Hypoxia: a master regulator of microRNA biogenesis and activity. Free Radic Biol Med 64:20–30

Nowak JS, Michlewski G (2013) miRNAs in development and pathogenesis of the nervous system. Biochem Soc Trans 41:815–820

Ortega-Saenz P, Pardal R, Levitsky K, Villadiego J, Munoz-Manchado AB, Duran R, Bonilla-Henao V, Arias-Mayenco I, Sobrino V, Ordonez A, Oliver M, Toledo-Aral JJ, Lopez-Barneo J (2013) Cellular properties and chemosensory responses of the human carotid body. J Physiol 591:6157–6173

Pepper DR, Landauer RC, Kumar P (1995) Postnatal development of CO2-O2 interaction in the rat carotid body in vitro. J Physiol 485(Pt 2):531–541

Poellinger L, Johnson RS (2004) HIF-1 and hypoxic response: the plot thickens. Curr Opin Genet Dev 14:81–85

Rodriguez A, Vigorito E, Clare S, Warren MV, Couttet P, Soond DR, van Dongen S, Grocock RJ, Das PP, Miska EA, Vetrie D, Okkenhaug K, Enright AJ, Dougan G, Turner M, Bradley A (2007) Requirement of bic/microRNA-155 for normal immune function. Science 316:608–611

Saldana MJ, Salem LE, Travezan R (1973) High altitude hypoxia and chemodectomas. Hum Pathol 4:251–263

Tili E, Michaille JJ, Cimino A, Costinean S, Dumitru CD, Adair B, Fabbri M, Alder H, Liu CG, Calin GA, Croce CM (2007) Modulation of miR-155 and miR-125b levels following lipopolysaccharide/TNF-alpha stimulation and their possible roles in regulating the response to endotoxin shock. J Immunol 179:5082–5089

Wilczynska A, Bushell M (2015) The complexity of miRNA-mediated repression. Cell Death Differ 22:22–33

Williams Z, Ben-Dov IZ, Elias R, Mihailovic A, Brown M, Rosenwaks Z, Tuschl T (2013) Comprehensive profiling of circulating microRNA via small RNA sequencing of cDNA libraries reveals biomarker potential and limitations. Proc Natl Acad Sci U S A 110:4255–4260

Zhu C, Zhou R, Zhou Q, Chang Y, Jiang M (2016) microRNA-539 suppresses tumor growth and tumorigenesis and overcomes arsenic trioxide resistance in hepatocellular carcinoma. Life Sci 166:34–40

TASK-1 ($K_{2P}3$) and TASK-3 ($K_{2P}9$) in Rabbit Carotid Body Glomus Cells

4

Dawon Kang, Jiaju Wang, James O. Hogan, and Donghee Kim

Abstract

Glomus cells isolated from rabbit and rat/mouse carotid bodies have been used for many years to study the role of ion channels in hypoxia sensing. Studies show that hypoxia inhibits the inactivating K^+ channels (Kv4) in rabbits, but inhibits TASK in rats/mice to elicit the hypoxic response. Because the role of TASK in rabbit glomus cells is not known, we isolated glomus cells from rabbits and studied the expression of TASK mRNA in the whole carotid body (CB), changes in $[Ca^{2+}]_i$ and TASK activity. RT-PCR showed that rabbit CB expressed mRNA for TASK-3 and several Kv (Kv2.1, Kv3.1 and Kv3.3). In rabbit glomus cells in which 20 mM KCl_o elevated $[Ca^{2+}]$, anoxia also elicited a strong rise in $[Ca^{2+}]$. In cell-attached patches with 140 mM KCl in the pipette, basal openings of ion channels with single-channel conductance levels of 16-pS, 34-pS, and 42-pS were present. TREK-like channels were also observed. In inside-out patches with high $[Ca^{2+}]_i$, BK was activated. The 42-pS channel opened spontaneously and briefly. The 16-pS and 34-pS channels showed properties similar to those of TASK-1 and TASK-3, respectively. TASK activity in cell-attached patches was lower than that in rat glomus cells under identical recording conditions. Hypoxia (~0.5%O_2) reduced TASK activity by ~52% and depolarized the cells by ~30 mV. Our results show that the O_2-sensitive TASK contributes to the hypoxic response in rabbit glomus cells.

Keywords

Rabbit · Carotid body · Hypoxia · Ca^{2+} · TASK · Kv

4.1 Introduction

Early studies using rabbit carotid body (CB) glomus cells have shown that hypoxia inhibits the voltage-dependent outward K^+ current, which then produces the hypoxic response such as cell depolarization, elevation of $[Ca^{2+}]_i$ and transmitter secretion (Biscoe et al. 1989; Chen et al. 2000; Delpiano and Hescheler 1989; Lopez-Barneo et al. 1988; Lopez-Lopez et al. 1989; Montoro et al. 1996). In rabbit glomus cells, the inactivating K^+ channels (Kv4) have been reported to be

D. Kang (✉)
Department of Physiology and Institute of Health Sciences, College of Medicine, Gyeongsang National University, Jinju, South Korea
e-mail: dawon@gnu.ac.kr

J. Wang · J. O. Hogan · D. Kim
Department of Physiology and Biophysics, Chicago Medical School, Rosalind Franklin University of Medicine and Science, North Chicago, IL, USA
e-mail: jiaju.wang@my.rfums.org; james.hogan@rosalindfranklin.edu; donghee.kim@rosalindfranklin.edu

E. B. Gauda et al. (eds.), *Arterial Chemoreceptors*, Advances in Experimental Medicine and Biology 1071, https://doi.org/10.1007/978-3-319-91137-3_4

the target of hypoxia (Perez-Garcia et al. 2000; Sanchez et al. 2002). In rat and mouse glomus cells, hypoxia inhibits TASK-1/3 ($K_{2P}3/9$), the background K^+ channels that are active at rest, to produce the hypoxic response (Buckler 2015; Buckler et al. 2000; Kim et al. 2009). In the human carotid body, O_2-sensitive ion channels have not yet been recorded, but mRNA transcripts of K^+ channels including TASK-1, BK and different Kv subtypes are expressed (Fagerlund et al. 2010; Mkrtchian et al. 2012). Because TASK is expressed in the carotid body from different species (rat, mouse, human), we asked if TASK is also present in the rabbit CB glomus cells and contributes to the hypoxic response.

To determine the expression and function of TASK in rabbit glomus cells, we recorded and characterized ion channel currents that were basally active in cell-attached patches. We also studied the expression of TASK-3 mRNA and hypoxia-induced changes in $[Ca^{2+}]_i$. The level of TASK activity in rabbit glomus cells was low compared to those in rat and mouse glomus cells. However, TASK inhibition by hypoxia in rabbit glomus cells was as strong as those observed in rat and mouse glomus cells. Our study shows that TASK-1 and TASK-3 (or TASK-1/3) provide part of the background K^+ conductance, and contribute to the hypoxic response in rabbit glomus cells.

4.2 Methods

4.2.1 Cell Isolation

The Animal Care and Use Committee of Rosalind Franklin University approved the use of animals. Rabbits (8–12 days old; oryctolagus cuniculus; European rabbit) were fully anesthetized with sodium pentabarbital (60 mg/kg, i.p.) and decapitated. Carotid artery bifurcation was dissected and placed in ice-cold, low Ca^{2+}, low Mg^{2+} phosphate buffered saline (low Ca^{2+}/Mg^{2+} PBS: 137 mM NaCl, 2.8 mM KCl, 2 mM KH_2PO_4, 0.07 mM $CaCl_2$, 0.05 mM $MgCl_2$, pH 7.4). CB was dissected out and digested in a solution containing trypsin (0.4 mg/ml) and collagenase (0.4 mg/ml) in low Ca^{2+}/Mg^{2+} PBS for 25 min at 37 °C. Following gentle trituration, the dispersed cells were separated by centrifugation (3000 rpm for 5 min) and suspended in CB growth medium (Ham's F-12, 10% fetal bovine serum, 23 mM glucose, 4 mM Glutamax-1 (L-alanyl glutamine), 10 kU penicillin, 10 kU streptomycin, and 300 μg/ml insulin). Cells were plated on poly-L-lysine-pretreated glass coverslips, incubated at 37 °C in a humidified atmosphere of 95% air/5% CO_2 and used within 6 hr. after plating.

4.2.2 Reverse Transcriptase (RT)-Polymerase Chain Reaction (PCR)

First-strand cDNAs were synthesized from total RNA isolated from rabbit carotid body using oligo dT (SuperScript™ First-Strand Synthesis System; Invitrogen, Carlsbad, CA) and used as template for PCR amplification using Taq polymerase. The PCR conditions were as follows: initial denaturation at 94 °C for 5 min; 30 cycles at 94 °C for 30 sec, 55 °C for 30 sec, and 72 °C for 30 sec; and a final extension step at 72 °C for 10 min. The products were electrophoresed on a 1.5% (w/v) agarose gel to verify the product size. The DNA fragments were directly sequenced with the ABI PRISM® 3100-Avant Genetic Analyzer (Applied Biosystems). Primer sequences (forward and reverse) are as follows: TASK-3 (tactgcttcatcacgctcacc and cctgtacgagatggactgcaa), Kv2.1 (ggatgaggacgacaccaagtt and aaattctccccgttcttctcc), Kv2.2 (agaagagcaccaaagagccag and agcaaagagtttctgcttggg), Kv3.1 (ggtgtggttcaccttcgagtt and gcctatcctctcggcgtagta) and Kv3.3 (gaagctgcccaagaagaagaa and ctcaatcacctcctcctgagc).

4.2.3 Electrophysiology

Ion channels were recorded using Axopatch 200B amplifier. Channel current was filtered at 2 kHz using an eight-pole Bessel filter (−3 dB) and transferred to a computer using the Digidata

1320 interface at a sampling rate of 20 kHz (Kang et al. 2014). Single-channel currents were analyzed with the pCLAMP software (versions 10). Channel openings were analyzed to obtain channel activity (NPo; N = number of channels in the patch, P_o = open probability). In experiments using cell-attached and inside-out patches, pipette solution contained (mM): 140 KCl, 1 $MgCl_2$, 5 EGTA, 10 HEPES and 11 glucose (pH 7.3). Extracellular bath (physiological) solution contained (mM): 117 NaCl, 5 KCl, 23 $NaHCO_3$, 1 $MgCl_2$, 1 $CaCl_2$, 10 HEPES and 11 glucose (adjusted to pH 7.3 with NaOH).

4.2.4 $[Ca^{2+}]_i$ Measurement

Isolated cells plated on glass coverslips were incubated with 2 μM fura-2 acetoxymethyl ester (fura-2 AM) for 30 min at 37 °C in culture medium. A coverslip with attached cells was placed in a recording chamber positioned on the stage of an inverted microscope (IX71; Olympus). Fura-2 was alternately excited at 340 and 380 nm, and the emitted fluorescence was filtered at 510 nm and recorded using a CCD (charge-coupled device)-based imaging system running SimplePCI software (Hamamatsu Corp.). The cells in the recording chamber were continuously perfused with a solution containing (mM): 117 NaCl, 5 KCl, 23 $NaHCO_3$, 1 $MgCl_2$, 1 $CaCl_2$, 10 HEPES and 11 glucose (pH 7.3). Ratiometric data were calibrated by applying experimentally determined constants to the equation: $[Ca^{2+}] = K_d \times \beta \times (R-R_{min})/(R_{max}-R)$ (Grynkiewicz et al. 1985). Values for Rmax (12.0), Rmin (0.1) and β (11.6) were determined *in vitro*, and a Kd value of 300 nM for fura-2 was assumed.

4.2.5 Hypoxia Studies

In our experiments that used a recording chamber that was open to the atmosphere, we generated two levels of hypoxic solutions (~0.5% and ~1.6% O_2). These levels of hypoxia were achieved by bubbling solutions with $0\%O_2/5\%CO_2/95\%N_2$ with 2 units/ml of glucose oxidase/100 units/ml of catalase, and $0\%O_2/5\%CO_2/95\%$ N_2 without enzymes. After a steady-state basal $[Ca^{2+}]_i$ was obtained with normoxic solution bubbled with $21\%O_2/5\%CO_2/74\%N_2$, the perfusion solution was switched to a desired hypoxic solution. The temperature of the perfusion solutions was ~35 °C.

4.2.6 Statistical Analysis

Student's t-test (for comparison of two sets of data) was used for data analysis using PRISM software. $p < 0.05$ was considered significant.

4.3 Results

4.3.1 Expression of TASK3 mRNA in Rabbit CB

We first assessed the expression of TASK-3 mRNA by RT-PCR using primer sequences designed to detect rabbit TASK-3 (GenBank Acc: XM002710554). TASK-3 primers amplified the expected region of the TASK-3 cDNA sequence, as shown by the 408 base pair band in Fig. 4.1 (first lane). Rabbit TASK-1 sequence was not available and therefore we did not test the expression of TASK-1 mRNA. Using primers specific for Kv channels, we also identified the expression of a number of Kv subtypes such as Kv2.1, Kv2.2, Kv3.1 and Kv3.2 (Fig. 4.1: top bands).

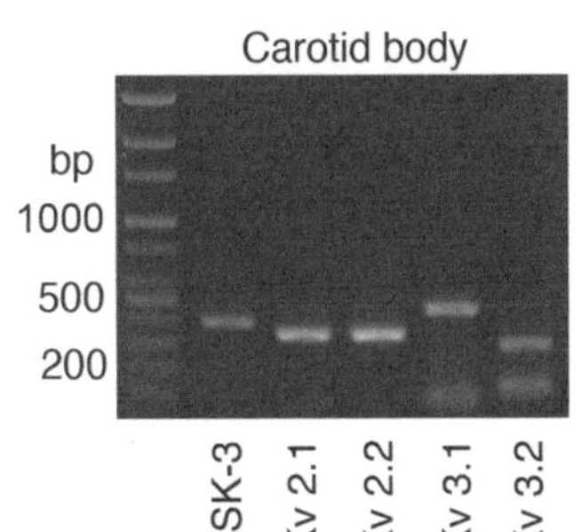

Fig. 4.1 Expression of TASK-3 and Kv in rabbit carotid body (adult male). The expected size of the PCR products for TASK-3, Kv2.1, Kv2.2, Kv3.1 and Kv3.2 are 408 bp, 354 bp, 352 bp, 466 bp, and 307 bp, respectively

Because whole CBs that consist of many different types of cells were used to generate cDNA, these K^+ channels could be expressed in glomus cells or non-glomus cells. Nevertheless, our results show that TASK3 mRNA is expressed in the rabbit CB. Expression of TASK-1 mRNA was not determined, as rabbit TASK-1 sequence was not available.

4.3.2 Intracellular Ca^{2+} Measurement

To be sure that cells that we isolated from rabbit CB contain O_2-sensitive chemoreceptor cells, we studied changes in $[Ca^{2+}]_i$ in response to high extracellular [KCl] and hypoxia. Fura-2-loaded cells attached to a glass coverslip were perfused with bicarbonate-buffered physiological solution. Under normoxic condition, the mean basal $[Ca^{2+}]_i$ was ~200 nM in this particular group of cells. Brief application of 20 mM KCl elicited a rapid reversible increase in $[Ca^{2+}]_i$ to ~500 nM (Fig. 4.2). Hypoxia (~1.6% O_2) caused a small but visible elevation of $[Ca^{2+}]_i$ whereas anoxia (~0.5% O_2) caused a strong elevation to a level similar to that produced by 20 mM KCl. These effects of hypoxia and anoxia showed that cells isolated from rabbit CB were indeed O_2-sensitive glomus cells. These types of cells were used subsequently for ion channel recording.

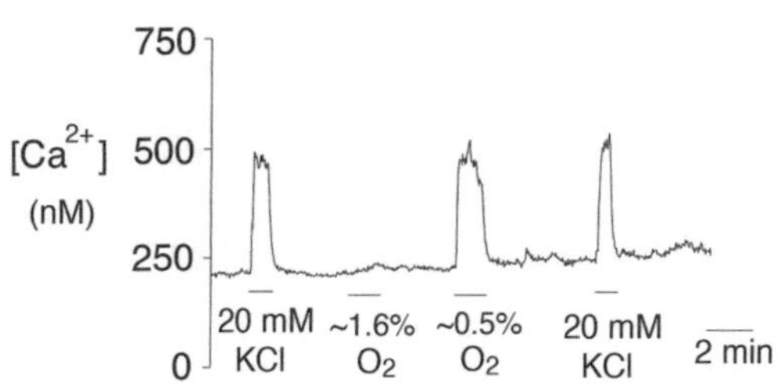

Fig. 4.2 Elevation of intracellular $[Ca^{2+}]$ by hypoxia in rabbit glomus cells. Averaged $[Ca^{2+}]$ from one coverslip of cells is shown

4.3.3 Expression of TASK-Like Channels

To help identify TASK-like channels, cell-attached patches were formed with 140 mM KCl in the pipette and perfused with bath solution containing 5 mM KCl with the pipette potential initially held at 0 mV. This experimental condition allows TASK-1 and TASK-3 to be detected if they are present in the patch membrane. Under this steady state condition, we observed openings of three channels in the inward current direction with single conductance levels of 16-pS, 34-pS and 42-pS (Fig. 4.3). The 16-pS channel showed a mildly inwardly rectifying current-voltage relationship with the current reversing at pipette potential of ~ −60 mV (Fig. 4.3a). The channel openings were brief, and the duration histogram showed that the mean open time was 0.6 ± 0.1 ms ($n = 3$). The kinetic properties of this channel were nearly identical to those of TASK-1 (Kim et al. 2009). The 34-pS channel also exhibited mildly inwardly rectifying current-voltage relationship and the current reversed at pipette potential of ~−60 mV (Fig. 4.3b). The channel openings were also brief in duration, with a mean open time of ~0.7 ± 0.1 ms ($n = 3$). The 34-pS channel was also basally active and not observed when KCl in the pipette was replaced with NaCl. The kinetic properties of the 34-pS channel are nearly identical to those of TASK-3 or TASK-/3 heteromer (Kang et al. 2004). These findings show that TASK-1 and TASK-3-like channels are basally active, and thus provide part of the background K^+ conductance in rabbit glomus cells.

The 42-pS channel openings in two different cell-attached patches are shown in Fig. 4.3c. This channel opened in a random fashion, showed long burst duration and many closings within each burst. We did not test its voltage-dependence, but this channel showed single-channel properties similar to the O_2-sensitive Kv that was previously described in rabbit glomus cells (Ganfornina and Lopez-Barneo 1991). We also observed opening of TREK-like channels in two patches, indicating that glomus cells from 1–2 week old rabbits express a TREK (Fig. 4.3d). As expected, applying negative pressure to the patch membrane

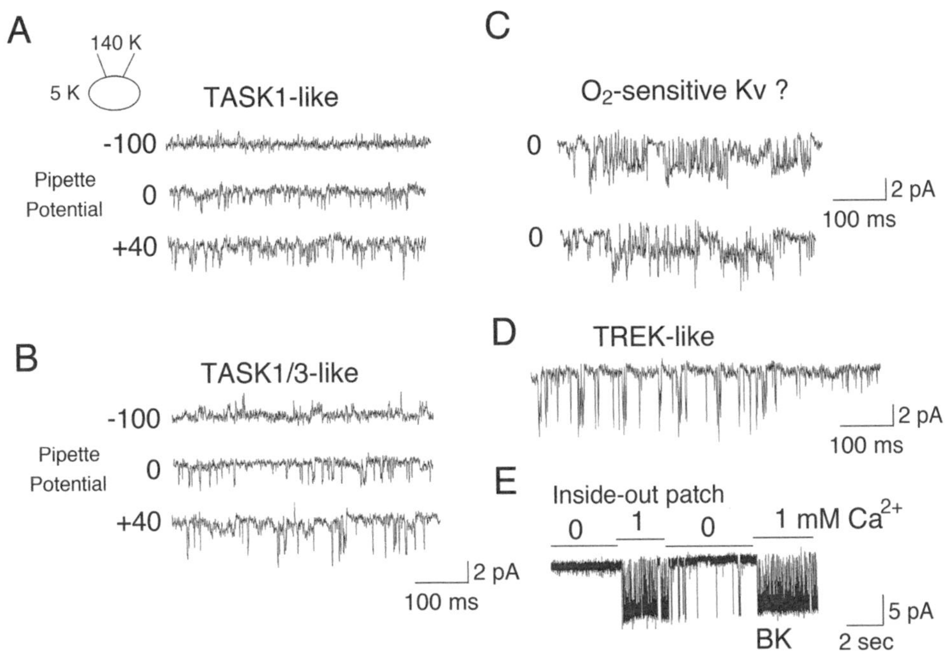

Fig. 4.3 Ion channels observed in cell-attached patches in the resting state (pipette potential is 0 mV) in rabbit glomus cells perfused with physiological solution. BK was observed only in inside-out patches

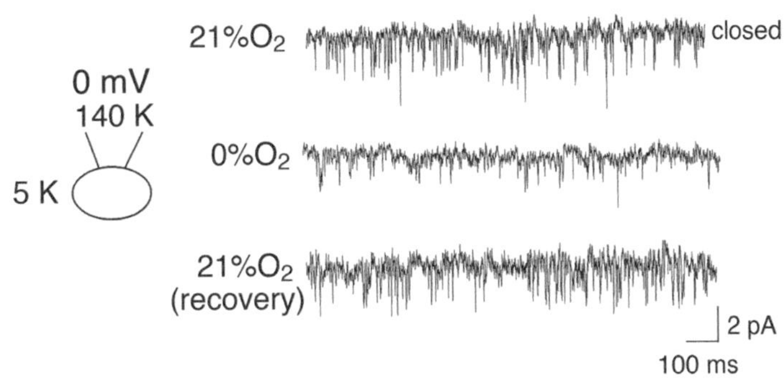

Fig. 4.4 Inhibition of TASK by hypoxia in rabbit glomus cells. Cell-attached patches are perfused with normoxic and hypoxic solutions. Hypoxia also reduced channel amplitude, indicating that cell depolarization has occurred.

increased the activity of the TREK-like channel. Such mechanosensitive TREKs have also been observed in glomus cells from 1–7 day old rat glomus cells (Kim 2013). In inside-out patches, opening of Ca^{2+}-sensitive BK was observed in all patches that expressed TASK-like channels (Fig. 4.3e).

4.3.4 Inhibition of TASK-Like Channels by Hypoxia

To determine the effect of hypoxia on TASK activity, cell-attached patches showing TASK-like channels were perfused with normoxic solution first and then with anoxic solution. Hypoxia produced a ~50% reduction of single channel amplitude, indicating that hypoxia caused ~30 mV depolarization if the resting cell Em is assumed to be ~ −60 mV. In three cell-attached patches, hypoxia also produced a significant decrease in TASK activity by 52 ± 8% (Fig. 4.4).

These results show that hypoxia also inhibits TASK in rabbit glomus cells, as it does in rat and mouse glomus cells, and therefore that TASK inhibition contributes to the hypoxic response in rabbit glomus cells.

4.4 Discussion

In rat and mouse glomus cells, TASK (TASK-1, TASK-3 and TASK-1/3 heteromer) is the primary background K^+ channel whose inhibition by hypoxia contributes to the excitatory response. The purpose of our study was to determine whether TASK is also functionally expressed in rabbit glomus cells and inhibited by hypoxia. Here we show evidence for the functional presence of TASK (TASK-1 and TASK-3) in rabbit glomus cells. We also show that hypoxia (0.5%O_2) inhibits TASK to levels similar to those found in rat and mouse glomus cells, indicating that TASK is involved in hypoxia sensing in rabbit glomus cells.

An earlier study showed that anandamide, an inhibitor of TASK, caused stimulation of the rabbit carotid sinus nerve activity to a level similar to that produced by hypoxia, which led the authors to suggest that TASK-1 is importantly involved in the hypoxic response (Kobayashi and Yamamoto 2010). However, anandamide and its analogues are highly non-specific and modulate many ion channels including TRP and Kv channels (Oliver et al. 2004). Thus, anandamide-mediated inhibition of Kv, rather than inhibition of TASK, could have elicited the hypoxic response by reducing the repolarization during each action potential firing. Using RT-PCR, we have identified mRNA expression of TASK-3 along with those of Kv2 and Kv3 in whole CB that contains many types of cell. Single-channel recording in cell-attached patches showed basal opening of channels with current-voltage relationships and single-channel kinetic properties similar to those of TASK-1 and TASK-3. Thus, rabbit glomus cells express TASK that is likely to provide some fraction of the background K^+ current.

On average, the basal TASK activity in cell-attached patches in rabbit glomus cells was lower than those in rat and mouse glomus cells under identical experimental conditions. Although the levels of TASK expression activity may be low in rabbit glomus cells, we could clearly identify TASK -like channels. It is unclear why TASK was not detected in earlier studies, but it could be that the studies focused only on voltage-gated K^+ channels (Ganfornina and Lopez-Barneo 1991, 1992). In addition to TASK, we observed frequent opening of ~42-pS channel that we have not characterized here. It seems likely that the ~42-pS channel is the O_2-sensitive inactivating Kv (K_{O2}) channel that was also observed in cell-attached patches in earlier studies (Ganfornina and Lopez-Barneo 1991, 1992). We also show the presence of TREK-like channel whose function in glomus cells remain to be determined. The mechanosensitive TREK could be involved in flow-mediated changes in glomus cell excitability (Schultz et al. 2013).

The ability of hypoxia to inhibit TASK in rabbit glomus cells suggests that TASK partly contributes to the hypoxic response. The overall contribution of TASK to the hypoxic response may be smaller in rabbit than in rat glomus cells. However, because TASK is active at rest, inhibition of TASK may be important for the initial hypoxic response, whereas inhibition of Kv contributes to the sustained hypoxic response as the frequency of spontaneous action potentials increases (Lopez-Lopez et al. 1989; Rocher et al. 2005).

Acknowledgements This work was supported by funds from NIH, Rosalind Franklin University, and the Ministry of Science, ICT and Future Planning (NRF-2015R1A5A2008833, Korea).

References

Biscoe TJ, Duchen MR, Eisner DA et al (1989) Measurements of intracellular Ca^{2+} in dissociated type I cells of the rabbit carotid body. J Physiol 416:421–434

Buckler KJ (2015) TASK channels in arterial chemoreceptors and their role in oxygen and acid sensing. Pflugers Arch 467(5):1013–1025

Buckler KJ, Williams BA, Honore E (2000) An oxygen-, acid- and anaesthetic-sensitive TASK-like background potassium channel in rat arterial chemoreceptor cells. J Physiol 525(Pt 1):135–142

Chen J, He L, Dinger B et al (2000) Cellular mechanisms involved in rabbit carotid body excitation elicited by endothelin peptides. Respir Physiol 121(1):13–23

Delpiano MA, Hescheler J (1989) Evidence for a PO_2-sensitive K^+ channel in the type-I cell of the rabbit carotid body. FEBS Lett 249(2):195–198

Fagerlund MJ, Kahlin J, Ebberyd A et al (2010) The human carotid body: expression of oxygen sensing and signaling genes of relevance for anesthesia. Anesthesiology 113(6):1270–1279

Ganfornina MD, Lopez-Barneo J (1991) Single K^+ channels in membrane patches of arterial chemoreceptor cells are modulated by O_2 tension. Proc Natl Acad Sci U S A 88(7):2927–2930

Ganfornina MD, Lopez-Barneo J (1992) Potassium channel types in arterial chemoreceptor cells and their selective modulation by oxygen. J Gen Physiol 100(3):401–426

Grynkiewicz G, Poenie M, Tsien RY (1985) A new generation of Ca^{2+} indicators with greatly improved fluorescence properties. J Biol Chem 260(6):3440–3450

Kang D, Han J, Talley EM et al (2004) Functional expression of TASK-1/TASK-3 heteromers in cerebellar granule cells. J Physiol 554(Pt 1):64–77

Kang D, Wang J, Hogan JO et al (2014) Increase in cytosolic Ca^{2+} produced by hypoxia and other depolarizing stimuli activates a non-selective cation channel in chemoreceptor cells of rat carotid body. J Physiol 592(Pt 9):1975–1992

Kim D (2013) K(+) channels in O(2) sensing and postnatal development of carotid body glomus cell response to hypoxia. Respir Physiol Neurobiol 185(1):44–56

Kim D, Cavanaugh EJ, Kim I et al (2009) Heteromeric TASK-1/TASK-3 is the major oxygen-sensitive background K+ channel in rat carotid body glomus cells. J Physiol 587(Pt 12):2963–2975

Kobayashi N, Yamamoto Y (2010) Hypoxic responses of arterial chemoreceptors in rabbits are primarily mediated by leak K channels. Adv Exp Med Biol 669:195–199

Lopez-Barneo J, Lopez-Lopez JR, Urena J et al (1988) Chemotransduction in the carotid body: K^+ current modulated by PO2 in type I chemoreceptor cells. Science 241(4865):580–582

Lopez-Lopez J, Gonzalez C, Urena J et al (1989) Low pO2 selectively inhibits K^+ channel activity in chemoreceptor cells of the mammalian carotid body. J Gen Physiol 93(5):1001–1015

Mkrtchian S, Kahlin J, Ebberyd A et al (2012) The human carotid body transcriptome with focus on oxygen sensing and inflammation--a comparative analysis. J Physiol 590(16):3807–3819

Montoro RJ, Urena J, Fernandez-Chacon R et al (1996) Oxygen sensing by ion channels and chemotransduction in single glomus cells. J Gen Physiol 107(1):133–143

Oliver D, Lien CC, Soom M et al (2004) Functional conversion between A-type and delayed rectifier K^+ channels by membrane lipids. Science 304(5668):265–270

Perez-Garcia MT, Lopez-Lopez JR, Riesco AM et al (2000) Viral gene transfer of dominant-negative Kv4 construct suppresses an O_2-sensitive K^+ current in chemoreceptor cells. J Neurosci 20(15):5689–5695

Rocher A, Geijo-Barrientos E, Caceres AI et al (2005) Role of voltage-dependent calcium channels in stimulus-secretion coupling in rabbit carotid body chemoreceptor cells. J Physiol 562(Pt 2):407–420

Sanchez D, Lopez-Lopez JR, Perez-Garcia MT et al (2002) Molecular identification of Kvalpha subunits that contribute to the oxygen-sensitive K^+ current of chemoreceptor cells of the rabbit carotid body. J Physiol 542(Pt 2):369–382

Schultz HD, Marcus NJ, Del Rio R (2013) Role of the carotid body in the pathophysiology of heart failure. Curr Hypertens Rep 15(4):356–362

Molecular Characterization of Equilibrative Nucleoside Transporters in the Rat Carotid Body and Their Regulation by Chronic Hypoxia

5

Shaima Salman and Colin A. Nurse

Abstract

The mammalian carotid body (CB) is the main peripheral arterial chemoreceptor organ that is excited by decreases in blood PO_2 (hypoxia) and increases in blood PCO_2/H^+. An increase in CB afferent carotid sinus nerve (CSN) discharge results in respiratory and cardiovascular reflex responses that help maintain homeostasis. The CB consists mainly of innervated clusters of the chemoreceptive type I (glomus) cells that are associated with the processes of glial-like type II cells. Extracellular ATP and adenosine (ADO) levels increase in response to acute hypoxia and there is evidence that during chronic sustained hypoxia ADO elevation plays a major role in regulating CB chemosensitivity and CSN discharge. We recently characterized the molecular identities of ectonucleotidase enzymes involved in regulating extracellular ATP hydrolysis to produce ADO in the rat CB. In the present study, we focus on a molecular characterization of the equilibrative nucleoside transporter (ENT) system that is known to regulate extracellular ADO concentrations in the rat CB based on pharmacological studies. Examination of ENT expression using *quantitative* PCR (qPCR) analysis revealed the expression of both ENT1 and ENT2 mRNAs in whole CB extracts from ~2-week-old juvenile rats. In dissociated rat CB cultures, both ENT1 and ENT2 immunoreactivity was localized to type I cell clusters. Furthermore, we show that ENT1 and ENT2 mRNA expression is downregulated in CBs isolated from rat pups exposed to chronic hypobaric hypoxia (~1 week). These findings reveal the molecular identities of the ENT system expressed in the rat CB and are consistent with the proposed shift to ADO signaling during chronic hypoxia.

Keywords

Equilibrative nucleoside transporter · Adenosine · Carotid body · Hypoxia

5.1 Introduction

Mammalian carotid bodies are the main peripheral arterial chemoreceptors that sense changes in arterial PO_2 and PCO_2/H^+ and elicit respiratory and cardiovascular reflexes to maintain homeostasis (Gonzalez et al. 1994; Kumar and Prabhakar 2012). The functional unit of the carotid body (CB) consists of innervated clusters of chemosensitive type I (glomus) cells that are ensheathed by the elongated processes of type II (glial-like) cells. The type I cells sense a decrease in blood O_2 (e.g. hypoxia) and elevated CO_2/H^+ (acid

S. Salman (✉) · C. A. Nurse
Department of Biology, McMaster University, Hamilton, ON, Canada
e-mail: ssalman7@jhmi.edu

E. B. Gauda et al. (eds.), *Arterial Chemoreceptors*, Advances in Experimental Medicine and Biology 1071, https://doi.org/10.1007/978-3-319-91137-3_5

hypercapnia) and elicit the release of neurochemicals that shape the carotid sinus nerve (CSN) discharge (Kumar and Prabhakar 2012; Lopez-Barneo et al. 2008; Nurse 2010; Nurse and Piskuric 2013; Peers and Buckler 1995).

ATP and adenosine, acting postsynaptically and/or presynaptically via purinergic receptors, play a major role in shaping the CSN discharge in response to chemostimuli (Bairam et al. 2013; Conde et al. 2012a; Kumar and Prabhakar 2012; Nurse 2014; Nurse and Piskuric 2013). Two sources of ATP release have been identified within the CB in response to hypoxia and hypercapnia. ATP released from type I cells acts postsynaptically on P2X2/3 receptors expressed on the CSN terminals (Prasad et al. 2001; Rong et al. 2003; Zhang et al. 2000), or on P2Y2 receptors expressed on glial-like type II cells (Tse et al. 2012; Xu et al. 2003; Zhang et al. 2012), inducing further ATP release through pannexin-1 (Panx-1) channels (Zhang et al. 2012). Furthermore, adenosine, produced via the hydrolysis of ATP by ectonucleotidases or released through the bidirectional equilibrative nucleoside transporters, may act on postsynaptic and presynaptic P1 receptors (Conde et al. 2009, 2012a; Holmes et al. 2017; Murali and Nurse 2016).

Extracellular adenosine may be produced by two major metabolic sources, i.e. hydrolysis of extracellular ATP by the combined action of ecto-nucleotidases and ecto-5′ nucleotidase, and/or release from type I cells via the Na^+ independent equilibrative nucleoside transport (ENT) system. Both sources have been described in the CB using pharmacological tools, with the latter being responsible for ~45% of the adenosine release during mild hypoxia (Conde et al. 2012b).

We recently provided molecular and immunohistochemical evidence for the presence of surface-located ectonucleotidases NTPDase1, 2 and 3, and ecto-5′ nucleotidase (E5′Nt) in the rat CB (Salman et al. 2017), responsible for the hydrolysis of tri-, di-, and monophosphate molecules to produce adenosine. However, though previous studies demonstrated pharmacological evidence for the presence of the equilibrative transporter system in the rat CB (Conde and Monteiro 2004), information on the molecular identity of the corresponding transporters is lacking. The equilibrative transporter family consists of three subtypes, equilibrative nucleoside transporter (ENT)-1-3 (Baldwin et al. 2004). ENT1 and ENT2 are trafficked to the plasma membrane while ENT3 is trafficked intracellularly to the mitochondria (Baldwin et al. 2005; Govindarajan et al. 2009). In this study, we report on the molecular characterization of ENT1 and ENT2 mRNAs using quantitative PCR, as well as protein localization using immunofluorescence labeling in the rat CB. We show that the rat CB expresses ENT1 and ENT2 mRNA and protein localized to type I cell clusters. In addition, we examined the regulation of ENT1 and ENT2 in CBs obtained from rat pups exposed to chronic hypobaric hypoxia for ~1 week.

5.2 Methods

See Salman et al. (2017). Mouse monoclonal human ENT1 (sc-377283) and ENT2 (sc-373871) antibodies were purchased from Santa Cruz.

5.3 Results

Quantitative real-time PCR (qPCR) analyses of ENT1 and ENT2 mRNA expression in the carotid body relative to brain and peripheral ganglia. Expression of the adenosine transporters, equilibrative nucleoside transported 1 (ENT1) and (ENT2) was determined in extracts of the whole rat carotid body (CB) and compared to the expression pattern in tissues/cells from the central and peripheral nervous system. Products of RT-PCR analysis confirmed mRNA expression of both ENT1 and ENT2 in the rat carotid body. Whole brain tissue extract was used as a positive control for ENT expression in central nervous system (Che et al. 1995). Whole tissue extracts from superior cervical ganglion (SCG) and petrosal ganglion (PG) were used as positive controls for ENT expression in peripheral nervous system. Expression of ENT1 and ENT2 mRNA was detected in the CB, SCG and PG. We also examined ENT expression in an immortalized adrenal

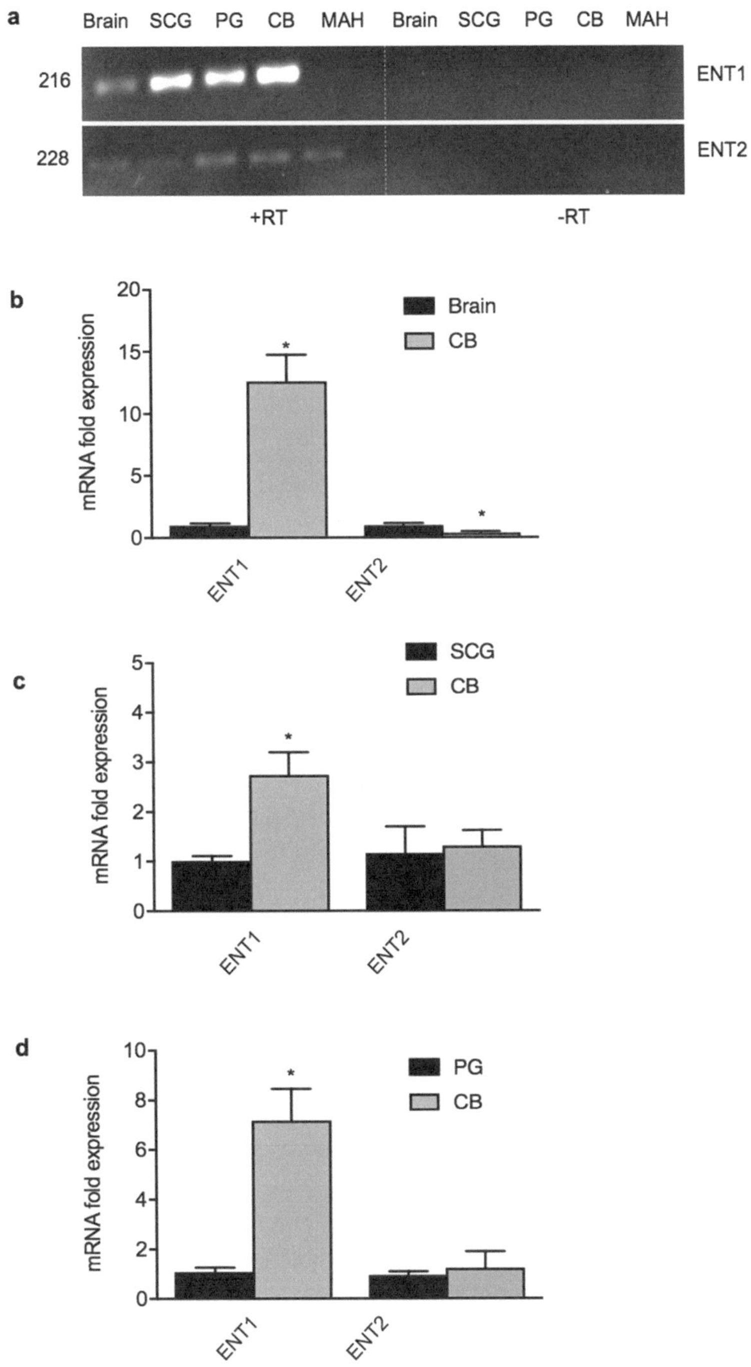

Fig. 5.1 Expression analysis of equilibrative nucleoside transporter 1 (ENT1) and 2 (ENT2) genes in rat carotid body (CB) and tissues/cells from central and peripheral nervous systems. Gel electrophoresis of RT-PCR products for ENT1 and 2 mRNA in rat brain, superior cervical ganglion (SCG), petrosal ganglion (PG), whole CB tissue, and the immortalized chromaffin cell line (MAH). No-Reverse Transcriptase (-RT) samples were used as a negative control (**a**, $n = 3$). *Quantitative* real-time PCR analysis of ENT1 and ENT2 mRNA expression in whole rat CB relative to brain (**b**), or superior cervical ganglion (SCG) (**c**), or petrosal ganglion (PG) (**d**). Experimental replicate represents expression in tissues pooled from individual rat litters (n). Data are presented as means ±SEM of ENT expression in CB tissue relative to specified groups; $n = 4$. Asterisks denote $P < 0.05$ vs. control tissue

chromaffin (MAH) cell line (see Salman et al. 2017). The expression pattern of ENT1 and ENT2 transcripts appeared quite variable among these tissues/cells as shown in Fig. 5.1a. Note the CB expresses transcripts of both ENT1 and ENT2. In contrast to chromaffin-derived PC12 cells which are known to express both ENT1 and ENT2 (Kobayashi et al. 2000), the MAH cell line appeared to express ENT1 at a lower level than ENT2 (Fig. 5.1a; $n = 3$).

When compared with ENT mRNA expression in the brain and SCG, which express adenosine

A1 receptors (Che et al. 1995; Connolly et al. 1993), we found the rat CB expresses significantly higher levels of ENT1 mRNA (Fig. 5.1b, c). Petrosal ganglia (PG), which relay chemosensory information from the CB to the central pattern generator in brainstem are known to express adenosine A_{2a}, A_{2b}, and A_1 receptors (Conde et al. 2012a; Gauda 2000; Nurse 2014; Nurse and Piskuric 2013). As shown in Fig. 5.1d, expression of ENT1 mRNA was significantly higher in the CB relative to the PG. By comparison, expression of ENT2 mRNA did not seem to be differentially expressed in the CB relative to SCG and PG, though there was an apparent lower expression in the CB relative to the brain (Fig. 5.1b). These data suggest that ENT1 is the dominantly expressed member of the nucleoside transporter family in the rat CB ($n = 4$).

Localization of ENT1 and ENT2 immunoreactivity in cell cultures of dissociated rat carotid body. Using immunofluorescence labeling, we localized ENT expression in permeabilized and non-permeabilized cultures of dissociated rat CB. These cultures consist of isolated type I cell clusters that are easily identified under phase contrast microscopy (Murali and Nurse 2016; Nurse 2010; Salman et al. 2017; Zhang et al. 2012). In permeabilized CB cultures, ENT1 and ENT2 immunoreactivity (green) was found associated with tyrosine hydroxylase (TH)-immunoreactive (ir; red) type I cells as shown in Figs. 5.2a–c and Figs. 5.2d–f, respectively ($n = 2$). Interestingly, ENT1 staining appeared to be mostly localized to the nuclei of TH-positive cells in permeabilized CB cultures as shown in Fig. 5.2a. In non-permeabilized CB cultures, type I cell clusters were also selectively immunopositive for both ENT1 (Fig. 5.2g–i, n = 2) and ENT2 (Fig. 5.2j–l, n = 2), suggesting an additional extracellular localization for both transporters. Note the confinement of ENT1-ir and ENT2-ir to type I cell clusters in Fig. 5.2i, l respectively.

Regulation of ENT1 and ENT2 mRNA expression in rat carotid bodies exposed to chronic hypoxia in vivo.

We recently demonstrated that chronic hypobaric hypoxia *in vivo* up-regulates expression of NTPDase3 and the adenosine producing enzyme, E5'Nt/CD73 in the rat CB (Salman et al. 2017), thereby favoring increased generation of extracellular adenosine. Previous studies showed that chronic hypoxia represses ENT expression in human umbilical vein endothelium (HUVEC) (Casanello et al. 2005) and human microvascular endothelial cells (HMEC) (Eltzschig et al. 2005). We therefore examined whether chronic hypoxia regulates ENT expression in CB, SCG, and PG relative to normoxic controls To this end, we exposed juvenile rat pups (P11-14) to chronic hypobaric hypoxia (60 kPa, simulating an altitude of approximately 4300 m) for 5–7 days as previously described (Salman et al. 2017). For comparison, *quantitative* real time-PCR (qPCR) data adapted from the latter study is plotted in Fig. 5.3, and reveals a significant upregulation in NTPDase3 and E5'Nt mRNAs in the CBs from chronically hypoxic (CHox) animals compared normoxic (Nox) controls in Fig. 5.3. By contrast, similar to previous studies (Casanello et al. 2005; Eltzschig et al. 2005), ENT1 expression was significantly downregulated in the CB and PG of CHox animals relative to Nox controls (Fig. 5.3). ENT2 mRNA was also downregulated and to a lesser extent, though this was specific to the CB tissue of CHox animals (Fig. 5.3). Taken together, these data suggest that chronic hypoxia differentially regulates NTPDase, E5'Nt, and ENT expression in the CB so as to elevate extracellular adenosine levels.

5.4 Discussion

In response to chemostimuli (e.g. hypoxia and acid hypercapnia), ATP is released from the carotid body (CB) and its rapid breakdown to adenosine activates a number of autocrine and paracrine signaling pathways (Buttigieg and Nurse 2004; Conde and Monteiro 2004; Conde et al. 2007) add refs (Murali and Nurse 2016; Nurse and Piskuric 2013; Tse et al. 2012). The main purpose of the present study was to explore mRNA expression pattern and localization of equilibrative nucleoside transporters (ENTs) in the CB and their potential role in adenosine clearance during chronic hypoxia.

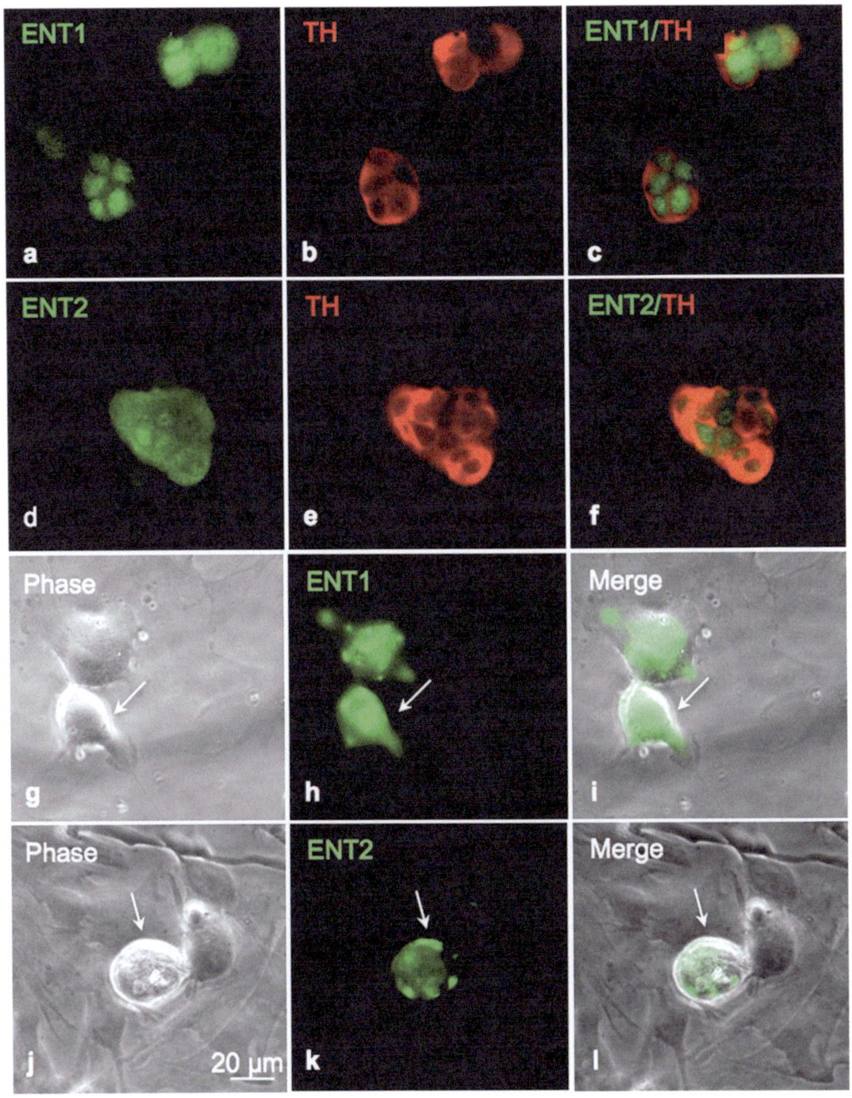

Fig. 5.2 Equilibrative nucleoside transporter1 (ENT1) and ENT2 immunoreactivity in cultures of dissociated rat carotid body. Immunofluorescence staining of permeabilized cultures showing ENT1- and ENT2-ir (green; **a** and **d** respectively) associated within TH-ir (red; **b** and **e**) type I cell clusters; merged image is shown in **c** and **f** ($n = 2$). *G-I* and *J-L,* ENT1 and ENT2 immunoreactivity in non-permeabilized CB cultures respectively. Note positive extracellular ENT1 and ENT2 immunoreactivity (**h** and **k**) respectively in type I cell clusters (arrow in phase contrast images, **g** and **j**) and merged images in **i** and **l**; note ENT1- and -2 negative background cells surrounding clusters in **i** and **l** (n = 2)

We found that the rat carotid body expresses mRNA and protein for both equilibrative transporter 1 (ENT1) and ENT2. Moreover, mRNA of ENT1 and ENT2 was downregulated in rat CB following *in vivo* exposure to chronic hypobaric hypoxia, and it is proposed that these mechanisms could lead to increased extracellular adenosine concentration at the chemosensory synapse.

Two major metabolic sources for the generation of extracellular adenosine have been identified during mild hypoxia, i.e. ATP catabolism by a series of ectonucleotidase enzymes, and export via the equilibrative nucleoside transport (ENT)

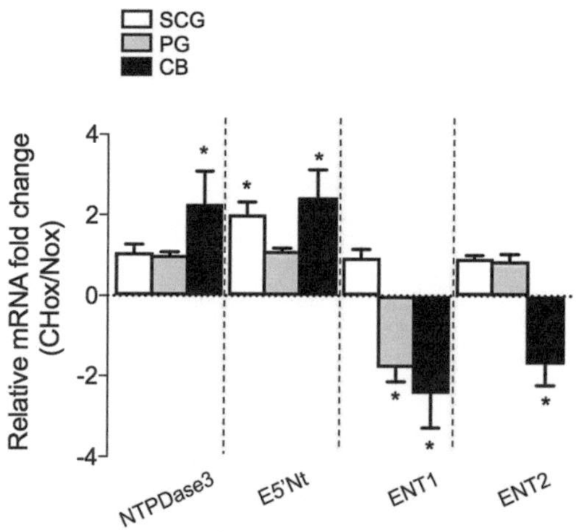

Fig. 5.3 Effects of chronic hypobaric hypoxia on ectonucleoside triphosphate diphosphohydrolase3 (NTPDase3), ecto-5′-nucleotidase (E5′Nt), ENT1 and ENT2 mRNA expression in rat carotid body. Rat pups (~2-week old) were exposed to normoxia or continuous hypobaric hypoxia (60 kPa, ~4300 m) for 5–7 days, starting at day 7. *Quantitative* real-time PCR (qPCR) analysis was used to compare mRNA fold change in chronic hypoxia relative to normoxia of NTPDase3, E5′Nt, ENT1 and ENT2 in SCG, PG and CB tissues. Data reveal a significant upregulation of NTPDase3 and E5′Nt concomitant with a downregulation in ENT1 and ENT2 expression ($n = 4$ litters). Results were normalized to TBP and shown as fold change relative to expression in normoxia. Each experimental replicate represents expression in tissues pooled from individual rat litters (n). Asterisks denote significant differences between normoxic and hypoxic groups, $P < 0.05$

system (Conde et al. 2009). A recent study from our laboratory identified expression of NTPDase1,2,3 and E5′Nt mRNAs in the rat CB, and NTPDase 2,3 and E5′Nt immunoreactivities were localized to type I cell clusters (Salman et al. 2017). The presence of an equilibrative nucleoside transporter (ENT) system in the CB was inferred from pharmacological studies using the ENT inhibitor NBTI (Conde et al. 2008; Conde and Monteiro 2004). Based on those studies it was concluded that ~45% of extracellular adenosine present during mild hypoxia (10% O_2) is trafficked though the ENT system (Conde et al. 2008, Conde and Monteiro 2004). In the present study, we uncovered the molecular identity of the ENT system in the rat CB by showing the rat CB expresses mRNA and protein of ENT1 and ENT2 subtypes. During episodes of high hypoxia intensity, adenosine release by ENT is significantly reduced and extracellular adenosine is preferentially produced by extracellular ATP hydrolysis via the enzymatic action of ectonucleotidases and ecto-5′ nucleotidase (Conde et al. 2012a). The remaining source of adenosine has been proposed to be generated from cAMP (Perez-Garcia et al. 1991).

Previous studies reported transcriptional repression of ENT expression in cardiac myocytes (Chaudary et al. 2004), vascular endothelia and mucosal epithelia (Casanello et al. 2005; Eltzschig et al. 2005) during chronic hypoxia. Given that chronic hypoxia can augment CB chemosensitivity (Kumar and Prabhakar 2012; Powell 2007), and the key role of purinergic signaling in driving the CB afferent discharge (Nurse and Piskuric 2013), it was of interest to contrast ENT expression in CBs isolated from animals exposed to chronic hypobaric hypoxia. Interestingly, we found that chronic hypoxia downregulates mRNA expression of ENT1 and ENT2 in the rat CB, consistent with previous studies reporting a similar downregulation of ENT mRNA in other tissues during chronic hypoxia (Eltzschig et al. 2005).

Adenosine, generated by hydrolysis of extracellular ATP by surface-located enzymes or released through bidirectional equilibrative transporters in type I cells, may activate presynaptic and postsynaptic P1 receptors in the CB (Conde et al. 2009, 2012a; Murali and Nurse 2016; Nurse and Piskuric 2013). In conclusion, the functional consequence of the increased capacity to produce extracellular adenosine from ATP by ectonucleotidases (Conde et al. 2009, 2012a; Gauda 2000; Murali and Nurse 2016; Salman et al. 2017), coupled with the decreased capacity to transport adenosine across type I cells by ENT1,2, predicts a major shift towards enhanced adenosine signaling at the CB chemosensory synapse during chronic hypoxia.

5.5 Statistical Analysis

Statistical analyses were performed using GraphPad Prism7 software. Data are expressed as means ± SE and compared using a non- parametric (Mann-Whitney) test. *n* value indicates the number of litters used in each experiment. Asterisks denote significant differences where $P < 0.05$ between groups.

Acknowledgments We wish to thank Cathy Vollmer for excellent technical assistance and Grant B. McClelland for providing the hypobaric chambers.

References

Bairam A, Niane LM, Joseph V (2013) Role of ATP and adenosine on carotid body function during development. Respir Physiol Neurobiol 185:57–66

Baldwin SA, Beal PR, Yao SY, King AE, Cass CE, Young JD (2004) The equilibrative nucleoside transporter family, SLC29. Pflugers Arch 447:735–743

Baldwin SA, Yao SY, Hyde RJ, Ng AM, Foppolo S et al (2005) Functional characterization of novel human and mouse equilibrative nucleoside transporters (hENT3 and mENT3) located in intracellular membranes. J Biol Chem 280:15880–15887

Buttigieg J, Nurse CA (2004) Detection of hypoxia-evoked ATP release from chemoreceptor cells of the rat carotid body. Biochem Biophys Res Commun 322:82–87

Casanello P, Torres A, Sanhueza F, Gonzalez M, Farias M et al (2005) Equilibrative nucleoside transporter 1 expression is downregulated by hypoxia in human umbilical vein endothelium. Circ Res 97:16–24

Chaudary N, Naydenova Z, Shuralyova I, Coe IR (2004) Hypoxia regulates the adenosine transporter, mENT1, in the murine cardiomyocyte cell line, HL-1. Cardiovasc Res 61:780–788

Che M, Ortiz DF, Arias IM (1995) Primary structure and functional expression of a cDNA encoding the bile canalicular, purine-specific Na(+)-nucleoside cotransporter. J Biol Chem 270:13596–13599

Conde SV, Monteiro EC (2004) Hypoxia induces adenosine release from the rat carotid body. J Neurochem 89:1148–1156

Conde SV, Obeso A, Gonzalez C (2007) Low glucose effects on rat carotid body chemoreceptor cells' secretory responses and action potential frequency in the carotid sinus nerve. J Physiol 585:721–730

Conde SV, Gonzalez C, Batuca JR, Monteiro EC, Obeso A (2008) An antagonistic interaction between A2B adenosine and D2 dopamine receptors modulates the function of rat carotid body chemoreceptor cells. J Neurochem 107:1369–1381

Conde SV, Monteiro EC, Obeso A, Gonzalez C (2009) Adenosine in peripheral chemoreception: new insights into a historically overlooked molecule--invited article. Adv Exp Med Biol 648:145–159

Conde SV, Monteiro EC, Rigual R, Obeso A, Gonzalez C (2012a) Hypoxic intensity: a determinant for the contribution of ATP and adenosine to the genesis of carotid body chemosensory activity. J Appl Physiol 1985(112):2002–2010

Conde SV, Ribeiro MJ, Obeso A, Rigual R, Monteiro EC, Gonzalez C (2012b) Chronic caffeine intake in adult rat inhibits carotid body sensitization produced by chronic sustained hypoxia but maintains intact chemoreflex output. Mol Pharmacol 82:1056–1065

Connolly GP, Stone TW, Brown F (1993) Characterization of the adenosine receptors of the rat superior cervical ganglion. Br J Pharmacol 110:854–860

Eltzschig HK, Abdulla P, Hoffman E, Hamilton KE, Daniels D et al (2005) HIF-1-dependent repression of equilibrative nucleoside transporter (ENT) in hypoxia. J Exp Med 202:1493–1505

Gauda EB (2000) Expression and localization of A2a and A1-adenosine receptor genes in the rat carotid body and petrosal ganglia. A2a and A1-adenosine receptor mRNAs in the rat carotid body. Adv Exp Med Biol 475:549–558

Gonzalez C, Almaraz L, Obeso A, Rigual R (1994) Carotid body chemoreceptors: from natural stimuli to sensory discharges. Physiol Rev 74:829–898

Govindarajan R, Leung GP, Zhou M, Tse CM, Wang J, Unadkat JD (2009) Facilitated mitochondrial import of antiviral and anticancer nucleoside drugs by human equilibrative nucleoside transporter-3. Am J Physiol Gastrointest Liver Physiol 296:G910–G922

Holmes AP, Ray CJ, Pearson SA, Coney AM, Kumar P (2017) Ecto-5′-nucleotidase (CD73) regulates peripheral chemoreceptor activity and cardiorespiratory responses to hypoxia. J Physiol

Kobayashi S, Zimmermann H, Millhorn DE (2000) Chronic hypoxia enhances adenosine release in rat PC12 cells by altering adenosine metabolism and membrane transport. J Neurochem 74:621–632

Kumar P, Prabhakar NR (2012) Peripheral chemoreceptors: function and plasticity of the carotid body. Compr Physiol 2:141–219

Lopez-Barneo J, Ortega-Saenz P, Pardal R, Pascual A, Piruat JI (2008) Carotid body oxygen sensing. Eur Respir J 32:1386–1398

Murali S, Nurse CA (2016) Purinergic signalling mediates bidirectional crosstalk between chemoreceptor type I and glial-like type II cells of the rat carotid body. J Physiol 594:391–406

Nurse CA (2010) Neurotransmitter and neuromodulatory mechanisms at peripheral arterial chemoreceptors. Exp Physiol 95:657–667

Nurse CA (2014) Synaptic and paracrine mechanisms at carotid body arterial chemoreceptors. J Physiol 592:3419–3426

Nurse CA, Piskuric NA (2013) Signal processing at mammalian carotid body chemoreceptors. Semin Cell Dev Biol 24:22–30

Peers C, Buckler KJ (1995) Transduction of chemostimuli by the type I carotid body cell. J Membr Biol 144:1–9

Perez-Garcia MT, Almaraz L, Gonzalez C (1991) Cyclic AMP modulates differentially the release of dopamine induced by hypoxia and other stimuli and increases dopamine synthesis in the rabbit carotid body. J Neurochem 57:1992–2000

Powell FL (2007) The influence of chronic hypoxia upon chemoreception. Respir Physiol Neurobiol 157:154–161

Prasad M, Fearon IM, Zhang M, Laing M, Vollmer C, Nurse CA (2001) Expression of P2X2 and P2X3 receptor subunits in rat carotid body afferent neurones: role in chemosensory signalling. J Physiol 537:667–677

Rong W, Gourine AV, Cockayne DA, Xiang Z, Ford AP et al (2003) Pivotal role of nucleotide P2X2 receptor subunit of the ATP-gated ion channel mediating ventilatory responses to hypoxia. J Neurosci 23:11315–11321

Salman S, Vollmer C, McClelland GB, Nurse CA (2017) Characterization of ectonucleotidase expression in the rat carotid body: regulation by chronic hypoxia. Am J Physiol Cell Physiol 313:C274–CC84

Tse A, Yan L, Lee AK, Tse FW (2012) Autocrine and paracrine actions of ATP in rat carotid body. Can J Physiol Pharmacol 90:705–711

Xu J, Tse FW, Tse A (2003) ATP triggers intracellular Ca2+ release in type II cells of the rat carotid body. J Physiol 549:739–747

Zhang M, Zhong H, Vollmer C, Nurse CA (2000) Co-release of ATP and ACh mediates hypoxic signalling at rat carotid body chemoreceptors. J Physiol 525(Pt 1):143–158

Zhang M, Piskuric NA, Vollmer C, Nurse CA (2012) P2Y2 receptor activation opens pannexin-1 channels in rat carotid body type II cells: potential role in amplifying the neurotransmitter ATP. J Physiol 590:4335–4350

6 Mitochondrial Complex I Dysfunction and Peripheral Chemoreflex Sensitivity in a FASTK-Deficient Mice Model

Angela Gomez-Niño, Inmaculada Docio, Jesus Prieto-Lloret, Maria Simarro, Miguel A. de la Fuente, and Asuncion Rocher

Abstract

The molecular mechanisms underlying O_2-sensing by carotid body (CB) chemoreceptors remain undetermined. Mitochondria have been implicated, due to the sensitivity of CB response to electron transport chain (ETC) blockers. ETC is one of the major sources of reactive oxygen species, proposed as mediators in oxygen sensing. Fas-activated serine/threonine phosphoprotein is a sensor of mitochondrial stress that modulates protein translation to promote survival of cells exposed to adverse conditions. A translational variant of Fas-activated serine/threonine kinase (FASTK) is required for the biogenesis of ND6 mRNA, the mitochondrial encoded subunit 6 of the NADH dehydrogenase complex (Complex I). Ablating FASTK expression reduced Complex I activity in vivo by about 50%. We have tested the hypothesis of Complex I participation in O_2-sensing structures by studying the effect of hypoxia in FASTK$^{-/-}$ knockout mice. Ventilatory response to acute hypoxia and hypercapnia tests showed similar sensitivity and CB catecholaminergic activity in knockout and wild type mice; hypoxic pulmonary vasoconstriction response also was similar. Pulmonary artery contractility in vitro, using small vessel myography, showed a significantly decreased relaxation to rotenone in knockout mice preconstricted vessels with $PGF_{2\alpha}$. In conclusion, FASTK$^{-/-}$ knockout mice maintain respiratory chemoreflex under hypoxia and hypercapnia stress suggesting that completely functional Complex I ND6 protein is not required for these responses.

Keywords

Mitochondria · Complex I. FASTK family proteins · Carotid body chemoreceptors

A. Gomez-Niño (✉)
Departamento de Biologia Celular, Histologia y Farmacologia/IBGM, Universidad de Valladolid-CSIC, Valladolid, Spain

CIBER de Enfermedades Respiratorias (CIBERES), Instituto de Salud Carlos III, Madrid, Spain
e-mail: angela@biocel.uva.es

I. Docio · J. Prieto-Lloret · A. Rocher
Departamento de Bioquimica, Biologia Molecular y Fisiologia/IBGM, Universidad de Valladolid-CSIC, Valladolid, Spain

CIBER de Enfermedades Respiratorias (CIBERES), Instituto de Salud Carlos III, Madrid, Spain

M. Simarro
Departamento de Enfermeria/IBGM, Universidad de Valladolid, Valladolid, Spain

M. A. de la Fuente
Departamento de Biologia Celular, Histologia y Farmacologia/IBGM, Universidad de Valladolid-CSIC, Valladolid, Spain

E. B. Gauda et al. (eds.), *Arterial Chemoreceptors*, Advances in Experimental Medicine and Biology 1071, https://doi.org/10.1007/978-3-319-91137-3_6

6.1 Introduction

Carotid body (CB) chemoreceptor cells and pulmonary artery smooth muscle cells (PASMC) detect hypoxia with a very high sensitivity (PO_2 threshold ~ 70–75 mmHg) generating homeostatic responses that guarantee a normal O_2 supply to cells, in spite of environmental O_2 deficit (Gonzalez et al. 1994). Homeostatic responses include CB triggered hyperventilation allowing arterial blood PO_2 to increase, and vasoconstriction of pulmonary arteries (PA), optimizing ventilation/perfusion matching and blood oxygenation (Richalet 1997). These two initial responses are immediate, involving ionic conductance changes in CB chemoreceptor cells and PASMC plasma membrane. In both cell types the hypoxic transduction cascade includes inhibition of membrane O_2-sensitive K^+ channels, activation of voltage operated Ca^{2+} channels, and Ca^{2+}-dependent responses (López-Barneo et al. 1988; Post et al. 1992). In CB chemoreceptor cells, the final output of the hypoxic transduction cascade is the release of neurotransmitters that, stimulating sensory nerves, triggers brain stem neuron activity and induces hyperventilation. In PASMC the final output is the PA vasoconstriction, redirecting blood flow from poorly, to well ventilated lung regions. The cascade of transduction of natural stimuli is incompletely understood and it has been proposed in both cell types that reactive oxygen species (ROS) may play a critical role linking low PO_2 to K^+ channel inhibition. Mitochondria have been related due to the sensitivity of CB to electron transport chain (ETC) blockers, one of the main cellular origins of ROS.

The participation of mitochondrial Complex I in acute oxygen-sensing by CB chemoreceptor cells has been recently reported by Fernández-Agüera et al. (2015) showing that conditional knockout mice lacking mitochondrial complex I genes lose the hyperventilatory response to hypoxia. The Fas-activated serine/threonine kinase proteins (FASTK) are mitochondria-associated proteins that promote the survival of cells exposed to adverse environmental conditions. FASTK have emerged as key post-transcriptional regulators of mitochondrial gene expression and function. A translational variant of FASTK regulates alternative splicing in the nucleus and co-localizes with two types of cytosolic RNA granules, stress granules and processing bodies. The mitochondrial isoform of FASTK (mitoFASTK) co-localizes with mitochondrial RNA granules and binds at multiple sites along the ND6 mRNA, the mitochondrial-encoded subunit 6 of the NADH dehydrogenase complex (Complex I). MitoFASTK and its precursors modulate degradosome activity to generate mature ND6 mRNA, essential for ND6 biogenesis and Complex I activity (Jourdain et al. 2015). Ablating FASTK expression in cultured cells and mice results specifically in loss of ND6 mRNA and reduced Complex I activity in vivo by 50% (Simarro et al. 2010).

We have tested the hypothesis of Complex I participation in O_2-sensing structures by studying the CB respiratory chemoreflex to acute hypoxia and hypercapnia tests, measured by plethysmography in a model of $FASTK^{-/-}$ knockout mice. Catecholamine (CA) synthesis and the effect of rotenone on the CA release by *in vitro* CB. The *in vivo* hypoxic pulmonary vasoconstriction and the *in vitro* contractility response of small PA to rotenone were also assessed. Data showed that $FASTK^{-/-}$ mice maintain respiratory chemoreflex and vasoconstriction responses under hypoxic stress implying that normally functional Complex I ND6 protein is not required for these responses.

6.2 Methods

6.2.1 Animals

The study was conducted on 12 wild type (WT) C57BL/6J and 12 $FASTK^{-/-}$ knockout male mice (10–12 weeks old). Animals had unrestricted access to water and regular chow diet, and were housed in standard conditions in a temperature-controlled room (24 ± 2 °C) with 12:12 h light–dark cycles.

All surgical procedures were done under general anesthesia with intraperitoneal injection of pentobarbital (30 mg/kg). After experiments, animals were euthanized by a cardiac overdose of sodium pentobarbital. The University of Valladolid Institutional Committee approved protocols for Animal Care and Use following international laws and policies (European Union Directive for Protection of Vertebrates Used for Experimental and Other Scientific Ends, 2010/63/EU).

6.2.2 Generation of FASTK$^{-/-}$ Mice

FASTK$^{-/-}$ mice were created as described in detail elsewhere (Simarro et al. 2010). Briefly, the entire Fastk gene was replaced by a loxP-flanked neomycin expression cassette. Mice carrying this mutation were crossed with EIIa-Cre transgenic mice (The Jackson Laboratory) to remove the neomycin cassette, and then backcrossed 12 times into the C57BL/6 background. Both the Fastk gene and its protein product were shown to be deleted in these mice. FASTK$^{-/-}$ mice were viable and fertile and lacked any gross abnormality.

6.2.3 Plethysmography

Ventilation was measured in freely-moving animals by whole body plethysmography (Gonzalez-Obeso et al. 2017) using methacrylate chambers (Emka Technologies, Paris, France; BUXCO Research Systems, Wilmington, NC, USA) continuously fluxed (2 l/min) with the desired gas mixture. Temperature was maintained within the thermo-neutral range (22–24 °C). Tidal volume (TV; ml/Kg), respiratory frequency (fr; breaths/min), minute ventilation (MV; ml/min/Kg), were assessed. Animals breathed room air until acquiring a standard resting behavior. Basal ventilatory parameters and acute hypoxic and hypercapnic test response (12% O_2, remainder N_2, and 5% CO_2 in air) were measured.

6.2.4 CB Synthesis and Evoked Release of CA

General procedures for studying the ^{3}H-CA synthesis rate and stimulus-evoked release have been previously described (Chen et al. 1997). Organs were incubated (37 °C; 2 h) in Tyrode solution (in mM: NaCl, 140; KCl, 5; $CaCl_2$, 2; $MgCl_2$, 1.1; HEPES, 10; glucose, 5; pH 7.40) containing 30 μM of 3,5-^{3}H-tyrosine (CA precursor; 6 Ci/mmol; Perkin Elmer) and 100 μM 6-methyltetrahydropterine and 1 mM ascorbic acid, cofactors for tyrosine hydroxylase and dopamine beta hydroxylase, respectively. Tissues were then washed in precursor-free Tyrode (4 °C; 5 min) and homogenized. ^{3}H-CA were quantified by scintillation counting. To evaluate stimuli-evoked secretory response, isolated CB were incubated 2 h with ^{3}H-tyrosine of high specific activity (40–50 Ci/mmol) and later transferred to vials containing the precursor-free Tyrode solution (in mM: NaCl, 116; KCl, 5; $CaCl_2$, 2; $MgCl_2$, 1.1; HEPES, 10; glucose, 5; $NaCO_3H$, 24). The solution was equilibrated with gas mixtures containing 5% CO_2 and 21% O_2 (pH 7.40). Specific mitochondrial Complex I blocker, rotenone (10 μM; Sigma, Madrid, Spain), was added to the incubating solution. Incubating solutions were renewed and collected every 10 min and ^{3}H-CA content measured by scintillation counter.

6.2.5 Myography

Mice were killed by lethal injection (i.p.) of sodium pentobarbital. The heart and lungs were excised and placed in cold physiological salt solution (PSS) containing (in mmol/L): 118 NaCl, 24 $NaHCO_3$, 1 $MgSO_4$, 0.435 NaH_2PO_4, 5.56 glucose, 1.8 $CaCl_2$, and 4 KCl. Rings of PA were cleaned of adventitia and parenchyma under a dissection microscope, mounted on an isometric small vessel wire myograph (Danish Myo Technology, Denmark), and stretched to a basal tension of ~5 mN. They were then equilibrated with three brief exposures (3 min) to PSS containing 80 mM KCl (80KPSS; isotonic replacement of NaCl by KCl). The pH of the solution was maintained at 7.4 by gassing with 20% O_2, 5% CO_2, balance N_2. Rotenone (10 μM) was applied during an ongoing stable contraction to $PGF_{2\alpha}$ (5 μM). The resulting relaxation was expressed as a percentage of the stable contraction to $PGF_{2\alpha}$.

6.2.6 Statistical Analysis

Data are presented as mean ± S.E.M. Statistical analysis were performed by paired or unpaired Student's t-test and by repeated measures, One-Way Analysis of Variance (ANOVA) with Dunnett's multiple comparison tests. Significance was taken as $p < 0.05$.

6.3 Results

6.3.1 Ventilatory Response to Acute Hypoxia and Hypercapnia Tests

To assess if the respiratory peripheral chemoreflex was modified in FASTK$^{-/-}$ mice, the ventilatory response to acute hypoxia (12% O_2; 10 min) and hypercapnia (5% CO_2 in air) tests were measured. Continuous recording of the respiratory frequency (fr; breaths/min) and tidal volume (TV; ml/kg) was made while switching from air to the gas mixture in 10 wild type (WT) and 9 knockout (KO) mice. Figure 6.1 shows that the respiratory response in normoxia was similar between KO and WT mice. Respiratory frequency, TV and MV increased also in the same way in response to hypoxia (1337 ± 205 vs 2906 ± 438 ml/min/Kg in WT and 1134 ± 107 vs 3050 ± 490 ml/min/Kg in KO), and hypercapnia (3166 ± 453 ml/min/Kg in WT and 2724 ± 149 ml/min/Kg in KO) tests. In summary, KO mice had respiratory response to hypoxia and hypercapnia very similar to WT mice.

6.3.2 CA Synthesis Rate and Secretory Response from CB

To investigate the CB functional properties in FASTK$^{-/-}$ mice, ^{3}H-CA synthesis was assessed. Fig. 6.2a shows the lack of differences in ^{3}H-CA synthesis rate in WT and KO CB mice (59.9 ± 6.2 in WT and 62.9 ± 12.5 fentomol/CB in KO; n = 6–8). Because an important part of CA content in the CB, mainly norepinephrine (NE), is contained in sympathetic nerve endings from the superior cervical ganglion (SCG; Chen et al. 1997), ^{3}H-CA synthesis was also measured in SCG under the same conditions (29.8 ± 4.45 and 23.7 ± 4.65 pmol/SCG; n = 6, Fig. 6.2b). No significant difference in CA synthesis from CB or SCG between WT and KO mice was found.

It is known that ETC blockers activate CB inducing the release of CA (Gomez-Niño et al. 2009). This activity index was used to measure CA secretion from WT and KO mice CB induced by adding the Complex I blocker rotenone (10 μM) to the incubating solution. The release of ^{3}H-CA expressed as % of tissue content was obtained from WT and KO CB incubated with a 20% O_2, 5% CO_2-equilibrated control solution, later exposed to rotenone in the incubation solution, and a post-stimulus period without the blocker. Figure 6.2c shows that rotenone evoked release of ^{3}H-CA was significantly higher in KO mice chemoreceptor cells (WT, 1.4 ± 0.4 vs 3.2 ± 0.4, KO n = 4–6; $p < 0.05$).

6.3.3 Contractile Response of Small Pulmonary Arteries

In a preliminary test, KO mice seemed to preserve the *in vivo* hypoxic pulmonary vasoconstriction (10% O_2; 2 min) with values of pulmonary artery pressure (PAP) similar to WT mice (ΔPAP 3.5 and 3.9 mm Hg in KO and WT mice, respectively; n = 2 animals of each group). To assess the effect of rotenone on PASMC,

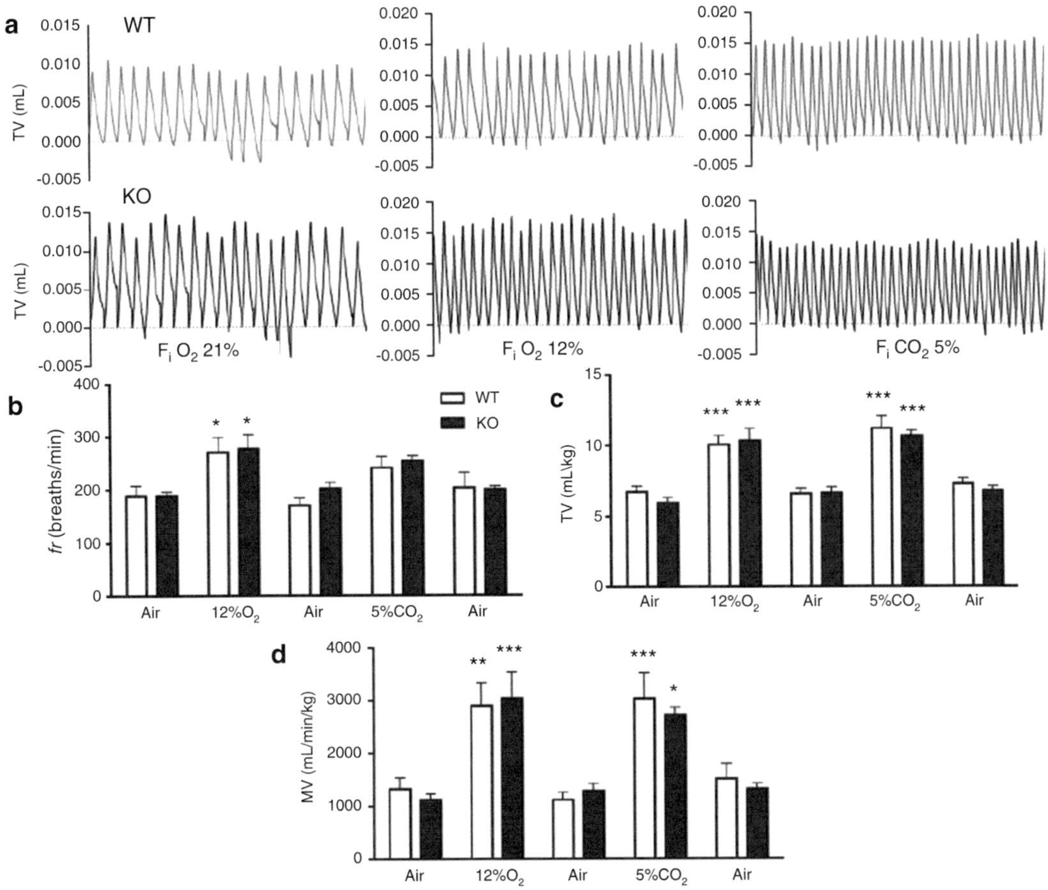

Fig. 6.1 Ventilatory parameters measured in freely moving wild type (WT) and knockout (KO) mice. (**a**) Representative plethysmographic recordings of TV showing the breathing patterns in air (FiO_2 21%), acute hypoxia (FiO_2 12%) and hypercapnia ($FiCO_2$ 5%) during 10 s. (**b–d**) show Breathing frequency (fr; breaths/min), Tidal volume (TV; ml/kg of body weight), and Minute ventilation (MV; ml/min/kg) in the same three conditions. Data are mean ± SEM obtained from 10 WT and 9 KO mice. $*p < 0.05$, $**p < 0.01$, $***p < 0.001$, acute hypoxia or hypercapnia vs air

myography was performed on the PA from 11 WT and 9 KO mice. No differences in PA contraction response to high extracellular K^+ (80 mM; 2.24 ± 0.26 mN from WT and 2.76 ± 0.29 mN from KO mice) were found (data not shown). Fig. 6.3c shows the significantly decreased relaxation response to Complex I blocker rotenone (10 μM) in KO mice (75.4 ± 5.6% in WT *vs* 35.8 ± 5.3% in KO; $p < 0.001$) in $PGF_{2\alpha}$ pre-constricted arteries.

6.4 Discussion

The high sensitivity of the CB to inhibitors of the mitochondria ETC has been one of the factors implying mitochondria in oxygen sensing by chemoreceptor cells. Changes in ROS production from mitochondrial origin imply their possible function as sensors of O_2 availability and as the origin of the low O_2 homeostatic response. Hypoxic inhibition of mitochondrial ETC results

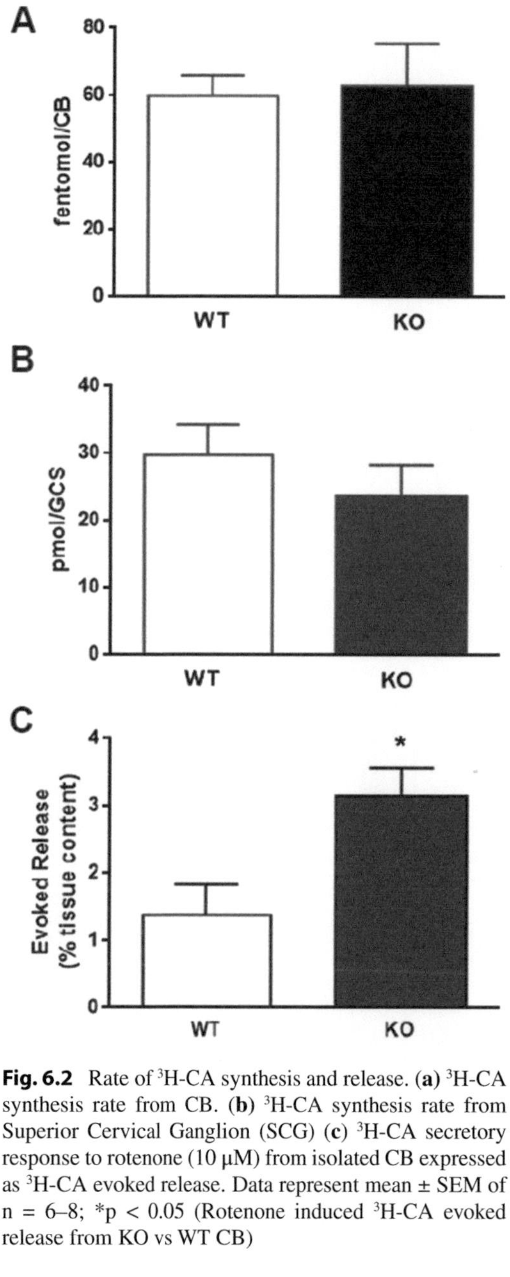

Fig. 6.2 Rate of ^{3}H-CA synthesis and release. (**a**) ^{3}H-CA synthesis rate from CB. (**b**) ^{3}H-CA synthesis rate from Superior Cervical Ganglion (SCG) (**c**) ^{3}H-CA secretory response to rotenone (10 μM) from isolated CB expressed as ^{3}H-CA evoked release. Data represent mean ± SEM of n = 6–8; *$p < 0.05$ (Rotenone induced ^{3}H-CA evoked release from KO vs WT CB)

in increased production of ROS from Complex I (Fernández-Agüera et al. 2015). This increase in ROS production is hypothesized to change the redox status of membrane ion channels and thus initiate excitation. The hypothesis that mitochondria participate in CB oxygen sensing has been tested in several studies using different mitochondrial protein deficient knockout mice. However, studies using KO mice lacking different mitochondrial Complex I proteins showed contradictory results. Lopez-Barneo's group (Fernández-Agüera et al. 2015) used conditional deletion of the Ndufs2 subunit of Complex I to study the mechanisms of O_2 transduction in the CB. They described that inactivating the Ndufs2 gene, which encodes a protein component of the ubiquinone-binding site of mitochondrial Complex I, prevented a glomus cells response to hypoxia. Loss of Ndufs2 also abolished the systemic hypoxic ventilatory response without disrupting the ventilatory response to hypercapnia, indicating that Complex I function was required for hypoxic sensitivity.

Conversely, Jain et al. (2016) deactivated the Ndufs4 gene, which codes for NADH ubiquinone oxidoreductase subunit S4, also a Complex I protein, in a mouse model that repeats many features of Leigh syndrome. They found that a hypoxic ventilatory response was triggered in KO mice similar to WT mice. Remarkably, chronic hypoxia prevented the development of many of the Leigh syndrome symptoms in the Ndufs4 KO mice without rescuing Complex I activity and significantly extended their survival, whereas chronic hyperoxia exacerbated the disease.

Archer et al. (1999) used mice lacking the gp91 subunit of NADPH oxidase (chronic granulomatous disease mice) to assess the hypothesis that NADPH oxidase is a pulmonary artery O_2-sensor. Deletion of gp91 phox did not alter p22 phox expression but severely inhibited ROS production. Nonetheless, hypoxia caused identical inhibition of whole-cell K^+ current in PASMC and hypoxic pulmonary vasoconstriction (in isolated lungs) from KO vs. WT mice. The rotenone effect was preserved in KO mice, consistent with a role for the mitochondrial ETC in O_2 sensing. Rotenone mimics hypoxia causing pulmonary vasoconstriction (Archer et al. 1993) and activating the CB. Inhibitors of the proximal mitochondrial ETC functionally block electron transport, leading to proximal accumulation of electron donors and shifting the redox state to a reduced state, as the redox hypothesis of oxygen sensing proposes. They conclude that NADPH oxidase, though a major source of lung ROS production is not the pulmonary vascular O_2 sensor in mice.

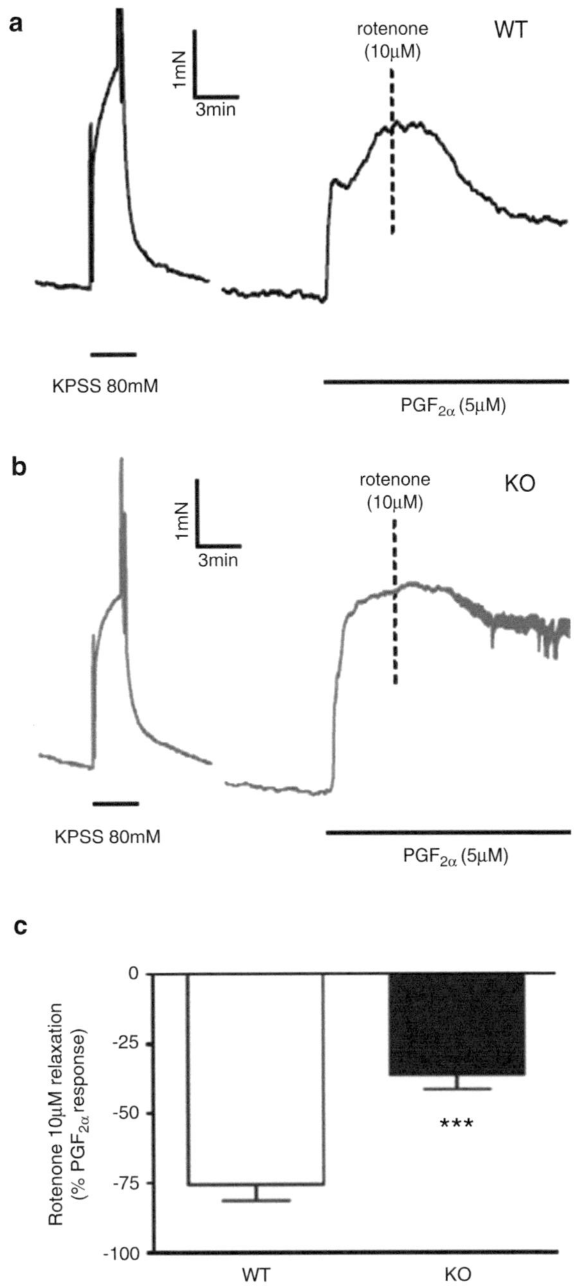

Fig. 6.3 Pulmonary artery contractile response to $PGF_{2\alpha}$ and rotenone. (**a**) Representative traces illustrating the relaxation induced by the Complex I blocker rotenone (10 μM) on PA rings pre-constricted with 5 μM $PGF_{2\alpha}$ in WT and KO mice. (**b**) Relaxation induced by 10 μM rotenone on pre-constricted arteries with $PGF_{2\alpha}$ in WT and KO mice. Data are expressed as mean ± SEM of n = 11 WT and 9 KO; $^{***}p < 0.001$)

Our results using $FASTK^{-/-}$ mice showed no significant differences of ventilatory response to acute hypoxia and hypercapnia, although a slightly increased sensitivity can be observed in TV and MV measurements. Similarly, in vitro data obtained from the isolated CB demonstrated no differences in the rate of CA synthesis. However, CA release from CB showed an increased release response from KO mice exposed to rotenone. Considering the effect of rotenone as a measure of ETC Complex I functionality, the CB release of CA is more sensitive to the Complex I blocker in KO mice. It has been reported that a moderate inhibition of the mitochondrial function can increase CB hypoxic sensitivity (Holmes et al. 2016) which is consistent with the augmented CA release in response to rotenone. The decreased functionality of FASTK$^{-/-}$ mice Complex I could explain the increased sensitivity to the Complex I blocker.

As it was found with the acute respiratory response to hypoxia, KO mice seem to preserve the *in vivo* hypoxic pulmonary vasoconstriction, showing similar pulmonary artery pressure values to WT mice. The PA response to rotenone was partially blocked in the FASTK$^{-/-}$. The reduced function of mitochondrial complex I in KO mice may explain the increased effect of the complex I blocker on PASMC, as it was also observed in the CB chemoreceptor cells CA release response to rotenone. It has been reported that inhibition of ETC in general and particularly rotenone, induce an increase of ROS generation and a decrease in ATP genesis in CB (Gomez-Niño et al. 2009). Mitochondrial inhibition is also known to reduce vascular contractility in the setting of unchanged global intracellular Ca^{2+} but with altered coding of cellular Ca^{2+}-waves (Rahman et al. 2013). Since signal transduction reactions are localized processes, it could be that specific redox or Ca^{2+} signaling might be linked to specific sensing in PASMC and CB chemoreceptor cells.

In summary, FASTK$^{-/-}$ and WT mice maintain a similar ventilatory response in normoxia and under hypoxic and hypercapnic stress, implying that there is not a direct effect of Complex I ND6 protein in CB oxygen sensing. The same lack of difference between both groups of mice appears to be in the case of HPV, although the number of experiments is not significant. The diverse results obtained from ablating different mitochondrial Complex I proteins show that hypoxia activates adaptive programs in KO mice that would allow survival with restrictive oxygen levels and that the normal functioning of mitochondrial Complex I is not strictly required for the hypoxic ventilatory chemoreflex response. Compensation for KO mice can occur, especially in the CB response, where it appears to have redundant mechanisms for oxygen sensing. An alternative explanation would take into consideration whether experiments using KO animals are an irrefutable evidence of the role of a protein in a specific process.

Acknowledgements This work was supported by Grants BFU2015-63706R (MINECO, FEDER-UE) and CIBER CB06/06/0050 from ISCiii (Spain). The Authors thank R. Cantalapiedra and R. Carretero for technical assistance.

References

Archer SL, Huang J, Henry T, Peterson D, Weir EK (1993) A redox-based O2 sensor in rat pulmonary vasculature. Circ Res 73(6):1100–1112

Archer SL, Reeve HL, Michelakis E, Puttagunta L, Waite R, Nelson DP, Dinauer MC, Weir EK (1999) O2 sensing is preserved in mice lacking the gp91 phox subunit of NADPH oxidase. Proc Natl Acad Sci U S A 96(14):7944–7949

Chen J, Gomez-Niño A, Gonzalez C, Dinger B, Fidone S (1997) Stimulus-specific mobilization of dopamine and norepinephrine stores in cat carotid body. J Auton Nerv Syst 67:109–113

Fernández-Agüera MC, Gao L, González-Rodríguez P, Pintado CO, Arias-Mayenco I, García-Flores P, García-Pergañeda A, Pascual A, Ortega-Sáenz P, López-Barneo J (2015) Oxygen sensing by arterial chemoreceptors depends on mitochondrial complex I signaling. Cell Metab 22(5):825–837

Gomez-Niño A, Agapito MT, Obeso A, Gonzalez C (2009) Effects of mitochondrial poisons on glutathione redox potential and carotidbody chemoreceptor activity. Respir Physiol Neurobiol 165:104–111

Gonzalez C, Almaraz L, Obeso A, Rigual R (1994) Carotid body chemoreceptors: from natural stimuli to sensory discharges. Physiol Rev 74(4):829–898

Gonzalez-Obeso E, Docio I, Olea E, Cogolludo A, Obeso A, Rocher A, Gomez-Niño A (2017) Guinea pig oxygen-sensing and carotid body functional properties. Front Physiol 8:285

Holmes AP, Turner PJ, Buckler KJ, Kumar P (2016) Moderate inhibition of mitochondrial function augments carotid body hypoxic sensitivity. Pflugers Arch 468(1):143–155

Jain IH, Zazzeron L, Goli R, Alexa K, Schatzman-Bone S, Dhillon H, Goldberger O, Peng J, Shalem O, Sanjana NE, Zhang F, Goessling W, Zapol WM, Mootha VK (2016) Hypoxia as a therapy for mitochondrial disease. Science 352(6281):54–61

Jourdain A, Koppen M, Rodley CD, Maundrell K, Gueguen N, Reynier P, Guaras AM, Enriquez JA, Anderson P, Simarro M, Martinou JC (2015) A mitochondria-specific isoform of FASTK is present in mitochondrial RNA granules and regulates gene expression and function. Cell Rep 10:1110–1121

López-Barneo J, López-López JR, Ureña J, González C (1988) Chemotransduction in the carotid body: K+ current modulated by PO2 in type I chemoreceptor cells. Science 241(4865):580–582

Post JM, Hume JR, Archer SL, Weir EK (1992) Direct role for potassium channel inhibition in hypoxic pulmonary vasoconstriction. Am J Phys 262(4 Pt 1):C882–C890

Rahman A, Ekman M, Shakirova Y, Andersson KE, Morgelin M, Erjefalt MJ, Brundin P, Li JL, Sward K (2013) Late onset vascular dysfunction in the R6/1 model of Huntington's disease. Eur J Pharmacol 698:345–353

Richalet JP (1997) Oxygen sensors in the organism: examples of regulation under altitude hypoxia in mammals. Comp Biochem Physiol A Physiol 118(1):9–14

Simarro M, Giannattasio G, De la Fuente MA, Benarafa C, Subramanian KK, Ishizawar R, Balestrieri B, Andersson EM, Luo HR, Orduña A, Boyce J, Anderson P (2010) Fas-activated serine/threonine phosphoprotein promotes immune-mediated pulmonary inflammation. J Immunol 184(9):5325–5332

7 Topical Application of Connexin43 Hemichannel Blocker Reduces Carotid Body-Mediated Chemoreflex Drive in Rats

David C. Andrade, Rodrigo Iturriaga, Camilo Toledo, Claudia M. Lucero, Hugo S. Díaz, Alexis Arce-Álvarez, Mauricio A. Retamal, Noah J. Marcus, Julio Alcayaga, and Rodrigo Del Rio

Abstract

The carotid body (CB) is the main arterial chemoreceptor involved in oxygen sensing. Upon hypoxic stimulation, CB chemoreceptor cells release neurotransmitters, which increase the frequency of action potentials in sensory nerve fibers of the carotid sinus nerve. The identity of the molecular entity responsible for oxygen sensing is still a matter of debate; however several ion channels have been shown to be involved in this process. Connexin-based ion channels are expressed in the CB; however a definitive role for these channels in mediating CB oxygen sensitivity has not been established. To address the role of these channels, we studied the effect of blockers of connexin-based ion channels on oxygen sensitivity of the CB. A connexin43 (Cx43) hemichannel blocking agent (CHBa) was applied topically to the CB and the CB-mediated hypoxic ventilatory response (F_iO_2 21, 15, 10 and 5%) was measured in adult male Sprague-Dawley rats (~250 g). In normoxic conditions, CHBa had no effect on tidal volume or respiratory rate, however Cx43 hemichannels inhibition by CHBa significantly impaired the CB-mediated

D. C. Andrade · C. Toledo · C. M. Lucero · H. S. Díaz
A. Arce-Álvarez
Laboratory of Cardiorespiratory Control, Department of Physiology, Faculty of Biological Sciences, Pontificia Universidad Católica de Chile, Santiago, Chile

R. Iturriaga
Laboratorio de Neurobiología, Facultad de Ciencias Biológicas, Pontificia Universidad Católica de Chile, Santiago, Chile

M. A. Retamal
Centro de Fisiología Celular e Integrativa, Clínica Alemana Universidad Del Desarrollo, Santiago, Chile

N. J. Marcus
Department of Physiology and Pharmacology, Des Moines University, Des Moines, IA, USA

J. Alcayaga
Laboratorio de Fisiología Celular, Universidad de Chile, Santiago, Chile

R. Del Rio (✉)
Centro de Envejecimiento y Regeneración (CARE), Pontificia Universidad Católica de Chile, Santiago, Chile

Centro de Excelencia en Biomedicina de Magallanes (CEBIMA), Universidad de Magallanes, Punta Arenas, Chile

Laboratory of Cardiorespiratory Control, Department of Physiology, Faculty of Biological Sciences, Pontificia Universidad Católica de Chile, Santiago, Chile
e-mail: rdelrio@bio.puc.cl

E. B. Gauda et al. (eds.), *Arterial Chemoreceptors*, Advances in Experimental Medicine and Biology 1071, https://doi.org/10.1007/978-3-319-91137-3_7

chemoreflex response to hypoxia. CHBa reduced both the gain of the hypoxic ventilatory response (HVR) and the maximum HVR by ~25% and ~50%, respectively. Our results suggest that connexin43 hemichannels contribute to the CB chemoreflex response to hypoxia in rats. Our results suggest that CB connexin43 hemichannels may be pharmacological targets in disease conditions characterized by CB hyperactivity.

Keywords

Carotid body · Chemoreflex · Connexin43 · Hypoxia

7.1 Introduction

The carotid body (CB) chemoreceptors are located in the bifurcation of the common carotid artery and have a primary role in sensing the partial pressure of O_2, CO_2 and pH in arterial blood (Iturriaga and Alcayaga 2004; Nurse and Piskuric 2013). The CB is composed of two primary cell types: the chemoreceptor cells (glomus) and the sustentacular cells (López-Barneo et al. 2008), which are believed to modulate glomus cell activity through a purinergic P2Y receptor-dependent pathway (Murali and Nurse 2016). Despite significant research devoted to understanding the function of this small but important organ, the molecular mechanisms by which glomus cells detect changes in PO_2 are not well understood. The most well-accepted theory is that reductions of P_{O2} (directly or indirectly) decrease K^+ channel (TASK) activity (Buckler et al. 2006), depolarizing the cell and leading to opening of voltage sensitive L-type Ca^{2+} channels, which in turn causes increases in intracellular Ca^{2+} which ultimately lead to the release of one or more neurotransmitters to the extracellular space (López-Barneo et al. 2016). Recent evidence also suggests a potentially important role of HIF-1α (Nurse 2017) and gaseous neurotransmitters such as CO and H_2S (Prabhakar and Peng 2017) in controlling CB oxygen sensing. Once activated by low oxygen partial pressure, glomus cells release several neurotransmitters including ATP (Zhang et al. 2000), acetylcholine (Nurse and Zhang 1999) and dopamine (Buerk et al. 1998). The release of these neurotransmitters impacts the rate of discharge of petrosal ganglion neurons (Smith and Mills 1976), which deliver "P_{O2} – mediated sensory information" to the nucleus of the tractus solitarii (Lipski et al. 1977; Iturriaga and Alcayaga 2004).

In addition to the chemical synapses within the CB, electrical coupling between glomus cells has also been observed (Monti-Bloch et al. 1993; Abudara and Eyzaguirre 1994). The structural basis of these electrical synapses are the connexins. These proteins are a family of transmembrane proteins which contain four transmembrane segments; two extracellular loops, one intracellular loop, and both the N- and C- terminals facing the cytoplasm (Retamal et al. 2016). Connexins are named according to their predicted molecular weight, thus connexin43 (Cx43) is predicted to have a molecular weight of ~43 kDa. In 1996, Kondo and Iwasa (1996) conducted an ultra-structural study of adult rat CBs which confirmed the presence of gap junction channels (GJC)-like structures between glomus cells, between glomus cells and sustentacular cells, and between glomus cells and petrosal neuron terminals. This observation was confirmed by immunofluorescent detection of Cx43 in the CB (Abudara et al. 1999). GJCs formed by Cx43 are regulated by cAMP and pH in cultured CB cells which induced an increase and decrease in cell coupling, respectively (for further details see Abudara et al. 2000, 2001). Other studies have demonstrated that sustained stimuli may also elicit changes in GJC-like expression, in which 2 weeks of hypobaric hypoxia increased expression of Cx43 in glomus cells and in petrosal neurons (Chen et al. 2002). Interestingly, it is well known that Cx43 can also form GJCs between cells and also hemichannels in cells *in vitro* as well as *in vivo* (Sáez et al. 2005). Hemichannels are half of a GJC and are found in non-appositional plasma membranes (Sáez et al. 2005). While GJCs allow the passive flow of ions and molecules between cells, hemichannels permit the exchange of ions and small molecules between the extracellular media and the cytoplasm (Retamal 2014). In spite of their low open probability (Contreras et al. 2003), Cx43 hemichannels allow the release of paracrine signaling molecules such as ATP and glutamate (Orellana and Stehberg 2014). Therefore, Cx43 hemichannels are of potential interest as potential modulators of CB chemosensory activity (Reyes et al. 2014). Despite the fact that numerous studies suggest a potential role of Cx43-based hemichannels in regulation of CB activity, this hypothesis has not been specifically tested. Thus, the aim of this work was to test the effect of a Cx43-specific hemichannel blocker (CHBa) on CB O_2 sensitivity *in vivo*.

7.2 Methods

7.2.1 Animals

Adult male Sprague-Dawley rats (n = 8), weighing between 250 and 300 g, were used in these experiments. All experiments were approved by the Bioethical Committee of the Facultad de Ciencias Biológicas, P. Universidad Católica de Chile and were carried out under the guidelines of the American Physiological Society, the Guía para el Cuidado y Uso de los Animales de Laboratorio from CONICYT and the National Institutes of Health Guide for the Care and Use of Laboratory Animals.

7.2.2 CB-Mediated Ventilatory Chemoreflex Function

Rats were anaesthetized with sodium pentobarbitone (40 mg kg^{-1}; ip) and placed in the supine position. The trachea was cannulated for airflow recording, and connected to a pneumotachograph to measure tidal volume (V_t), respiratory frequency (R_f), and minute inspiratory volume (V_E). During the experiment, rectal temperature was measured and maintained at 38.0 ± 0.5 °C with a regulated heating pad. Supplemental doses of sodium pentobarbitone were administered as necessary to maintain a surgical plane of anesthesia. Signals were acquired with an analogue-digital system PowerLAB 8SP (ADInstruments Pty Ltd., Castle Hill, Australia), calibrated and analyzed with the Labchart 7-Pro software (ADInstruments Pty Ltd., Australia). CB-mediated chemoreflex function was estimated as previously described (Del Rio et al. 2010, 2013, 2014). Briefly, rats breathed hypoxic gas mixtures (Fi_{O2} 5%,10%,15%) until a semi-steady state response was achieved (20–25 s). All recordings were made at an ambient temperature of 25 ± 2 °C. During chemoreflex testing, respiratory variables (RF and V_t) were averaged over 10 consecutive breaths.

7.2.3 Acute Blockade of Connexin 43 Hemichannels

A Cx43-selective hemichannel-blocking agent (CHBa) was used as previously described (Stehberg et al. 2012; Wang et al. 2013). CHBa (based on an 11 amino acid sequence targeting the second extracellular loop of the Cx43 protein) was dissolved in Pluronic F-127 at 20 °C, for a final concentration of 25 mM. The CBs were visualized and isolated from the surrounding tissue. Then, 20 μl of CHBa solution was topically applied to the CBs. Ten minutes were allowed for Pluronic F-127 to polymerize before ventilatory recordings were initiated.

7.2.4 Cx43 Expression

CB Cx43 protein expression was assessed using immunoblot as previously described (Del Rio et al. 2017). Briefly, after euthanizing the rats (pentobarbital 100 mg/kg i.p.), the carotid sinus region containing the CBs was quickly removed and snap frozen on dry ice and stored at −80 °C. The tissue was lysed in 200 μL of RIPA buffer with fresh protease inhibitor cocktail (Sigma Aldrich, St. Louis, MO, USA) and total protein concentration was assessed using a Thermo Scientific BCA Assay kit (Waltham, MA). Samples were loaded onto a 10% SDS-PAGE gel (50 μg/20 μl per well) and electrophoresis was performed. Then, proteins were transferred to a PVDF membrane (Millipore, USA) and the membrane was probed overnight with the following antibodies: rabbit anti- Cx43 (1:750, Cat. #PPS046, Novus Biological), and mouse anti- β-actin (1:500–1:1000, Cat. #sc-47,778, Santa Cruz, USA). Following washes with PBST, the appropriate secondary antibodies were added to each membrane. Blots were developed using a ECL kit (ThermoFischer, USA) scanner, and quantitative analysis of band densitometry was performed using Li-Cor Imaging software (Li-cor, USA).

7.2.5 Statistical Analysis

Data was expressed as means ± S.E.M. The differences in hypoxic ventilatory response was analyzed by paired t-test. A p value of <.05 was considered statistically significant.

7.3 Results

7.3.1 Expression of Cx43 in the Carotid Body

Figure 7.1 illustrates representative immunoblots of Cx43 expression in homogenates of rat CBs. Two bands close to 43 kDa were observed, however, bands with higher electrophoretic mobility were more abundant, indicating that in the CB, Cx43 is mainly in the dephosphorylated form.

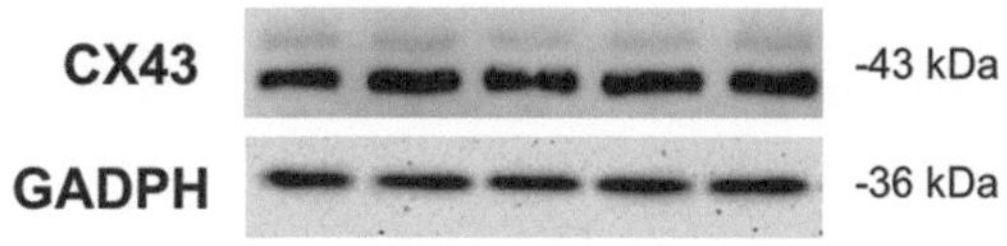

Fig. 7.1 Cx43 is highly expressed in the rat carotid body. Representative immunoblots from CBs excised from adult male Sprague-Dawley rats showing protein expression of Cx43. GADPH was used as the loading control

Table 7.1 Resting ventilatory parameters before and after Cx43 hemichannel blockade in the carotid bodies

	Before (n = 4)	After (n = 4)
V_t (ml)	4.80 ± 0.82	5.05 ± 0.29
R_f (breaths/min)	81.80 ± 3.34	80.66 ± 4.64
V_E (ml/min)	400.99 ± 83.51	409.36 ± 40.33

Data are showed as mean ± SEM. V_t tidal volume, R_f respiratory frequency, V_E minute ventilation

7.3.2 Effects of CHBa on Resting Breathing Parameters

The effects of CHBa on resting V_t, R_f and V_E are summarized in Table 7.1. CHBa did not induce significant changes in V_E when compared to values obtained before CHBa exposure ($p > 0.05$). However, we did observe a small trend for increasing V_t values following CHBa (Table 7.1).

7.3.3 Effects of Cx43 Hemichannel Blockade on the Hypoxic Ventilatory Response

In response to acute hypoxic stimulation, rats increased both Vt and Rf (Fig. 7.2). We observed a maximal response of >200% increase in V_E following exposure to F_iO_2 5% (Fig. 7.3). Remarkably, CHBa significantly reduced the hypoxic ventilatory response (HVR) in rats (Fig. 7.2). As shown in Fig. 7.3, CHBa markedly attenuated the V_t and R_f responses to several levels of hypoxia when compared to the values obtained before CHBa administration. Furthermore, CHBa reduced both the HVR gain and the maximum HVR by ~25% and ~50%, respectively. No effects on hyperoxic ventilatory responses were found (Fig. 7.3).

7.4 Discussion

The major findings of these studies are: (i) Cx43 is constitutively expressed in CB tissue and (ii) specific blockade of Cx43 hemichannels within the CB reduces the ventilatory response to hypoxia without changing resting breathing parameters. Taken together, our results strongly suggest that Cx43 hemichannels contribute to/

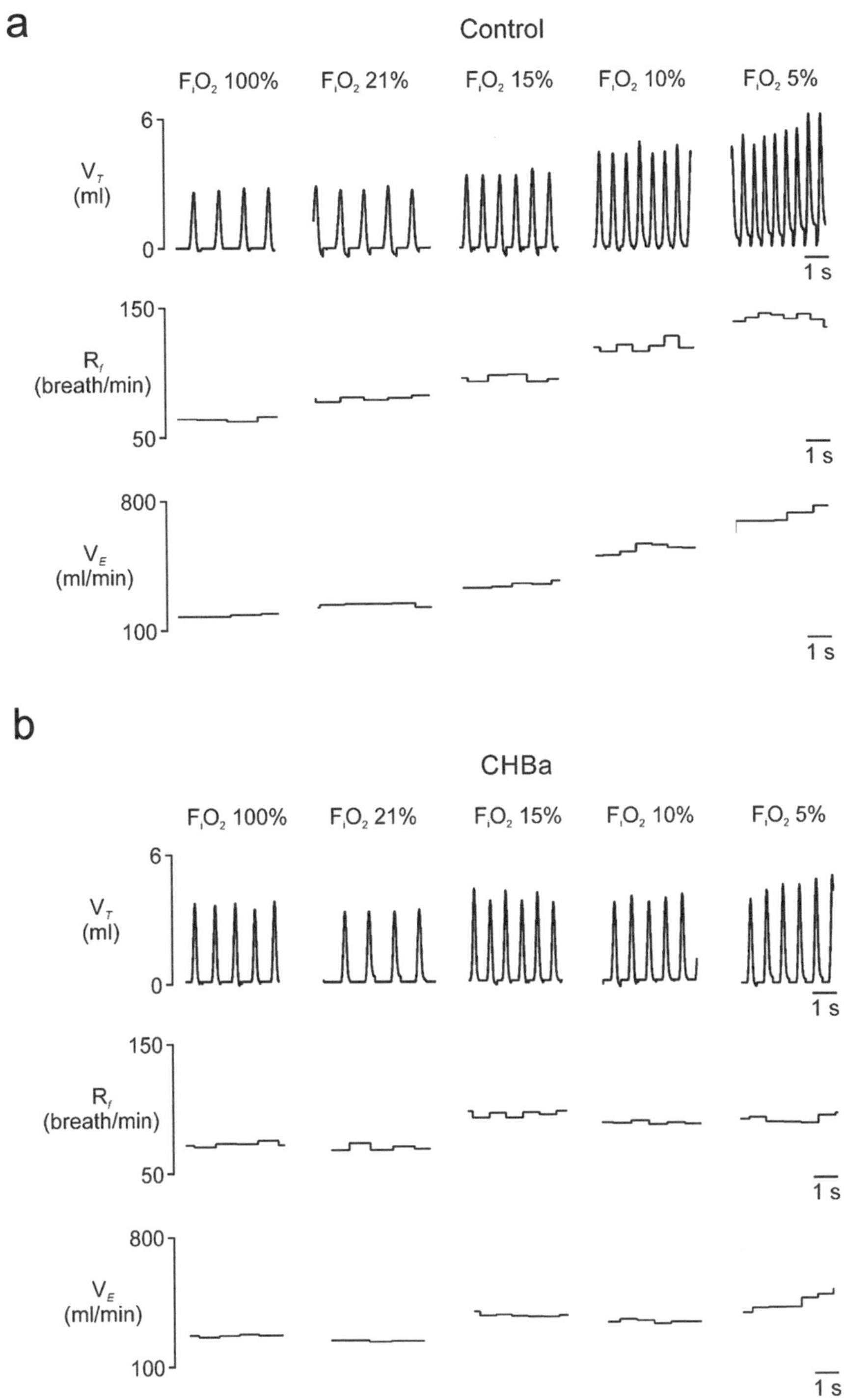

Fig. 7.2 Topical application of a connexin hemichannel blocking agent (CHBa) reduce the hypoxic ventilatory response. (**a**) Representative traces of tidal volume (V_t), respiratory frequency (R_f) and minute ventilation (V_E) during FiO_2 100, 21, 15, 10 and 5% in one Control animal. (**b**) Representative traces of tidal volume (V_t), respiratory frequency (R_f) and minute ventilation (V_E) during FiO_2 100, 21, 15, 10 and 5% after CHBa. Note that the ventilatory response to hypoxia is significantly decreased compared to the Control condition

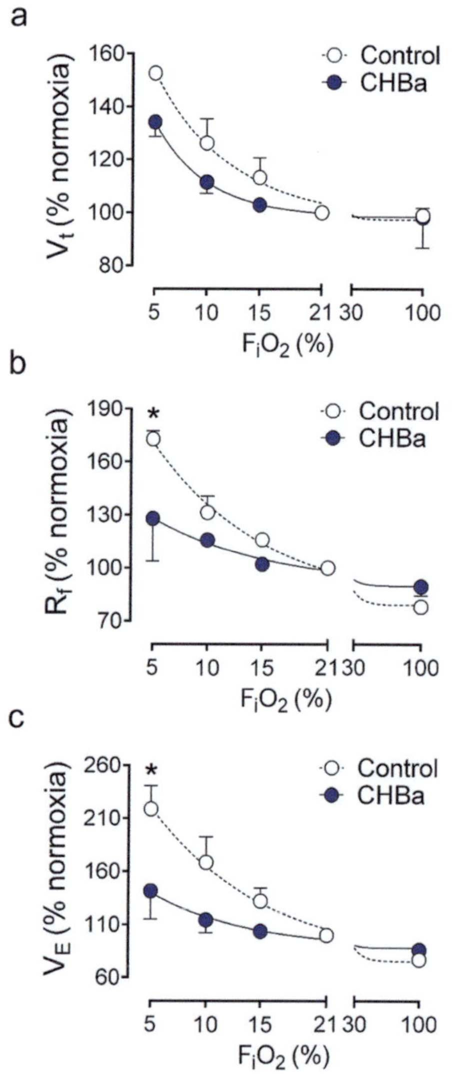

Fig. 7.3 Summary data showing the effect of CHBa on the CB-mediated ventilator chemoreflex response to hypoxia. (**a**) Effects of CHBa on Vt. Note that V_t responses in hypoxia were slightly reduced after CHBa application. (**b**) Effects of CHBa application on hypoxic R_f response. Note that CHBa markedly attenuate the R_f response to hypoxia. (**c**) After CHBa application the V_E response to hypoxia was significantly reduced. Holm-Sidak posthoc after two ways ANOVA. *, $p < 0.05$ vs. CHBa FiO_2 5%, n = 4

modulate the acute CB-mediated ventilatory responses to hypoxia.

Previous studies indicate that CB glomus cells express gap junction-like structures (Kondo and Iwasa 1996), and the work of Abudara et al. (1999) showed that CBs express Cx43. Our results confirm these previous observations and indicate a functional role for Cx43 in the CB. Activated Cx43 hemichannels allow the exchange of small molecules (~1 kDa) from extracellular media to the cytoplasm (Sáez et al. 2005). It is worth noting that our data showed that resting breathing parameters were not affected by Cx43 hemichannel blockade, which suggests that these hemichannels are not normally open in the CB during normoxic conditions. This finding is consistent with the observations of Contreras and co-workers in 2003 who found that Cx43 hemichannels have a very low open probability under physiological conditions (Contreras et al. 2003). In contrast to the lack of an effect observed in normoxia, our results show that CHBa markedly reduced the CB-mediated ventilatory response to hypoxia. This effect strongly suggests that Cx43 hemichannels are activated at low oxygen partial pressures. Cx43 hemichannel opening has been observed under hypoxic conditions or metabolic stress in other experimental models (John et al. 1999; Wang et al. 2014). Interestingly, it has been shown that stimulation of CB glomus cells with hypoxia induces the release of several small molecules to the extracellular milieu (Zhang et al. 2000). Thus we postulate that Cx43 hemichannels may participate in this process. Future studies are needed to address the biophysical properties of Cx43 hemichannels in the CB under normoxic and hypoxic conditions.

One of the most critical adaptive physiological responses to hypoxic challenges is an increase in pulmonary ventilation (Moore et al. 1986). Importantly, this ventilatory response is almost entirely dependent on functional CBs as denervation of the CBs totally abolishes the hypoxic ventilatory response (Del Rio et al. 2013). Despite this appreciation of the CBs as the primary mediator of the hypoxic ventilatory response, the precise molecular mechanism(s) underpinning oxygen sensing in the CB chemoreceptors is still debated. In these studies, we show that topical application of a Cx43 hemichannel blocker to the CB, significantly reduces the hypoxic ventilatory response. These results strongly suggest that Cx43 hemichannels modulate CB chemotransduction of hypoxic stimuli. Considering that these hemichannels allow the exchange of ions and other small molecules (i.e. ATP and glutamate) between the extracellular media and the cytoplasm (Sáez et al. 2005), and that CB chemoreceptor cells release ATP to the extracellular milieu (Zhang et al. 2000), it is likely that Cx43 hemichannels are activated by low oxygen partial pressure and that this results in the release of one or several neuromodulators within the CB. Further studies are needed to establish a causal relationship between Cx43 hemichannel function and CB hypoxic sensitivity.

In summary, our data indicate that topical application of a Cx43 hemichannel blocking agent (CHBa) in the CB significantly attenuates the hypoxic ventilatory response. Taking into account that Cx43 protein is expressed in the CB and that upon activation Cx43 hemichannels allow the release of several molecules from glomus cells, it is plausible that one or more neuromodulators are released by CB glomus cells through hemichannels during hypoxic stimulation. Future studies should focus on the functional role of connexin-based hemichannels in CB oxygen sensing and mediation of the hypoxic ventilatory response.

Acknowledgments This study was supported by Fondecyt 1140275 and 1180172 (RDR) and Fondecyt 1160227 (MAR). NJM is supported by a grant from the National Institutes of Health (HL-138600-01).

References

Abudara V, Eyzaguirre C (1994) Electrical coupling between cultured glomus cells of the rat carotid body: observations with current and voltage clamping. Brain Res 664:257–265

Abudara V, Garcés G, Sáez JC (1999) Cells of the carotid body express connexin43 which is up-regulated by cAMP. Brain Res 849:25–33

Abudara V, Eyzaguirre C, Sáez JC (2000) Short- and long-term regulation of rat carotid body gap junctions by cAMP. Identification of connexin43, a gap junction subunit. Adv Exp Med Biol 475:359–369

Abudara V, Jiang RG, Eyzaguirre C (2001) Acidic regulation of junction channels between glomus cells in the rat carotid body. Possible role of [Ca(2+)]. Brain Res 916:50–60

Buckler KJ, Williams BA, Orozco RV, Wyatt CN (2006) The role of TASK-like K+ channels in oxygen sensing in the carotid body. Novartis Found Symp 272:73–85

Buerk DG, Osanai S, Mokashi A, Lahiri S (1998) Dopamine, sensory discharge, and stimulus interaction with CO2 and O2 in cat carotid body. J Appl Physiol 85:1719–1726

Chen J, He L, Dinger B, Stensaas L, Fidone S (2002) Chronic hypoxia upregulates connexin43 expression in rat carotid body and petrosal ganglion. J Appl Physiol 92:1480–1486

Contreras JE, Sáez JC, Bukauskas FF, Bennett MV (2003) Gating and regulation of connexin 43 (Cx43) hemichannels. Proc Natl Acad Sci U S A 100(20):11388–11393

Del Rio R, Moya EA, Iturriaga R (2010) Carotid body and cardiorespiratory alterations in intermittent hypoxia: the oxidative link. Eur Respir J 36(1):143–150

Del Rio R, Marcus NJ, Schultz HD (2013) Carotid chemoreceptor ablation improves survival in heart failure: rescuing autonomic control of cardiorespiratory function. J Am Coll Cardiol 62(25):2422–2430

Del Rio R, Moya EA, Iturriaga R (2014) Carotid body potentiation during chronic intermittent hypoxia: implication for hypertension. Front Physiol 5:434

Del Rio R, Andrade DC, Toledo C et al (2017) Carotid body-mediated Chemoreflex drive in the setting of low and high output heart failure. Sci Rep 7:8035

Iturriaga R, Alcayaga J (2004) Neurotransmission in the carotid body: transmitters and modulators between glomus cells and petrosal ganglion nerve terminals. Brain Res Rev 46:46–53

John SA, Kondo R, Wang SY, Goldhaber JI, Weiss JN (1999) Connexin-43 hemichannels opened by metabolic inhibition. J Biol Chem 274:236–240

Kondo H, Iwasa H (1996) Re-examination of the carotid body ultrastructure with special attention to intercellular membrane appositions. Adv Exp Med Biol 410:45–50

Lipski J, McAllen RM, Spyer KM (1977) The carotid chemoreceptor input to the respiratory neurones of the nucleus of tractus solitarus. J Physiol 269:797–810

López-Barneo J, Ortega-Sáenz P, Pardal R, Pascual A, Piruat JI (2008) Carotid body oxygen sensing. Eur Respir J 32:1386–1398

López-Barneo J, González-Rodríguez P, Gao L, Fernández-Agüera MC, Pardal R, Ortega-Sáenz P (2016) Oxygen sensing by the carotid body: mechanisms and role in adaptation to hipoxia. Am J Physiol Cell Physiol 310:629–642

Monti-Bloch L, Abudara V, Eyzaguirre C (1993) Electrical communication between glomus cells of the rat carotid body. Brain Res 622:119–131

Moore LG, Harrison GL, McCullough RE, McCullough RG, Micco AJ, Tucker A, Weil JV, Reeves JT (1986) Low acute hypoxic ventilatory response and hypoxic depression in acute altitude sickness. J Appl Physiol 60:1407–1412

Murali S, Nurse CA (2016) Purinergic signalling mediates bidirectional crosstalk between chemoreceptor type I and glial-like type II cells of the rat carotid body. J Physiol 594:391–406

Nurse CA (2017) A sensible approach to making sense of oxygen sensing. J Physiol 595:6087–6088

Nurse CA, Piskuric NA (2013) Signal processing at mammalian carotid body chemoreceptors. Semin Cell Dev Biol 24:22–30

Nurse CA, Zhang M (1999) Acetylcholine contributes to hypoxic chemotransmission in co cultures of rat type 1 cells and petrosal neurons. Respir Physiol 115:189–199

Orellana JA, Stehberg J (2014) Hemichannels: new roles in astroglial function. Front Physiol 17:193

Prabhakar NR, Peng YJ (2017) Oxygen sensing by the carotid body: past and present. Adv Exp Med Biol 977:3–8

Retamal MA (2014) Connexin and Pannexin hemichannels are regulated by redox potential. Front Physiol 25:80

Retamal MA, García IE, Pinto BI, Pupo A, Báez D, Stehberg J, Del Rio R, González C (2016) Extracellular cysteine in connexins: role as redox sensors. Front Physiol 28:1

Reyes EP, Cerpa V, Corvalán L, Retamal MA (2014) Cxs and Panx- hemichannels in peripheral and central chemosensing in mammals. Front Cell Neurosci 8:123

Sáez JC, Retamal MA, Basilio D, Bukauskas FF, Bennett MV (2005) Connexin-based gap junction hemichannels: gating mechanisms. Biochim Biophys Acta 1711:215–224

Smith PG, Mills E (1976) Autoradiographic identification of the terminations of petrosal ganglion neurons in the cat carotid body. Brain Res 113:174–178

Stehberg J, Moraga-Amaro R, Salazar C, Becerra A, Echeverría C, Orellana JA, Bultynck G, Ponsaerts R, Leybaert L, Simon F, Sáez JC, Retamal MA (2012) Release of gliotransmitters through astroglial connexin 43 hemichannels is necessary for fear memory consolidation in the basolateral amygdala. FASEB J 26:3649–3657

Wang N, De Bock M, Decrock E, Bol M, Gadicherla A, Bultynck G, Leybaert L (2013) Connexin targeting peptides as inhibitors of voltage- and intracellular Ca2+–triggered Cx43hemichannel opening. Neuropharmacology 75:506–516

Wang J, Ma A, Xi J, Wang Y, Zhao B (2014) Connexin 43 and its hemichannels mediate hypoxia-ischemia-induced cell death in neonatal rats. Child Neurol Open 1:2329048x14544955

Zhang M, Zhong H, Vollmer C, Nurse CA (2000) Co-release of ATP and ACh mediates hypoxic signalling at rat carotid body chemoreceptors. J Physiol 525:143–158

Proinflammatory Cytokines in the Nucleus of the Solitary Tract of Hypertensive Rats Exposed to Chronic Intermittent Hypoxia

8

María Paz Oyarce and Rodrigo Iturriaga

Abstract

Obstructive sleep apnea (OSA) is characterized by chronic intermittent hypoxia (CIH), which is considered the main factor for developing hypertension. Sympathetic overflow, oxidative stress and inflammation have been associated with the CIH-induced hypertension. In rats exposed to CIH mimicking OSA, intermittent hypoxia enhanced carotid body (CB) chemosensory discharge, leading to an increase in arterial blood pressure in 3–5 days. In addition, CIH increases the CB levels of proinflammatory cytokines IL-1β, IL-6 and TNF-α in the CB. Proinflammatory molecules have been also involved in neurogenic hypertension acting on brain cardiovascular centers, like the nucleus of the solitary tract (NTS), which is the primary site for afferent CB inputs. Accordingly, we aim to study if proinflammatory cytokines in the NTS may play a role in the hypertension induced by CIH. Male Sprague-Dawley rats 250 g were exposed to CIH (5% O_2, 12 times/h, 8 h/day) for 7–28 days. Brains were removed and processed to measure IL-1β, IL-6 and TNF-α in the NTS using qPCR and immunofluorescence. The mRNA levels were significantly augmented in the NTS of rats exposed during 21 days to CIH compared with control animals. In addition, a significant increase of IL-1β, IL-6 and TNF-α immunofluorescence was found in the NTS at day 28 of CIH exposure compared with control rats. Present results suggest that proinflammatory cytokines in the NTS may contribute to the maintenance of hypertension in CIH-exposed animals.

Keywords

Obstructive sleep apnea · Carotid body · Intermittent hypoxia · Proinflammatory cytokines · Nucleus of the tractus solitary · Hypertension

M. P. Oyarce (✉) · R. Iturriaga
Laboratorio de Neurobiología, Facultad de Ciencias Biológicas, Pontificia Universidad Católica de Chile, Santiago, Chile
e-mail: mpoyarce@uc.cl; riturriaga@bio.puc.cl

8.1 Introduction

Obstructive sleep apnea (OSA) is a sleep-breathing disorder recognized as an independent risk factor for systemic hypertension and stroke (Dempsey et al. 2010; Somers et al. 2008). Chronic intermittent hypoxia (CIH) is considered the main factor for the progression of the hypertension in OSA patients (Dempsey et al. 2010; Iturriaga et al. 2016; Somers et al. 2008). It has been proposed that CIH produces oxidative stress, inflammation, and sympathetic hyperactivity, leading to endothelial dysfunction and

E. B. Gauda et al. (eds.), *Arterial Chemoreceptors*, Advances in Experimental Medicine and Biology 1071, https://doi.org/10.1007/978-3-319-91137-3_8

hypertension (Iturriaga et al. 2016, 2017; Dempsey et al. 2010; Garvey et al. 2009; Somers et al. 2008). In rats exposed to CIH mimicking OSA, intermittent hypoxia produces a potentiation of the carotid body (CB) chemosensory discharges in normoxic and hypoxic conditions, leading to sympathetic overactivity that contributes to increase arterial blood pressure (~10 mmHg) in 3–5 days (Del Rio et al. 2016; Iturriaga et al. 2017).

The arterial blood pressure regulatory systems, such as the renin-angiotensin system and sympathetic nervous system are activated by proinflammatory cytokines like IL-6 and TNF-α (Granger 2006). Indeed, proinflammatory molecules acting on brainstem cardiovascular centers have been involved in the maintenance of high arterial pressure in experimental models of neurogenic hypertension (Waki et al. 2008; McBryde et al. 2013). In the CB, proinflammatory cytokines have been proposed as mediators of the chemosensory potentiation induced by CIH (Lam et al. 2012; Iturriaga et al. 2009; Del Rio et al. 2011, 2012). Indeed, Del Rio et al. (2012) studied the effects of the anti-inflammatory agent ibuprofen on the TNF-α and IL-1β levels in the rat CB, the potentiation of the CB chemosensory responses and the development of systemic hypertension following 21 days of CIH. They found that ibuprofen prevented the overexpression of the cytokines in the CB and the hypertension, but failed to block the enhanced CB chemosensory responses to hypoxia. Thus, it is likely that proinflammatory cytokines levels may contribute to the CIH-induced hypertension acting at other levels of the chemoreflex pathway. The systemic treatment with ibuprofen also prevents the increased number of c-fos positive neurons in the caudal portion of the NTS in rats exposed to CIH for 21 days (Del Rio et al. 2012). Accordingly, we studied if CIH may increase proinflammatory cytokines in the NTS, the primary site for the afferent input from the CB, playing a role in the maintenance of the hypertension induced by CIH.

8.2 Methods

8.2.1 Animals and Exposure to Intermittent Hypoxia

Experiment were performed on male Sprague-Dawley rats (250 g). Rats were fed with standard diet *ad libitum* and kept on a 12:12-h light dark cycle. Unrestrained, freely moving rats housed in individual chambers were exposed to 5% of inspired O_2 for 20 s followed by room air for 280 s applied 12 episodes/h; 8 h/day for 7, 14, 21 and 28 days (CIH- rats). A computerized system based on solenoid valves control the alternating cycles of N_2 to produce hypoxic episodes followed by normoxic conditions. The hypoxic pattern was applied from 08:30 to 16:30 h. Control rats were exposed to room air. Room temperature was kept at 22–23 °C. The protocol was approved by the Bioethical Committee of the Biological Science Faculty of the Pontificia Universidad Católica de Chile, Santiago.

8.2.2 Gene Expression Measurements of Proinflammatory Cytokines

In both Control and CIH-rats we measured the expression of mRNA for IL-1β, IL-6, and TNF-α in biopsies of the NTS using quantitative PCR (qPCR). Rats were euthanized with sodium pentobarbitone (100 mg/kg) and their brainstems were removed and immediately frozen in dry ice and stored at −80 °C. The brains were cut in coronal sections in a cryostat at −20 °C selecting a portion around 300 μm thick of brainstem with the caudal NTS according to the coordinates of the rat brain atlas of Paxinos and Watson (2004). The NTS biopsies were excised with a chilled micro-punch needle of 500 μm of inner diameter and placed in a 1.5 mL Eppendorf tube. The mRNA was extracted according to the manufacturer's protocol (RNAqueous Total RNA, AM1912, USA). The concentration of mRNA was determined with a spectrophotomer (Epoch 2). The cDNA synthesis was made with a reverse transcript kit (Invitrogen SuperScript III First-Strand Synthesis System for

RT-PCR). The mRNA levels were analyzed using SYBR Green Power SYBR Green (Applied Biosystems, USA) in a qPCR equipment (Applied Biosystems, 7500 fast). We used the same primer sequences for IL-1β, IL-6 and TNF-α employed by Popa et al. (2011). Rat β-2 microglobulin was used as housekeeping gene.

8.2.3 Immunofluorescence of Proinflammatory Cytokines

Separated experimental series were used to measure cytokines. Control and CIH-rats were anaesthetized with sodium pentobarbitone (60 mg/kg) and perfused transcardially with saline solution followed by 4% paraformaldehyde (PFA; Sigma-Aldrich, Rockville, MD, USA). The brains were removed and post fixed 2 h by immersion and then placed in gradient of 30% of sucrose in 0.01 M of phosphate buffered saline (PBS 1X) at 4 °C. Frozen tissues were serially sectioned at ~20 μm in a cryostat at −20 °C. The NTS sections were identified according to the coordinates of the rat brain atlas of Paxinos and Watson at −13.80 mm of Bregma (Paxinos and Watson 2004). Sections were treated during 1 h with blocking solution (0.5% Triton X-100 and 5% of gelatin from cold water fish skin [G7765, SIGMA] dissolved in PBS 1X). Antibodies against PIC TNF-α, IL-1β and IL-6 (1:50; goat anti-rabbit Santa Cruz, USA) were used overnight a 4 °C. Then sections were treated with Alexa 594 (donkey anti-goat, 1:250) 1 h at room temperature. Finally, sections were incubated during 20 min with Hoechst (Molecular Probes, USA, 1:500), covered with coverslip and were viewed under an epifluorescence microscope. The analysis of fluorescence measurement was performed with the ImageJ (NIH, Bethesda, USA).

8.2.4 Data Analysis

All the data are reported as mean ± SEM. Differences between two groups was assessed by Student t-test. Differences between more groups was assessed with one-way ANOVA and when significant interactions were detected ($p < 0.05$), a post hoc Dunnet's test was used to determine which means were significantly different ($p < 0.05$) from each other. Statistical analysis was performed using GraphPad Prisma 6.0 (GraphPad Software, Inc., San Diego, CA, USA).

8.3 Results

8.3.1 CIH Increased mRNA Levels of Proinflammatory Cytokines in the NTS of Rats

The analysis of the results showed that IL-1β, IL-6, and TNF-α mRNA levels progressively increased reaching a significant difference at 21 days of CIH exposure compared with the Control group ($p < 0.01$, one-way ANOVA followed by Dunnett's test). No significant changes were found in proinflammatory cytokine levels at 7 or 14 days of CIH exposure related to Control rats (Fig. 8.1).

8.3.2 CIH Increased Proinflammatory Cytokines Immunoreactivity in the NTS of Rats

We found a significant increase of TNF-α, IL-1β and IL-6 in the NTS following 28 days of CIH exposure in the NTS compared to the control group (* $p < 0.01$; ** $p < 0.002$, t Student's). No significant changes were found in proinflammatory cytokines levels at 7 or 14 days of CIH exposure related to control rats (Fig. 8.2).

8.4 Discussion

Our results showed that CIH exposure produced a significant increase of TNF-α, IL-1β and IL-6 levels in the NTS, the first integration nucleus in the brainstem for chemo and baroreceptor afferent inputs, suggesting that proinflammatory cytokines may be involved in CIH-induced

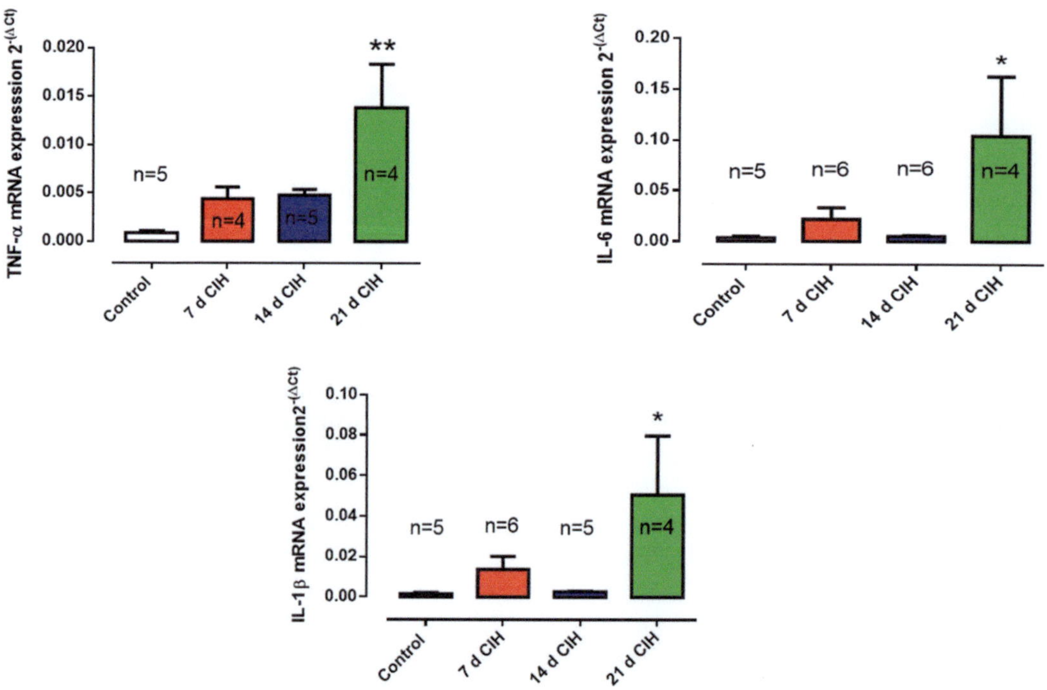

Fig. 8.1 mRNA levels of proinflammatory cytokines in the NTS of rats exposed to CIH. Bars showed qPCR mRNA levels of TNF-α, IL-1β and IL-6 in the NTS of rats exposed to 7, 14 and 21 days to CIH compared to Control rats ($p < 0.01$, one-way ANOVA followed by Dunnett's test)

hypertension. We found a progressive increase of TNF-α, IL-1β and IL-6 levels in the NTS, which reached a statistical significance at 21 days of CIH. Previously, we reported that CIH increased TNF-α and IL-1β immunoreactive levels in the rat CB, with a similar time-course (Del Rio et al. 2011; Del Rio et al. 2012). Similarly, Lam et al. (2012) reported that CIH increased the mRNA levels of IL-1β, TNF-α and IL-6 in the CB. Thus, the available data suggested that proinflammatory cytokines may be involved in the CB chemosensory potentiation induced by CIH. Accordingly, we studied if the CIH-induced increase of TNF-α and IL-1β immunoreactivity in the rat CB may contribute to enhance CB chemosensory discharge (Del Rio et al. 2012). We found that ibuprofen systemically administrated by osmotic pumps during CIH exposure for 21 days prevented the increase of cytokine immunoreactivity in the rat CB as well as the systemic hypertension, but was unable to prevent the potentiation of CB chemosensory responses to hypoxia. Thus, these results indicated that the CB chemosensory potentiation was independent of the increased TNF-α and IL-1β immunoreactivity. However, these results suggest that proinflammatory cytokines may act on other neural structures of the hypoxic chemoreflex pathway to maintain the hypertension. Indeed, Del Rio et al. (2011) found that ibuprofen reduced the CIH-induced activation of c-fos-positive neurons in the NTS. Taken together the available data show that proinflammatory cytokines may play a role in the activation of NTS neurons induced by CIH.

The increase of proinflammatory cytokines levels in the NTS started at day 21 of CIH exposure and remained elevated following 28 days. However, it is worth to note that the arterial blood pressure in rats increases after 3–5 days of the onset of the CIH stimulation (Del Rio et al. 2016). Thus, our results support a plausible contribution of TNF-α, IL-1β and IL-6 in the NTS on the maintenance, but not in the onset of the hypertension in CIH-exposed rats.

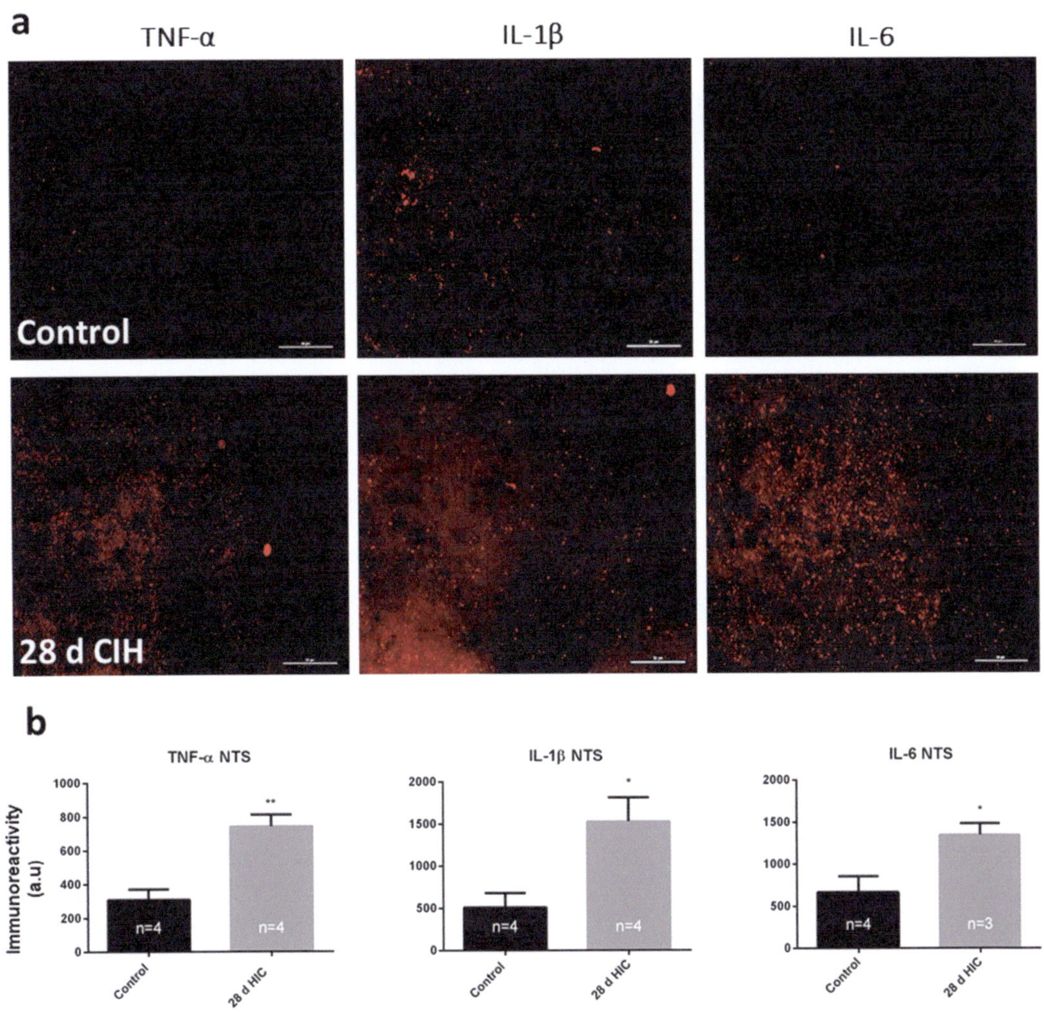

Fig. 8.2 Immunoreactivity levels of proinflammatory cytokines in the cNTS of rats exposed to 28 days of CIH. (**a**), Representative images of immunofluorescence of TNF-α, IL-1β and IL-6 in the NTS of Control rats (upper panel) and rats exposed during 28 days to CIH (lower panel). Bar scale: 50 μm. (**b**), Quantifications of immunofluorescence for proinflammatory cytokines as arbitrary units (a.u) * $p < 0.01$; ** $p < 0.002$, t Student's. Bar scale: 50 μm

Acknowledgements This work was supported by grant FONDECYT 1150040 from the National Fund for Scientific and Technological Development of Chile.

References

Del Rio R, Moya EA, Iturriaga R (2011) Differential expression of proinflammatory cytokines, endothelin-1 and nitric oxide synthases in the rat carotid body exposed to intermittent hypoxia. Brain Res 1395:74–85

Del Rio R, Moya EA, Parga MJ, Madrid C, Iturriaga R (2012) Carotid body inflammation and cardiorespiratory alterations in intermittent hypoxia. Eur Respir J 39:1492–1500

Del Rio R, Andrade D, Lucero C, Arias P, Iturriaga R (2016) Carotid body ablation abrogates hypertension and autonomic alterations induced by intermittent hypoxia in rats. Hypertension 68:436–445

Dempsey JA, Veasey SC, Morgan BJ, O'donnell CP (2010) Pathophysiology of sleep apnea. Physiol Rev 90:47–112

Garvey JF, Taylor CT, McNicholas WT (2009) Cardiovascular disease in obstructive sleep apnoea

syndrome: the role of intermittent hypoxia and inflammation. Eur Respir J 33:1195–1205
Granger JP (2006) An emerging role for inflammatory cytokines in hypertension. Am J Physiol Heart Circ Physiol 290:H923–H924
Iturriaga R, Moya EA, Del Rio R (2009) Carotid body potentiation induced by intermittent hypoxia: implications for cardiorespiratory changes induced by sleep apnoea. Clin Exp Pharmacol Physiol 36:1197–1204
Iturriaga R, Del Rio R, Idiáquez J, Somers VK (2016) Carotid body chemoreceptors, sympathetic neural activation, and cardiometabolic disease. Biol Res 49:13
Iturriaga R, Oyarce MP, Dias ACR (2017) Role of carotid body in intermittent hypoxia-related hypertension. Curr Hypertens Rep 19:38
Lam SY, Liu Y, Ng KM, Lau CF, Liong EC, Tipoe GL, Fung ML (2012) Chronic intermittent hypoxia induces local inflammation of the rat carotid body via functional upregulation of proinflammatory cytokine pathways. Histochem Cell Biol 137:303–317
Mcbryde FD, Abdala AP, Hendy EB, Pijacka W, Marvar P, Moraes DJA, Sobotka PA, Paton JFR (2013) The carotid body as a putative therapeutic target for the treatment of neurogenic hypertension. Nat Commun 4:2395
Paxinos G, Watson C (2004) The Rat brain in stereotaxic coordinates: the new coronal set, 5th edn. Elsiever/Academic, Amsterdam/Boston
Popa D, Fu X, Go A, Powell FL (2011) Ibuprofen blocks time-dependent increases in hypoxic ventilation in rats. Respir Physiol Neurobiol 178:381–386
Somers VK, White DP, Amin R, Abraham WT, Costa F, Culebras A, Daniels S, Floras JS, Hunt CE, Olson LJ, Pickering TG, Russell R, Woo M, Young T (2008) Sleep apnea and cardiovascular disease. J Am Coll Cardiol 52:686–717
Waki H, Gouraud SS, Maeda M, Paton JFR (2008) Specific inflammatory condition in nucleus tractus solitarii of the SHR: Novel insight for neurogenic hypertension? Auton Neurosci Basic Clin 42:25–31

9 Central Hypoxia Elicits Long-Term Expression of the Lung Motor Pattern in Pre-metamorphic *Lithobates Catesbeianus*

Tara A. Janes and Richard Kinkead

Abstract

During vertebrate development, the neural networks underlying air-breathing undergo changes in connectivity and functionality, allowing lung ventilation to emerge. Yet, the factors regulating development of these critical homeostatic networks remain unresolved. In amphibians, air-breathing occurs sporadically prior to metamorphosis. However, in tadpoles of *Lithobates catesbeianus* (American bullfrog), hypoxia stimulates gill and lung ventilation during early development. Because accelerated metamorphosis is a useful strategy to escape deterioration of the milieu, we hypothesized that central hypoxia would elicit long-term expression of the lung motor command for air breathing in pre-metamorphic tadpoles (TK stages VI–XIII). To do this, we recorded respiratory activity from cranial nerves V and VII in isolated brainstems before, during, and up to 2 h after exposure to 15 min of mild (PwO_2 range: 114–152 Torr) or moderate (PwO_2 range: 38–76 Torr) hypoxia. To test for stage-dependent effects, data were compared between early (VI–IX) and mid (X–XIII) stages. Early stages responded strongly during moderate hypoxia with increased lung burst frequency (167%). Mild and moderate hypoxia increased lung burst frequency during the 2 h re-oxygenation period in early stage brainstems (136%, 497%, respectively), but produced only marginal effects on mid stage brainstems (39%, 31%, respectively). In contrast, hypoxia was not an important factor controlling fictive buccal burst frequency, which drives continuous gill ventilation in tadpoles prior to metamorphosis (all stages showed <25% increase). These preliminary results suggest that central hypoxia elicits long-term increases in lung burst frequency in a severity- and stage-dependent manner.

Keywords

Hypoxia · Chemoreflex · Respiration · Neurodevelopment · Amphibian

9.1 Introduction

The emergence of lung ventilation in air breathing vertebrates occurs during early development and requires a myriad of changes within the underlying neural networks. Environmental perturbations can influence respiratory development

T. A. Janes (✉) · R. Kinkead
Department of Pediatrics, Université Laval, Québec, QC, Canada

Institut Universitaire de Cardiologie et de Pneumologie de Québec, Québec, QC, Canada
e-mail: tara-adele.janes.1@ulaval.ca; richard.kinkead@fmed.ulaval.ca

E. B. Gauda et al. (eds.), *Arterial Chemoreceptors*, Advances in Experimental Medicine and Biology 1071, https://doi.org/10.1007/978-3-319-91137-3_9

(see reviews by Bavis and Mitchell 2008; Burggren and Reyna 2011). In particular, exposure of the developing brain to hypoxia can be readily encountered in diverse environments (aquatic, altitude, nesting, in utero). Hypoxia, acting via central chemoreflexes, alters the activity of respiratory neurons (Horn and Waldrop 1997; Solomon et al. 2000) and may exert significant consequences for the developing respiratory neural networks. However, the neurological processes by which air breathing develops, and hypoxia's impacts upon these processes remain significant gaps-in-knowledge.

Over the years, our lab has helped to establish an amphibian model of respiratory development and identified fundamentally conserved mechanisms giving rise to the air-breathing phenotype at adulthood (Fournier et al. 2007; Fournier and Kinkead 2008; Rousseau et al. 2016). The time course of amphibian development, from hatching through metamorphosis, is heavily influenced by environmental factors. Accordingly, amphibian development is not defined by time since hatching, but is categorized as physical "stages" based on progressive hind- and fore-limb development, tail regression, etc. (Taylor and Kollros 1946). During aquatic tadpole stages (Taylor-Kollros stages I–XVII, without forelimbs), respiratory demands are met by cutaneous gas exchange and gill ventilation. The latter behaviour is driven by a continuous low amplitude motor pattern engaging the buccal musculature (similar to fish; Kogo et al. 1994). In response to environmental cues, hormonal signalling cascades trigger metamorphosis of tadpoles into terrestrial frogs (Denver 1997). During metamorphosis (Taylor-Kollros stages XVIII–XXV, with forelimbs), the gills regress and lungs emerge as the primary gas exchange surface. This physical transformation is accompanied by emergence of a high amplitude, low frequency motor pattern driving more forceful lung ventilation (Kogo et al. 1994). In the absence of a diaphragm, positive-pressure lung ventilation is accomplished, as with gill ventilation, via the buccal musculature. Although the lungs of pre-metamorphic tadpoles are rudimentary, sporadic fictive air breathing is observed in intact animals and brainstem preparations (Burggren and Doyle 1986; Fournier and Kinkead 2008). Thus, the neural circuits producing air breathing are present in early development, but their activity is inhibited prior to metamorphosis (Straus et al. 2000).

Hypoxia has been shown to alter expression of air breathing in both intact and *in vitro* amphibian preparations. During chronic hypoxia, air-breath frequency increases in intact tadpoles (Burggren and Doyle 1986), arguing for a direct effect of hypoxia capable of revealing the lung motor pattern in pre-metamorphic stages. Similarly, our lab has previously shown that central hypoxia increases fictive lung burst frequency when applied to pre-metamorphic brainstems, but the same stimulus is inhibitory in adult frogs (Fournier et al. 2007; Fournier and Kinkead 2008). The central oxygen chemoreflex of amphibians underlying the hypoxic response shows a developmental shift from excitatory (tadpoles) to inhibitory (adult frogs; Fournier and Kinkead 2008). Therefore, the effects of central hypoxia may be contingent on developmental stage and studies of hypoxia-induced respiration in freely-behaving tadpoles support this contention (Burggren and Doyle 1986; Winmill et al. 2005). These considerations led us to hypothesize that exposure of tadpole brainstems to acute central hypoxia would elicit long-term increases in lung burst frequency that are dependent on both developmental stage and hypoxic severity.

9.2 Materials and Methods

9.2.1 Isolated Brainstem Preparations

Experiments were performed on *Lithobates catesbeianus* tadpoles between TK stages VI–XIII. Isolated brainstem preparations were made as previously described (Fournier et al. 2007). All experiments complied with the guidelines of the Canadian Council on Animal Care and were approved by the institutional animal care committee. Briefly, tadpoles were anesthetized (MS-222, 0.06 g l^{-1}) and the cranium opened to permit dissection of the brainstem and cranial nerves (CN's). During dissection, neural tissue was irrigated continuously with cold (0–5 °C) artificial cerebrospinal fluid (aCSF) prepared as follows: (in mM): 104 NaCl, 4 KCl, 1.4 $MgCl_2$, 2.4 $CaCl_2$, 25 $NaHCO_3$, 10 D-glucose (Sigma Aldrich) and equilibrated with a mixture of 1.8% CO_2 + 98.2% O_2 to pH 7.90 ± 0.08.

9.2.2 Electrophysiological Recordings and Hypoxia Protocol

Brainstem preparations recovered for 40 mins in oxygenated aCSF (1.8% CO_2 + 98.2% O_2) before experiments commenced. Fictive motor activity was recorded simultaneously from CN's V/ VII and X using suction electrodes (Fournier et al. 2007). Once a stable neural signal was achieved, 20 min of baseline activity was recorded during perfusion with oxygenated aCSF (5–7 mL/min). The perfusion was then switched to hypoxic aCSF (1.8% CO_2 + 98.2% N_2) at a speed of 3 mL/min (mild) or 10 mL/min (moderate), for 15 min. Finally, the preparation was perfused with oxygenated aCSF for 2 h. Time-matched controls received 3 h of oxygenated aCSF. All solutions were maintained at room temperature (21–23 °C) with a pH of 7.90 ± 0.08. Oxygen levels within the recording chamber were monitored using a MI-730 microelectrode and OM-4 Oxygen Meter (Microelectrodes Inc. NH, USA). Mild hypoxia ranged from 114–152 Torr while moderate hypoxia was measured to be 38–76 Torr.

9.2.3 Data Analysis

Tadpole brainstems produce two patterns of respiratory-related neural bursting (Torgerson et al. 1998; Fig. 9.1). Buccal and lung bursts occurring in CN's V and VII were classified as previously described (Sakakibara 1984a), and lung bursts were verified by referencing simultaneous activity from CN X. The frequency of buccal (F_B) and lung bursts (F_L) were analyzed during the following time points: 20 min baseline, last 10 min hypoxia, and during re-oxygenation: 1 h–1 h 20 min, 2 h–2 h 20 min. The F_L was measured during 20 min (10 mins for hypoxia), F_B was measured for the last min of each time point, and both parameters were expressed as frequency min^{-1}. Data for each replicate were normalized to their baseline in order to be expressed as absolute change (Δ). Results are summarized as mean Δ frequency ± SEM.

9.3 Results

Hypoxic stimulation of lung ventilation is stage-dependent in freely-behaving tadpoles; becoming apparent at stage IV and significant by stage IX. In contrast, hypoxic enhancement of gill ventilation observed in intact animals at very early stages becomes non-existent by stage XVI (Burggren and Doyle 1986). Using these reports as a guide, data were analyzed according to the following classification: stage VI–IX tadpoles comprised the "early stage" group, while stages X–XIII were considered as "mid stages". Figure 9.1 presents representative neurograms of integrated fictive respiratory motor activity recorded from the cranial nerves during our protocol. Each respiratory "burst" represents the summation of action potentials generated by underlying motoneuron nuclei. Under baseline conditions (Fig. 9.1a–d, *Baseline*), motor activity was dominated by a high frequency, low amplitude bursting pattern underlying buccal ventilation in intact tadpoles (*buccal bursts*, min-max range: 31–53 min^{-1} for early stages and 30–58 min^{-1} for mid stages). At baseline, high amplitude events consistent with fictive lung bursting occurred with low frequency (Fig. 9.1b–d, Baseline, *lung bursts*) across most brainstem preparations (min-max range: 0.1–2.8 min^{-1} for early stages and 0.1–1.6 min^{-1} for mid stages). To control for potential time-dependent changes in respiratory activity over the 3 h protocol, time control experiments were performed. The control preparation shown in Fig. 9.1c did not exhibit changes in F_L or F_B during 3 h perfusion with oxygenated aCSF. This result was consistent for controls at all stages.

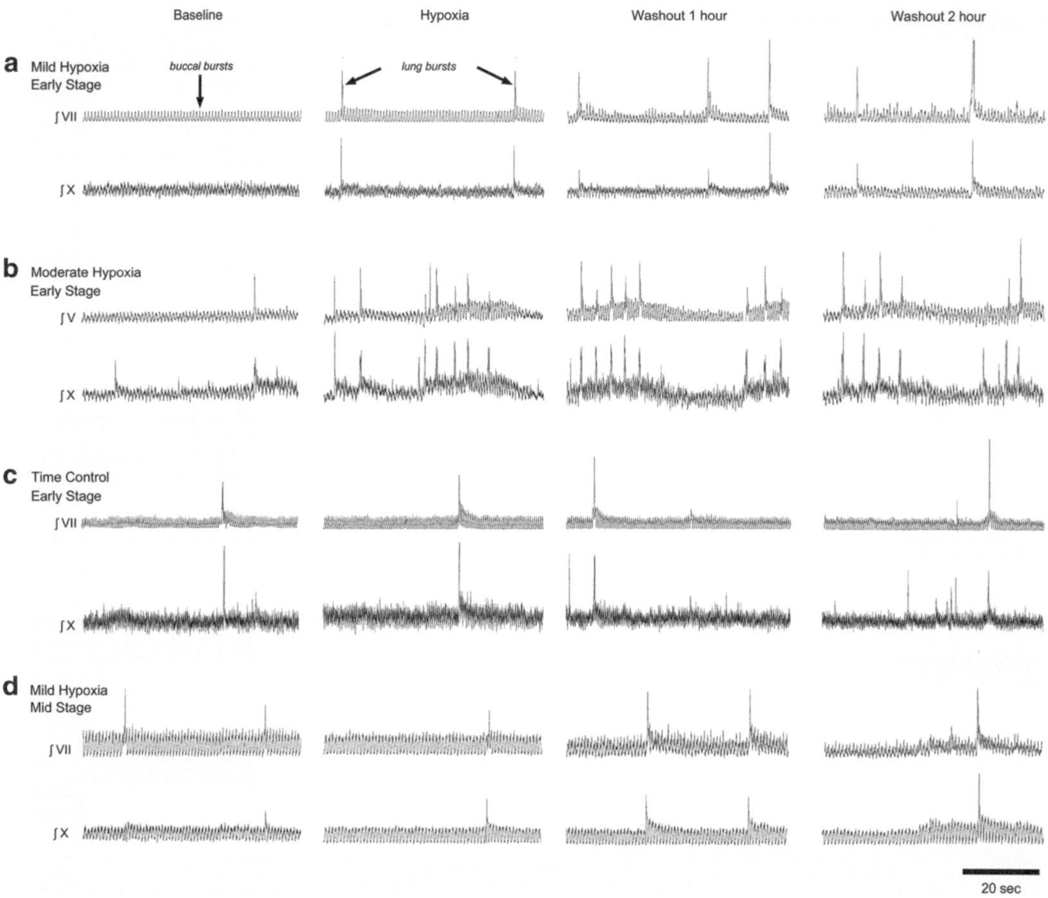

Fig. 9.1 Representative neurograms comparing fictive respiratory activity recorded from isolated brainstems in early and mid stage tadpoles during baseline, central hypoxia and 1 h–2 h washout (re-oxygenation). Integrated neurograms show fictive respiratory bursting underlying the buccal (*buccal bursts*) and lung (*lung bursts*) motor rhythms. As shown, mild (**a**) and moderate (**b**) hypoxia increased F_L, but not F_B, for early stage brainstems during the 2 h re-oxygenation period (*Washout 1 h, Washout 2 h*). All data were compared against time-matched controls (**c**). In contrast, neither mild (**d**) nor moderate (not shown) hypoxia elicited changes in F_L or F_B in mid stage tadpoles

9.3.1 Central Hypoxia Induces Long-Term Augmentation of Lung Bursting in Early Stage Tadpoles

During exposure to moderate central hypoxia, the F_L in early stage brainstems increased acutely (*versus* time controls). Interestingly, this acute hypoxic response was absent in the same stage group exposed to mild hypoxia. However, both mild and moderate hypoxia elicited long-term increases in F_L during the subsequent washout period. Figure 9.1a, b illustrates examples of the progressive augmentation of F_L (*Washout 1 h, 2 h*). Analysis of F_L data from mildly hypoxic brainstems revealed an increase $>\Delta 1\ \text{min}^{-1}$ in 2/4 preparations and a collective mean greater than controls after 1 h of re-oxygenation, but values returned to control levels after 2 h (Fig. 9.2a, Mild). The effect of moderate hypoxia was stronger and more persistent, with 2/2 preparations exhibiting an $F_L > \Delta 1\ \text{min}^{-1}$ after 2 h of

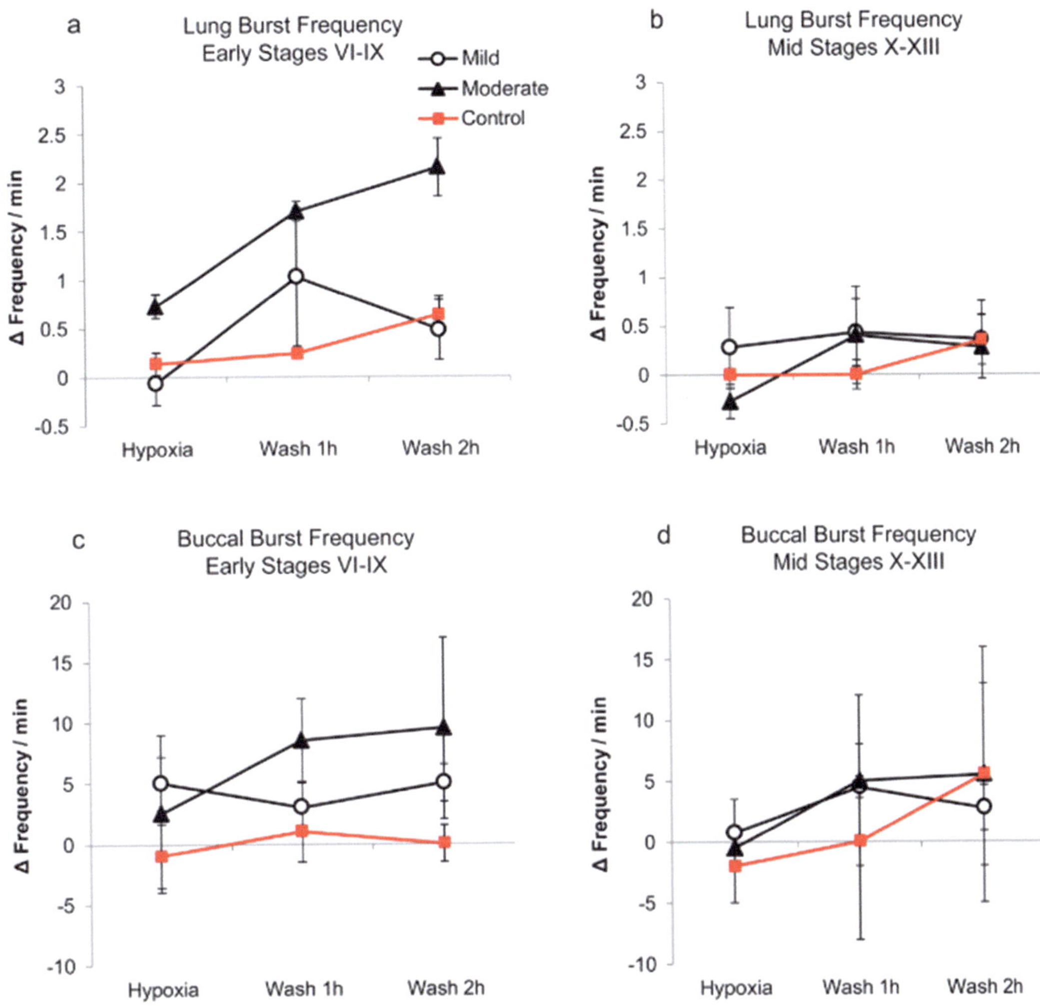

Fig. 9.2 Central hypoxia elicits severity-dependent, long-term augmentation of F_L in early stage tadpole brainstems. (**a**) Treatment of early stages with moderate hypoxia elicited an acute hypoxic response (*Hypoxia*); mild and moderate hypoxia induced severity-dependent long-term augmentation of F_L during re-oxygenation (*Wash 1 h, Wash 2 h*). (**b**) Neither mild nor moderate hypoxia affected F_L recorded from mid stage brainstems as compared to time-matched controls. The F_B was only marginally increased by hypoxia in early stage brainstems (**c**), and exhibited no change in mid stages (**d**) suggesting that the central hypoxic stimulus was specific for F_L in pre-metamorphic tadpole brainstems

re-oxygenation (Fig. 9.2a, Moderate). In contrast to the notable long-term effects on F_L, mild and moderate hypoxia had only marginal effects on the F_B of early stage brainstems after 2 h of re-oxygenation (Fig. 9.2c). Following moderate hypoxia, 2/2 preparations exhibited a $F_B > \Delta 5\ min^{-1}$ (Fig. 9.2c, *Wash 2 h*), but due to the high baseline F_B mean (43 ± 5), this change equated to only a 24% frequency increase.

9.3.2 Mid Stage Brainstems Do Not Respond to Central Hypoxia with Increased Lung Frequency

Central hypoxia applied to mid stage brainstems failed to elicit increased F_L during hypoxic perfusion, or during the subsequent re-oxygenation period (Figs. 9.1d and 9.2b). After 2 h of re-oxygenation, F_L increased (>Δ1 min^{-1}) in only 1/4 of preparations exposed to mild hypoxia, and 0/2 moderately hypoxic brainstems. The mean change in F_L after 2 h of washout from mild or moderate hypoxia was similar to that observed for time controls (Fig. 9.2b). Similarly, the F_B in mid stage brainstems was consistently unaffected after 2 h of re-oxygenation (Fig. 9.2d). Specifically, 0/4 brainstems exposed to mild hypoxia, and 0/2 exposed to moderate hypoxia exhibited any change (+ or −) in F_B greater than time controls after 2 h of washout.

9.4 Discussion

The present study provides preliminary evidence that acute central hypoxia is sufficient to elicit long-term expression of the motor pattern for lung ventilation in developing tadpoles. Further, these data collectively indicate that the ability of central hypoxia to elicit long-term augmentation of F_L is both stage- and severity-dependent. We suggest that during early stages of development, the neural networks underlying pulmonary ventilation enter a 'sensitive' period when hypoxia can impact their long-term motor activity.

The most important and novel results to come out of the present study were that brief exposure of early stage brainstems to hypoxia elicited augmentation of F_L which persisted beyond the hypoxic stimulus. Previous studies support an acute excitatory effect of hypoxia on F_L (during the hypoxic challenge) for early stage tadpoles and their isolated brainstems (Burggren and Doyle 1986; Fournier et al. 2007). Our results build upon these previous studies by demonstrating hypoxic augmentation of F_L persisting into the re-oxygenation period for early stage brainstems. In contrast, previous studies exposing isolated brainstems of stage VIII–XVI tadpoles (i.e. mainly mid stages) to chronic severe hypoxia (0 Torr) reported no change in F_L, further supporting our results (Winmill et al. 2005). F_B was not augmented by central hypoxia for any developmental stage (changes were marginal). This is not surprising given that the pattern generators for buccal and lung ventilation are spatially distinct (Wilson et al. 2002), and suggests that hypoxic signaling pathways can independently modulate buccal vs. lung motor activity. For example, peripheral chemoreceptors might modulate F_B during early development, thus explaining the increased F_B observed for intact pre-metamorphic tadpoles (Straus et al. 2001). Moreover, the specificity of hypoxia for F_L (and not F_B) precludes non-specific effects of tissue deterioration following hypoxic exposure.

Underlying the acute hypoxic response of isolated brainstems is the central chemoreflex. In vertebrates, the locus coeruleus (LC) is a chemosensitive region of the pons rostral to CN V containing neurons involved in respiratory control (Noronha-de-Souza et al. 2006). While the LC neurons of mammals are considered to be primarily CO_2-sensitive, an intact LC is essential for the acute hypoxic response of pre-metamorphic tadpoles (Fournier and Kinkead 2008). Therefore, this structure may contribute to the observed long-term F_L augmentation described herein. The amphibian central oxygen chemoreflex shows a developmental shift from excitatory in early stages to inhibitory in post-metamorphic tadpoles and adult frogs (Fournier and Kinkead 2008). This functional shift is thought to be a consequence of maturing GABA-ergic systems. The inability of central hypoxia to elicit augmentation of F_L in mid stage brainstems may reflect this chemoreflex maturation, suggesting that the developmental shift from excitation towards inhibition is initiated around stage X in *L. catesbienus*.

In the present study, we were primarily interested in the expression of respiratory motor activity during the re-oxygenation period. Our results demonstrate that even brief periods of central hypoxia can elicit long-term F_L, suggesting that mechanisms of neuroplasticity are invoked. While the precise nature of these mechanisms remains to be elucidated, we speculate that the excitatory central chemoreflex permits long-term plasticity of F_L in early stage tadpoles. The F_L observed under baseline conditions is low in early stages and increases as tadpoles approach metamorphosis. Interestingly, Straus et al. (2000) showed that central $GABA_B$ receptor antagonism in pre-metamorphic tadpoles increased F_L. Therefore, the neural networks sub serving lung ventilation are functional in early stage tadpoles, but their activity is normally inhibited until metamorphosis. As a consequence of this developmental inhibition, there may be only weak activity in the neural pathways for the lung motor pattern under baseline conditions, and an excitatory stimulus (i.e. hypoxia) engaging these pathways would be predicted to induce plasticity. Whether hypoxia provides excitatory drive to overcome inhibition or also weakens GABAergic pathways remains unresolved. However, transition of the central chemoreflex to inhibition may end the time period during which hypoxia elicits long-term augmentation of F_L.

A growing body of evidence demonstrates conservation between the respiratory control centers of vertebrate taxa (Wilson et al. 2006). In fetal and neonatal mammals, research supports the existence of a central oxygen chemoreflex modulating respiratory activity and requiring intact pontine structures (Dawes et al. 1983; Koos et al. 1992). However, the precise cellular identity, mechanisms and development of vertebrate central oxygen chemoreceptors remain as open issues (Gourine and Funk 2017). Our comparative amphibian model aims to garner a fundamental understanding of central hypoxia's effects on long-term respiratory activity during early neurodevelopment, and suggests involvement from developing chemoreceptive pathways which have yet to be fully defined. Amphibian brainstem preparations are robust and cranial nerve activity is a strong indicator of respiration at the behavioural level (Sakakibara 1984b). One limitation of the present study was that elimination of peripheral sensory inputs, while informative, warrants caution as peripheral and central chemoreflexes interact (Smith et al. 2010). However, our results correspond well with behavioural studies, and agree with mechanistic findings, giving us confidence in our conclusions. Although preliminary, these data are thus highly promising and current experiments aiming to substantiate these results are underway. The long-term augmentation of F_L we noted for early stage tadpoles in response to moderate hypoxia may depend upon specific changes in gene expression and protein synthesis and these mechanistic insights constitute an area for future study.

Acknowledgements This work was supported by a "Discovery Grant" and "Research Tools and Instruments" Grant from the National Sciences and Engineering Research Council of Canada awarded to R.K.

References

Bavis RW, Mitchell GS (2008) Long-term effects of the perinatal environment on respiratory control. J Appl Physiol Bethesda Md 1985 104:1220–1229

Burggren W, Doyle M (1986) Ontogeny of regulation of gill and lung ventilation in the bullfrog, Rana catesbeiana. Respir Physiol 66:279–291

Burggren WW, Reyna KS (2011) Developmental trajectories, critical windows and phenotypic alteration during cardio-respiratory development. Respir Physiol Neurobiol 178:13–21

Dawes GS, Gardner WN, Johnston BM, Walker DW (1983) Breathing in fetal lambs: the effect of brain stem section. J Physiol 335:535–553

Denver RJ (1997) Environmental stress as a developmental cue: corticotropin-releasing hormone is a proximate mediator of adaptive phenotypic plasticity in amphibian metamorphosis. Horm Behav 31:169–179

Fournier S, Kinkead R (2008) Role of pontine neurons in central O_2 chemoreflex during development in bullfrogs (Lithobates catesbeiana). Neuroscience 155:983–996

Fournier S, Allard M, Roussin S, Kinkead R (2007) Developmental changes in central O_2 chemoreflex in Rana catesbeiana: the role of noradrenergic modulation. J Exp Biol 210:3015–3026

Gourine AV, Funk GD (2017) On the existence of a central respiratory oxygen sensor. J Appl Physiol 123:1344–1349

Horn EM, Waldrop TG (1997) Oxygen-sensing neurons in the caudal hypothalamus and their role in cardiorespiratory control. Respir Physiol 110:219–228

Kogo N, Perry SF, Remmers JE (1994) Neural organization of the ventilatory activity in the frog, Rana catesbeiana. I. J Neurobiol 25:1067–1079

Koos BJ, Chao A, Doany W (1992) Adenosine stimulates breathing in fetal sheep with brain stem section. J Appl Physiol 72:94–99

Noronha-de-Souza CR, Bícego KC, Michel G, Glass ML, Branco LGS, Gargaglioni LH (2006) Locus coeruleus is a central chemoreceptive site in toads. Am J Physiol Regul Integr Comp Physiol 291:R997–R1006

Rousseau J-P, Bairam A, Kinkead R (2016) Aldosterone, corticosterone, and thyroid hormone and their influence on respiratory control development in Lithobates catesbeianus: an in vitro study. Respir Physiol Neurobiol 224:104–113

Sakakibara Y (1984a) The pattern of respiratory nerve activity in the bullfrog. Jpn J Physiol 34:269–282

Sakakibara Y (1984b) Trigeminal nerve activity and buccal pressure as an index of total inspiratory activity in the bullfrog. Jpn J Physiol 34:827–838

Smith CA, Forster HV, Blain GM, Dempsey JA (2010) An interdependent model of central/peripheral chemoreception: evidence and implications for ventilatory control. Respir Physiol Neurobiol 173:288–297

Solomon IC, Edelman NH, Neubauer JA (2000) Pre-Bötzinger complex functions as a central hypoxia chemosensor for respiration in vivo. J Neurophysiol 83:2854–2868

Straus C, Wilson RJ, Tezenas du Montcel S, Remmers JE (2000) Baclofen eliminates cluster lung breathing of the tadpole brainstem, in vitro. Neurosci Lett 292:13–16

Straus C, Wilson RJ, Remmers JE (2001) Oxygen sensitive chemoreceptors in the first gill arch of the tadpole, Rana catesbeiana. Can J Physiol Pharmacol 79:959–962

Taylor AC, Kollros JJ (1946) Stages in the normal development of Rana pipiens larvae. Anat Rec 94:7–13

Torgerson CS, Gdovin MJ, Remmers JE (1998) Fictive gill and lung ventilation in the pre- and postmetamorphic tadpole brain stem. J Neurophysiol 80:2015–2022

Wilson RJA, Vasilakos K, Harris MB, Straus C, Remmers JE (2002) Evidence that ventilatory rhythmogenesis in the frog involves two distinct neuronal oscillators. J Physiol 540:557–570

Wilson RJA, Vasilakos K, Remmers JE (2006) Phylogeny of vertebrate respiratory rhythm generators: the Oscillator Homology Hypothesis. Respir Physiol Neurobiol 154:47–60

Winmill RE, Chen AK, Hedrick MS (2005) Development of the respiratory response to hypoxia in the isolated brainstem of the bullfrog Rana catesbeiana. J Exp Biol 208:213–222

10 Cysteine Oxidative Dynamics Underlies Hypertension and Kidney Dysfunction Induced by Chronic Intermittent Hypoxia

Nuno R. Coelho, Clara G. Dias, M. João Correia, Patrícia Grácio, Jacinta Serpa, Emília C. Monteiro, Lucília N. Diogo, and Sofia A. Pereira

Abstract

Previous data showed the lack of efficacy of an adrenoceptor antagonist to revert hypertension induced by chronic intermittent hypoxia (CIH). We hypothesized that, in addition to sympathetic activation, CIH may change the availability and dynamics of cysteine. Temporal variation in total cysteine and its fractions, free reduced, free oxidized and protein-bound (CysSSP), were measured in homogenates of kidney cortex and medulla of Wistar rats. Animals were exposed to CIH for 14, 21 and 60 days and cysteine fractions and fibronectin gene expression were assessed at these time-points. Two different phases in cysteine dynamics were identified. An early phase (14d) characterized by an increase in cysteine oxidation and CysSSP forms. Late events (>21d) were characterized by a global reduction in cysteine, minimum level of CysSSP and maximum overexpression of fibronectin in kidney cortex. In conclusion, cysteine dynamics is influenced by the duration of CIH exposure: first there is a cysteine disulfide stress-like adaptive response followed by a progressive loss of cysteine availability and a decrease in CysSSP fraction. Kidney fibrosis associated to an unbalance in cysteine dynamics might contribute to the inefficacy of available antihypertensive drugs in patients with delayed diagnosis of sleep apnea.

Keywords

Cysteine · Protein *S*-cysteinylation · Disulfide stress · Kidney fibrosis · Systemic hypertension · Antihypertensive drug response

10.1 Introduction

We have previously reported the lack of efficacy of carvedilol (CVDL) in reversing hypertension induced by chronic intermittent hypoxia (CIH) in rats (Diogo et al. 2015). CVDL is an unselective beta-blocker with intrinsic anti-α1-adrenergic and antioxidant activities. Neither dose, bioavailability, nor absence of beta-blocking activity were responsible for our findings (Diogo et al. 2015). A plausible explanation is that CVDL would be unable to stop oxidative stress and kidney inflam-

N. R. Coelho · C. G. Dias · M. João Correia
P. Grácio · E. C. Monteiro (✉) · L. N. Diogo
S. A. Pereira
CEDOC, Centro de Estudos Doenças Crónicas, Nova Medical School, Faculdade de Ciências Médicas, Universidade Nova de Lisboa, Lisbon, Portugal
e-mail: emilia.monteiro@nms.unl.pt

J. Serpa
CEDOC, Centro de Estudos Doenças Crónicas, Nova Medical School, Faculdade de Ciências Médicas, Universidade Nova de Lisboa, Lisbon, Portugal

Instituto Português de Oncologia de Lisboa Francisco Gentil (IPOLFG), Lisbon, Portugal

E. B. Gauda et al. (eds.), *Arterial Chemoreceptors*, Advances in Experimental Medicine and Biology 1071, https://doi.org/10.1007/978-3-319-91137-3_10

mation in response to CIH. More, the chronicity of intermittent hypoxia promotes progressive cardiometabolic disease and irreversible renal damage (Wu et al. 2015, Sun et al. 2012) that compromise the antihypertensive effect of drugs, mainly when treatment is prescribed too late. The temporal variation in factors involved in kidney oxidative stress and inflammation during CIH is unknown. Additionally, the contribution of cysteine to the mechanisms underlying the regulation of blood pressure and kidney function has been supported by several reports (Vasdev et al. 2009). Evidence on cysteine effects in fibrosis is scarce (Horie et al. 2003). Mechanisms have been mostly investigated through N-acetylcysteine and cysteine supplementation (Vasdev et al. 2009), supporting that its low availability is related to kidney function, particularly in early phases of injury.

Cysteine availability is a net contribution of three pools which can dynamically exchange one- to-one, depending on the redox status of tissue involved. The protein-bound fraction (*S*- cysteinylated proteins, CysSSP) is generated by a reversible post-translational modification through a disulfide bound between free reduced low molecular weight cysteine and cysteine residues of proteins (Rossi et al. 2009). The free cysteine pool includes the reduced (CysSH) and oxidized cysteine fractions (CysSSX). The oxidized free cysteine pool contains mixed disulfides, like cysteinyl-glutathione or the combination of two cysteines (cystine).

Herein, we hypothesized that the duration of intermittent hypoxia exposure changes the dynamics of cysteine in the kidney and decreases its availability, favoring kidney fibrosis and consequently compromising the management of hypertension.

10.2 Methods

10.2.1 Animals

Thirty-five male Wistar Crl:WI (Han) (*Rattus norvegicus* L.), aged 10–13 weeks, with mean body weights of 293 ± 7 g were housed in polycarbonate cages with wire lids and maintained under standard laboratory conditions as follows: artificial 12 h light/dark cycles, room temperature (22 ± 2.0 °C) and relative humidity of 60 ± 10%. Rats were given reverse osmosis water *ad libitum* and standard laboratory diet (SDS diets RM1). This diet has 2.2 g/Kg of methionine and 2.4 g/kg of cystine (SDS, Special Diets Services, UK).

All applicable institutional and governmental regulations concerning the ethical use of animals were followed and the present study was also approved by the Institutional Ethics Committee of the NOVA Medical School for animal care and use in research.

10.2.2 Chronic Intermittent Hypoxia (CIH) Exposure

Rats were randomly assigned and divided into six groups: Group 1 – CIH 14 days (n = 6), Group 2 – normoxic conditions (Nx) 14 days (n = 5), Group 3 – CIH 21 days (n = 6), Group 4 – Nx 21 days (n = 6), Group 5 – CIH 60 days (n = 6) and Group 6 – Nx 60 days (n = 6). A period of 2 days was provided for chamber acclimatization under normoxic conditions (21% O_2 + 79% N_2) before the exposure to CIH. Normoxic rats were exposed for 14, 21 or 60 days, respectively, to 21% O_2 and 79% N_2 in the same room as the CIH animal groups. Animals were kept in a eucapnic atmosphere inside medium A-chambers (Biospherix Ltd., NY, USA). O_2 concentration inside the chambers was controlled using 100% nitrogen (N_2) and 100% O_2 via electronically regulated solenoid switches, which gradually lowered the O_2 in the chamber from 21% to 5% O_2 (OxyCycler AT series, Biospherix Ltd). Chambers were infused with 100% N_2 for 3.5 min to quickly reduce the O_2 concentration to 5%; then the chambers were infused with 100% O_2 for 7 min to restore O_2 to ambient levels of 21%, until the next CIH cycle. Each CIH cycle lasted 10.5 min (normoxic period: 3.5 min; hypoxic period: 7 min) and the rats were exposed during their sleep period (light phase of light/dark cycle) to 5.6 CIH cycles/h, 10.5 h/day (9:30–8:00 pm), for 14, 21 or 60 days. During the remaining time, the chambers were ventilated with room air (21% of O_2).

10.2.3 Gene Expression Analysis

Left kidney cortex and medulla samples were collected and homogenized in Trizol® (Life Technologies) using a tissue homogenizer (Heidolph DIAX 900). After total RNA extraction, cDNA was synthesized from 1 μg of RNA. Quantitative real time PCR (qPCR) was carried out in a final volume of 8.3 μL, with 4 μL of SYBR® Green Master Mix, 10 μM of each primer, plus 1 μL of cDNA for each sample. Relative quantification (arbitrary units) of fibronectin gene expression was assessed in kidney samples. Rat specific primers were used for the housekeeping gene *β-actin*. Fold change was calculated in CIH-exposed rats in relation to the normoxic group for each time-course.

10.2.4 Determination of Cysteine Fractions

Cysteine fractions were quantified in right kidney cortex and medulla homogenates according to Grilo and co-authors by high performance liquid chromatography with fluorescence detection (Grilo et al. 2017).

10.2.5 Statistical Analysis

Data are presented as the ratio of CIH/Nx groups or as percentage of effect whenever applicable and as mean values with their standard deviations. Differences were considered significant at $p < 0.05$ and obtained using *Two-way* ANOVA with *Bonferroni* multiple comparison test or Student *t*-test whenever applicable. GraphPad Prism version 5 (GraphPad Software Inc., San Diego, CA, USA) was used to perform all the statistical analysis.

10.3 Results

We have previously characterized the time-course of blood pressure evolution with the current paradigm of CIH (Diogo et al. 2015). Under this paradigm, hypertension is established at day 14, the plateau of increased blood pressure is attained at day 21 and carvedilol was administered from day 35 until day 60.

To clarify the degree of irreversible compromise of the kidney, we assessed fibronectin gene expression at day 21 and day 60 (Fig. 10.1). At day 21 there was a slight increase in fibronectin in both cortex and medulla ($p < 0.05$) that was remarkably higher at day 60 in kidney cortex ($p < 0.001$).

Renal cortex and renal medulla of animals not exposed to CIH presented a total cysteine content of 5.73 ± 1.16 and 4.83 ± 1.01 μM/mg tissue, respectively. In these tissues the major fraction was represented by reduced cysteine (2.37 ± 1.14 vs. 2.77 ± 0.64 μM/mg tissue) followed by free oxidized (2.10 ± 0.95 vs. 1.1 ± 0.73 μM/mg tissue) and CysSSP (1.26 ± 0.92 vs. 0.97 ± 0.5 μM/mg tissue).

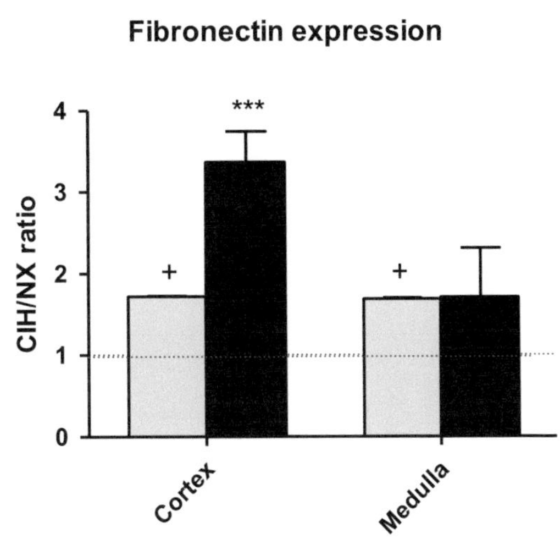

Fig. 10.1 Fibronectin gene expression upon exposure to CIH for 21 and 60 days. Gene expression was normalized with β-actin and quantified by qPCR in renal tissue homogenates. Data are presented as the ratio of chronic intermittent hypoxia (CIH) by normoxia (Nx) groups. (*) *Two-way* ANOVA with *Bonferroni* post-test *vs.* day 21; (+) Student *t*-test Nx *vs.* CIH at each represented time-point. Grey bar – 21 days; black bar – 60 days.

The total content of cysteine (free + protein bound) was quantified in order to investigate the impact of CIH exposure on cysteine availability. At day 14 the total cysteine availability in both renal cortex and medulla was similar to normoxia. It decreased at day 21, attaining its minimum in kidney medulla and remaining low until day 60 (Fig. 10.2a, $p < 0.001$).

Further, we addressed if changes in cysteine availability were accompanied by variation in cysteine dynamics. Although no loss in total cysteine availability was noted at day 14, it was already apparent an increase in CysSSP fraction at this time point (Fig. 10.2b, $p < 0.05$). This initial increase in CysSSP was maximal at day 14 of CIH and observed in both cortex ($p < 0.05$) and medulla ($p < 0.05$). Further, this cysteine fraction decreased over-time in both kidney cortex and kidney medulla, reaching its minimum at day 60 at kidney cortex ($p < 0.001$). Also, at both kidney cortex and medulla at days 14 and 21 the redox couple CysSH/CysSSX shifted towards an oxidized status in CIH-group and recovered to control levels at day 60 (Fig. 10.2c).

10.4 Discussion

Our data shows that CIH influences both availability and dynamics of cysteine in kidney Tissues in a time-dependent and biphasic manner. Increased cysteine oxidation is an early event in this time-course, which is followed by a latter gradual decrease in protein *S*- cysteinylation and simultaneously a recovery of the cysteine redox couple to normoxic levels. Our hypothesis-oriented research was motivated by the fact that the kidney is exposed to relatively high concentrations of cysteine (Stipanuk et al. 2002) and the availability of cysteine highly increases in developing kidney from term to adult life (Foreman and Segal 1987). In fact, the kidney is vital in

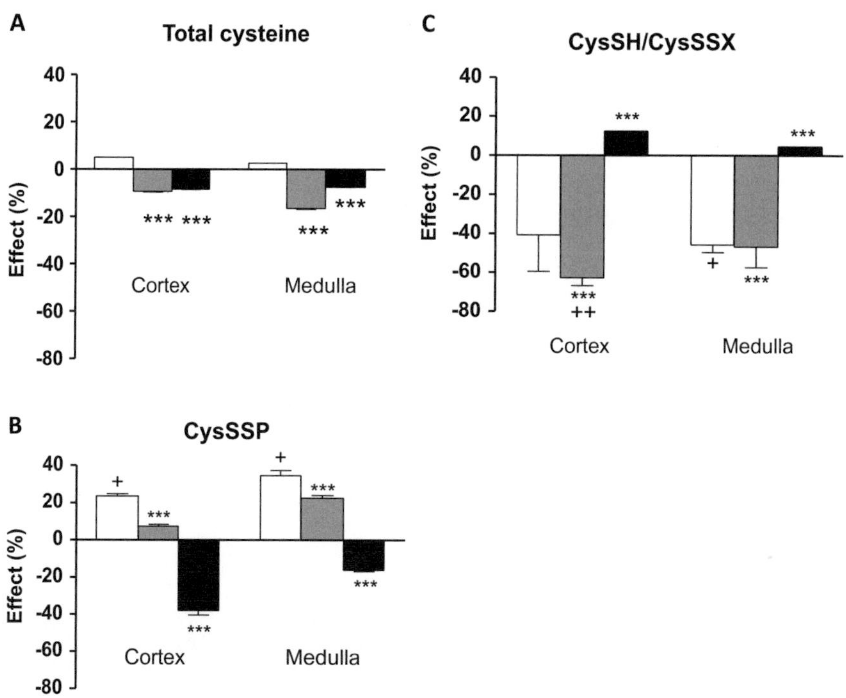

Fig. 10.2 Cysteine dynamics in kidney cortex and medulla upon 14, 21 and 60 days of CIH exposure. (**a**) Total content of cysteine in kidney cortex and medulla. (**b**) *S*-cysteinylation (CysSSP) of proteins in kidney homogenates. (**c**) CysSH/CysSSX in kidney homogenates. Data are presented as percentage of effect from normoxia (Nx) upon CIH exposure. (*) *Two-way* ANOVA with *Bonferroni* post-test *vs.* day 14. (+) Student *t*-test Nx *vs.* CIH at each timepoint. White bar – 14 days; grey bar – 21 days; black bar – 60 days

preventing the loss of cysteine from the organism (Kowalczyk-Pachel et al. 2016) and is an important source of this amino acid with antioxidant, anti-inflammatory and anti-fibrotic properties (Vasdev et al. 2009, Horie et al. 2003). Also, cysteine has been described as a blood pressure regulator and a renoprotective factor (Vasdev et al. 2009). Cysteine availability in kidney also regulates cysteine metabolism (Stipanuk et al. 1992).

We must highlight that in the present work we measured the net content of cysteine and its fractions that, to our best knowledge, was investigated for the first time in kidney and particularly in CIH model. However, total cysteine content was of the same magnitude of that reported by others (Stipanuk et al. 2002). This allowed us to distinguish cysteine dynamics and relate it to time-course establishment of hypertension and progression to kidney fibrosis. For instance, fibrosis is concomitant with a decrease in cysteine availability and proceeded by early events in its dynamics that were not reflected by changes in the total amount of cysteine.

Early after CIH exposure (day 14), the CysSH/CysSSX ratio, which is a pivotal node for redox signaling (Jones et al. 2004), shifted towards a more oxidized state, accompanied by an increase in CysSSP. Scarce information exists on the physiological role of CysSSP (Rossi et al. 2009). Nevertheless, *S*-thiolation events like CysSSP, have a key role regulating several signaling pathways, which may lead to adaptive responses in order to surpass kidney injury (Eaton et al. 2003).

Increases in cysteine free oxidized and in CysSSP fractions were previously described as "disulfide stress" in an acute inflammation model of pancreatitis (Moreno et al. 2014). This disulfide stress would promote redox signaling (Jones et al. 2004) with several targets including ribonuclease inhibitor, APE1/Ref1 (Naganuma et al. 2010), Keap1(Miyazaki et al. 2014), phosphatases and protein disulfide isomerases (Wang and Asghar 2017) that have important roles in kidney and blood pressure regulation.

On the other hand, maximal fibrosis was coincident with minimum values of CysSSP. A plausible hypothesis is that longer exposure to intermittent hypoxia would inhibit the enzymes involved in *S*-thiolation process (Wouters, Fan, and Haworth 2010). Besides its redox regulatory function (Banks et al. 2008), CysSSP can ensure protection of proteins against irreversible oxidation (Auclair et al. 2013), which is often associated with permanent loss of protein function and the accumulation of insoluble aggregates (Rossi et al. 2009, Dalle-Donne et al. 2007, Grilo et al. 2017), suggesting that at later stages proteins are more vulnerable to this pro-fibrotic process (Li et al. 2007).

Our findings show that when treatment with carvedilol was initiated to revert hypertension induced by CIH a significant degree of fibrosis was already established, and the variations in cysteine redox dynamics might not be targeted by this adrenergic blocker.

Funding

iNOVA4Health – UID/Multi/04462/2013 (Ref: 201601-02-021). FCT – PD/BD/114257/2016(NRC), PD/BD/105892/2014 (CGD) and SFRH/BD/130911/2017 (MJC).

References

Auclair JR, Johnson JL, Liu Q, Salisbury JP, Rotunno MS, Petsko GA, Ringe D, Brown RH Jr, Bosco DA, Agar JN (2013) Post-translational modification by cysteine protects Cu/Zn-superoxide dismutase from oxidative damage. Biochemistry 52(36):6137–6144

Banks DD, Gadgil HS, Pipes GD, Bondarenko PV, Hobbs V, Scavezze JL, Kim J, Jiang XR, Mukku V, Dillon TM (2008) Removal of cysteinylation from an unpaired sulfhydryl in the variable region of a recombinant monoclonal IgG1 antibody improves homogeneity, stability, and biological activity. J Pharm Sci 97(2):775–790

Dalle-Donne I, Rossi R, Giustarini D, Colombo R, Milzani A (2007) S-glutathionylation in protein redox regulation. Free Radic Biol Med 43(6):883–898

Diogo LN, Pereira SA, Nunes AR, Afonso RA, Santos AI, Monteiro EC (2015) Efficacy of carvedilol in reversing hypertension induced by chronic intermittent hypoxia in rats. Eur J Pharmacol 765:58–67

Eaton P, Jones ME, McGregor E, Dunn MJ, Leeds N, Byers HL, Leung K-Y, Ward MA, Pratt JR, Shattock MJ (2003) Reversible cysteine- targeted oxidation of proteins during renal oxidative stress. J Am Soc Nephrol 14(suppl 3):S290–S296

Foreman JW, Segal S (1987) Cysteine and glutathione levels in developing rat kidney and liver. Pediatr Res 22(5):605–608

Grilo NM, Joao Correia M, Miranda JP, Cipriano M, Serpa J, Matilde Marques M, Monteiro EC, Antunes AMM, Diogo LN, Pereira SA (2017) Unmasking efavirenz neurotoxicity: time matters to the underlying mechanisms. Eur J Pharm Sci 105:47–54

Horie T, Sakaida I, Yokoya F, Nakajo M, Sonaka I, Okita K (2003) L-cysteine administration prevents liver fibrosis by suppressing hepatic stellate cell proliferation and activation. Biochem Biophys Res Commun 305(1):94–100

Jones DP, Go YM, Anderson CL, Ziegler TR, Kinkade JM Jr, Kirlin WG (2004) Cysteine/cystine couple is a newly recognized node in the circuitry for biologic redox signaling and control. FASEB J 18(11):1246–1248

Kowalczyk-Pachel D, Iciek M, Wydra K, Nowak E, Górny M, Filip M, Włodek L, Lorenc-Koci E (2016) Cysteine metabolism and oxidative processes in the rat liver and kidney after acute and repeated cocaine treatment. PLoS One 11(1):e0147238

Li HY, Hou FF, Zhang X, Chen PY, Liu SX, Feng JX, Liu ZQ, Shan YX, Wang GB, Zhou ZM (2007) Advanced oxidation protein products accelerate renal fibrosis in a remnant kidney model. J Am Soc Nephrol 18(2):528–538

Miyazaki Y, Shimizu A, Pastan I, Taguchi K, Naganuma E, Suzuki T, Hosoya T, Yokoo T, Saito A, Miyata T (2014) Keap1 inhibition attenuates glomerulosclerosis. Nephrol Dial Transplant 29(4):783–791

Moreno M-L, Escobar J, Izquierdo-Álvarez A, Gil A, Pérez S, Pereda J, Zapico I, Vento M, Sabater L, Marina A (2014) Disulfide stress: a novel type of oxidative stress in acute pancreatitis. Free Radic Biol Med 70:265–277

Naganuma T, Nakayama T, Sato N, Zhenyan F, Soma M, Yamaguchi M, Shimodaira M, Aoi N, Usami R (2010) Haplotype-based case–control study on human apurinic/apyrimidinic endonuclease 1/redox effector factor-1 gene and essential hypertension. Am J Hypertens 23(2):186–191

Rossi R, Giustarini D, Milzani A, Dalle-Donne I (2009) Cysteinylation and homocysteinylation of plasma protein thiols during ageing of healthy human beings. J Cell Mol Med 13(9B):3131–3140

Stipanuk MH, Coloso RM, Garcia RA, Banks MF (1992) Cysteine concentration regulates cysteine metabolism to glutathione, sulfate and taurine in rat hepatocytes. J Nutr 122(3):420–427

Stipanuk MH, Londono M, Lee J-I, Mindy H, Anthony FY (2002) Enzymes and metabolites of cysteine metabolism in nonhepatic tissues of rats show little response to changes in dietary protein or sulfur amino acid levels. J Nutr 132(11):3369–3378

Sun W, Yin X, Wang Y, Tan Y, Cai L, Wang B, Cai J, Fu Y (2012) Intermittent hypoxia-induced renal antioxidants and oxidative damage in male mice: hormetic dose response. Dose-Response 11(3):385–400

Vasdev S, Singal P, Gill V (2009) The antihypertensive effect of cysteine. Int J Angiol 18(1):7–21

Wang X, Asghar M (2017) Protein disulfide isomerase regulates renal AT1 receptor function and blood pressure in rats. Am J Physiol Renal Physiol 313(2):F461–F466

Wouters MA, Fan SW, Haworth NL (2010) Disulfides as redox switches: from molecular mechanisms to functional significance. Antioxid Redox Signal 12(1):53–91

Wu H, Zhou S, Kong L, Chen J, Feng W, Cai J, Miao L, Tan Y (2015) Metallothionein deletion exacerbates intermittent hypoxia-induced renal injury in mice. Toxicol Lett 232(2):340–348

11 Adenosine Mediates Hypercapnic Response in the Rat Carotid Body via A_{2A} and A_{2B} Receptors

Joana F. Sacramento, Bernardete F. Melo, and Sílvia V. Conde

Abstract

Adenosine is one of the key neurotransmitters involved in hypoxic signaling in the carotid body (CB), and it was recently found to have a modulatory role in mediating hypercapnic sensitivity in the CB. Herein we have investigated the contribution of adenosine to the hypercapnic response in the rat CB and studied the adenosine receptors responsible for this effect. Experiments were performed in *Wistar* rats. Adenosine release in normoxia (21% O_2) and in response to hypercapnia (10% CO_2) was quantified by HPLC. Carotid sinus nerve (CSN) chemosensory activity was evaluated in response to hypercapnia in the absence and presence of ZM241385 (300 nM), an A_2 antagonist, and SCH58261 (20 nM), a selective A_{2A} antagonist. Hypercapnia increased the extracellular concentrations of adenosine by 50.01%. Both, ZM241385 and SCH58261, did not modify significantly the basal frequency of discharges of the CSN. Also, ZM241385 and SCH58261 did not modify the latency time and the time to peak in CSN chemosensory activity. CSN activity evoked by hypercapnia decreased by 58.82 and 33.59% in response to ZM241385 and to SCH58261, respectively. In conclusion, the effect of adenosine in mediating the hypercapnic response in the rat CB involves an effect on A_{2A} and A_{2B} adenosine receptors.

Keywords

Adenosine · Adenosine receptors · Carotid body · Carotid sinus nerve activity · Hypercapnia

11.1 Introduction

The carotid body (CB) is an arterial chemoreceptor activated by decreases in blood PaO_2 and pH or increases in blood $PaCO_2$ (Gonzalez et al. 1994). The CB plays a homeostatic role in response to high $PaCO_2$ and low pH and, it seems that approximately to 30–50% of the respiratory drive produced by arterial hypercapnia is mediated by CB (Heeringa et al. 1979; Rodman et al. 2001), with the remaining contribution arising from central chemoreceptors (Nattie 1999). Also, CB afferent activity is essential to set hypercapnic sensitivity in central chemoreceptors (Blain et al. 2010; Smith et al. 2015). While the transduction mechanisms and the neurotransmitters involved in the CB response to hypoxia are well studied, the CB chemotransduction in response to hypercapnia has received less attention. In 2015, Holmes and collaborators have

J. F. Sacramento · B. F. Melo · S. V. Conde (✉)
CEDOC, Centro de Estudos Doenças Crónicas, Nova Medical School, Faculdade de Ciências Médicas, Universidade Nova de Lisboa, Lisbon, Portugal
e-mail: silvia.conde@fcm.unl.pt

E. B. Gauda et al. (eds.), *Arterial Chemoreceptors*, Advances in Experimental Medicine and Biology 1071, https://doi.org/10.1007/978-3-319-91137-3_11

demonstrated that adenosine have a modulatory role in mediating hypercapnic sensitivity in the CB. Herein, we have investigated the role of adenosine, a key excitatory neurotransmitter involved in hypoxic signalling in the CB (Conde et al. 2009, 2012), on the CB response to hypercapnia. The role of adenosine was studied by testing ZM241385, an A_2 antagonist, and SCH5826, a selective A_{2A} antagonist on basal CSN chemosensory activity and in response to hypercapnia. We also evaluated the CB release of adenosine in response to hypercapnia. We found that hypercapnia induced the CB release of adenosine and that this mediator is involved in the hypercapnic response in the rat CB via both A_{2A} and A_{2B} adenosine receptors.

11.2 Methods

Experiments were performed in *Wistar* adult rats (300–380 g) obtained from *vivarium* of the NOVA Medical School|Faculdade de Ciências Médicas (NMS|UNL), Universidade Nova de Lisboa. Animals were kept under temperature and humidity control (21 ± 1 °C; 55 ± 10% humidity) with a 12 h light/12 h dark cycle. Animals were anaesthetized with sodium pentobarbital (60 mg/kg, i.p.), tracheostomized and the carotid arteries were dissected past the carotid bifurcation. To quantify adenosine at the CB, the CBs were cleaned free of CSN nearby connective tissues under dissection microscope and incubated in Tyrode solution (Conde et al. 2012). For CSN recordings, the preparation CB-CSN was identified and processed for CSN recordings as previously described by Rigual et al. (2002) and Conde et al. (2012). In all instances animals were killed by intracardiac overdoses of sodium pentobarbital. Principles of laboratory care were followed in accordance with the European Union Directive for Protection of Vertebrates Used for Experimental and Other Scientific Ends (2010/63/EU). Experimental protocols were approved by the ethics committee of the NMS|UNL.

11.2.1 CB Adenosine Release in Response to Hypercapnia

CBs were cleaned free of CSN nearby connective tissues under dissection microscope and incubated in Tyrode solution (Conde et al. 2012). To evaluate the release of adenosine from CBs in response to hypercapnia (10% CO_2) in control animals, the CBs were incubated in 500 mL of Tyrode bicarbonate solution with 2.5 μM of EHNA (an inhibitor of adenosine deaminase) and 5 μM of S-(p-nitrobenzyl)-6-thioinosine (NBTI, an inhibitor of equilibrative nucleoside transport system). Solutions were kept at 37 °C and continuously bubbled with normoxia (20%O_2 + 5%CO_2 + balanced N_2), except when hypercapnic stimuli was applied (20%O_2 + 10%CO_2 + balanced N_2). Protocol for adenosine release included: 2 periods of 10 min normoxic incubation, followed by 10 min incubation in hypercapnia and 2 post-hypercapnia incubations in normoxia during 10 min. The solutions were renewed at each fixed time, and all fractions were collected and quantified by HPLC as previously described (Conde et al. 2012). At the end of the experiment, the CBs have been placed in PCA 3 M and processed for the quantification of adenosine content.

11.2.2 CSN Electrophysiological Recordings

Extracellular recording from single or few fibers of CSN were performed using a suction electrode. The pipette potential was amplified (Neurolog Digimiter, Hertfordshire, UK), filtered with low pass (5 kHz) and high pass (10 Hz) filters, digitized at 5 kHz (Axonscope, Axon Instruments, Molecular Devices, Workingham, UK) and stored on a computer. Chemoreceptor activity was identified (spontaneous generation of action potentials at irregular intervals) and confirmed by its increase in response to hypoxia (0%O2 + 5% CO_2 + balanced N_2), which corresponds approximately to 20 mmHg, (Conde et al. 2012).

CSN unit activity was converted to logic pulses, which were summed every second and converted in a voltage proportional to the sum. The effect of ZM241385 (300 nM), an A_2 antagonist, and SCH58261 (20 nM), a selective A_{2A} antagonist, on the CSN activity has been investigated while perfusing the preparations with normoxic (20% O_2-equilibrated) and hypercapnic (10% CO_2-equilibrated) solutions.

11.2.3 Statistical Analysis

Data were evaluated using GraphPad Prism Software, version 6 and was presented as the mean ± SEM. The significance of the differences between the means was calculated by unpaired t test and one-way analysis of variance (ANOVA) with Bonferoni's multiple comparison tests. Differences were considered significant at $p < 0.05$.

11.3 Results

Figure 11.1 shows the effect of hypercapnia on adenosine release from the rat CB. Hyperca-pnia (10% CO_2) increased significantly the extracellular concentrations of adenosine with the release reaching 105.70 ± 11.45 (pmol/mg tissue), when compared with the basal normoxic release (before hypercapnia = 70.42 ± 8.83 pmol/mg tissue; after hypercapnia = 82.89 ± 22.13 pmol/mg tissue).

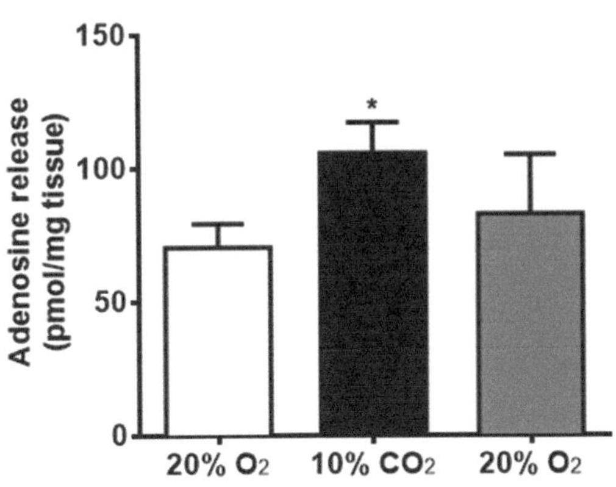

Fig. 11.1 Effect of hypercapnia (10% CO_2) on adenosine release from the rat CB. Protocol included 2 periods of 10 min in normoxia (20%O_2 + 5%CO_2), followed by 10 min of hypercapnia (20%O_2 + 10% CO_2) followed by two normoxic period (n = 6). Data are means ± SEM. Unpaired t test, $*p < 0.05$

Figure 11.2 depicts the effect of ZM241385 (300 nM), an A_2 antagonist, and SCH58261 (20 nM), a selective A_{2A} antagonist, on the CSN chemosensory activity in response to hypercapnia. Basal CSN chemosensory activity was not modified by the application of both adenosine receptors antagonists (Fig. 11.2a). The application of ZM241385 and SCH58261 showed a non-significant tendency to increase the onset of the response (latency time) (Fig. 11.2b). However, none of the drugs tested modified significantly the time to reach the maximal activity (time to peak, Fig. 11.2c).

Figure 11.2d represents typical recordings of CSN chemosensory activity in response to hypercapnia in control conditions, in the absence of any drugs, and in the presence of ZM241385 and SCH58261. The application of both drugs decreased the CSN activity in response to hypercapnia. However, ZM241385 was more effective in inhibiting CSN activity than SCH58261 (Fig. 11.2d). This result was confirmed when we plot the area under the curve (AUC) for the CSN chemosensory activity elicited by hypercapnia in control conditions and in the presence of ZM241385 and SCH58261 (Fig. 11.2e). ZM241385 decreased the AUC by 58.82% when compared with the effect of hypercapnia in the absence of the drugs, while SCH58261 decreased the AUC by 33.59% (Fig. 11.2e; AUC control = 2150.0 ± 194.2; AUC ZM241385 = 885.2 ± 153.2; AUC SCH58261 = 1428 ± 93.2 spikes/s).

11.4 Discussion

The present study demonstrates that adenosine mediates the CSN chemosensory activity evoked by hypercapnia, an effect that is mediated by A_2 adenosine receptors. We have showed that both adenosine receptor antagonists tested did not modify basal CSN chemosensory activity, the latency time and the time to peak. Additionally, both ZM241385 and SCH58261 decreased the

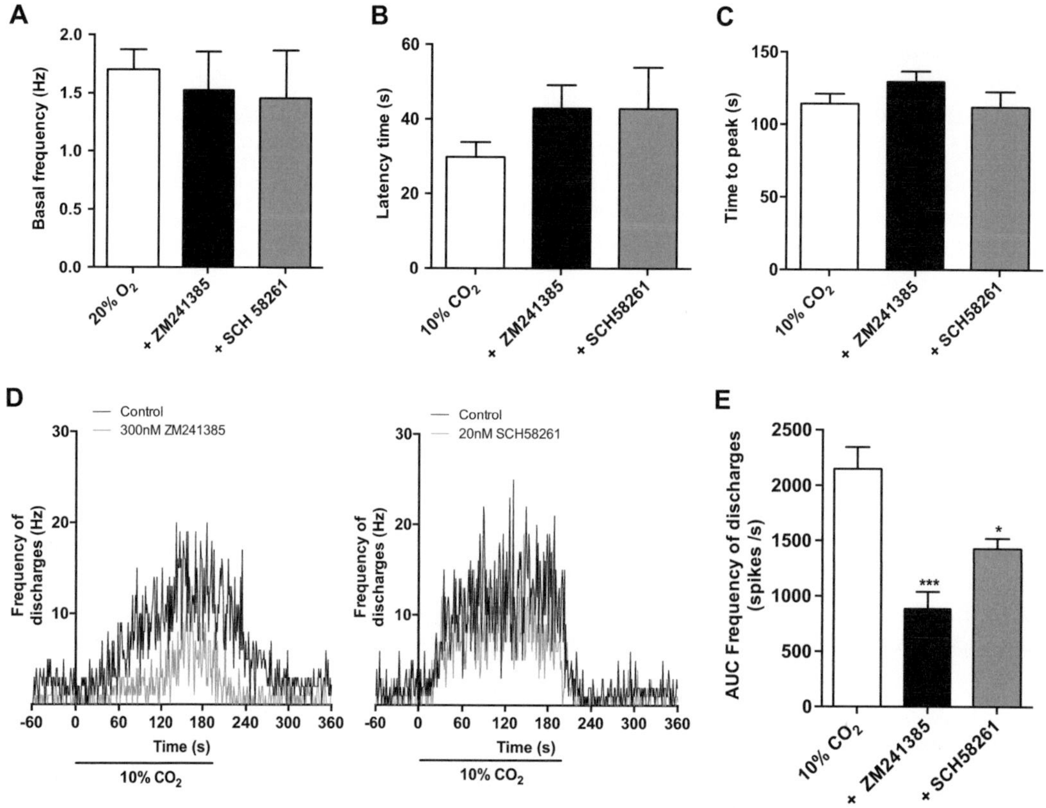

Fig. 11.2 Effect of ZM241385 (300 nM), an A_2 adenosine receptor antagonist, and SCH58261 (20 nM), a selective A_{2A} antagonist, on carotid sinus nerve (CSN) chemosensory activity elicited by hypercapnia (10%CO_2). (**a**) Mean basal frequencies. (**b**) Latency time (the onset of the response). (**c**) Time required to reach maximal activity (time to peak). (**d**) Typical recordings of the effect of ZM241385 and SCH58261 on the frequency of action potentials of CSN during superfusion with a solution equilibrated with 10%CO_2, respectively. (**e**) Area under the curve (AUC) obtained from the analysis of the curves plotted in D. Data represent means ± SEM of 5–7 animals. One-way ANOVA with Bonferroni multiple comparison tests, $*p < 0.05$ and $***p < 0.001$ *vs* control values

hypercapnia-evoked CSN chemosensory activity, being the effect more pronounced when ZM241385 was applied, suggesting that apart from A_{2A} receptors, A_{2B} adenosine receptors also contribute to this effect.

We have observed that hypercapnia induced the release of adenosine from the CB, as it happens in response to hypoxia (Conde and Monteiro 2004). In the present study we did not evaluate the source of the extracellular adenosine, however we can postulate the main origin of extracellular adenosine induced by hypercapnia could be the extracellular ATP catabolism, since Holmes et al. (2015) showed that the application of an 5′-ectonucleotidase (CD73) inhibitor, α,β-methylene ADP (AOPCP), dramatically (98%) decreased the CB sensitivity to hypercapnia.

In accordance with our previous findings we have observed herein that ZM241385 and SCH58261 did not modify the basal CSN activity (Conde et al. 2012), which suggest that adenosine did not contribute significantly to fix the basal activity of the CB in control conditions. Moreover, the application of ZM231485 and SCH58261 decreased the CSN activity elicited by hypercapnia, meaning that the effect of hypercapnia in the CB is mediated by A_2 adenosine receptors. Additionally, the fact that ZM241385, the non-selective A_2 antagonist decreased more the response to hypercapnia than SCH58261, the selective A_{2A} antagonist, indicated that both A_{2A} and A_{2B} adenosine receptors contribute to the effect of hypercapnia on the CB. In fact, from the observation of the Fig. 11.2d, A_{2A} seems to contribute with approximately 30% for the CSN activity in response to hypercapnia and A_{2B} with approximately 25%. These values are quite similar with those observed for the contribution of adenosine A_{2A} and A_{2B} receptors for the responses to moderate hypoxia in the CB (Conde et al. 2006). However, since we have not tested more intense hypercapnic stimuli we cannot anticipate if these contributions are maintained with higher or lower hypercapnic intensities. Both A_{2A} and A_{2B} adenosine receptors are Gs-coupled receptors and therefore adenosine action through these receptors involves the increase in cAMP levels (Sebastião and Ribeiro 2009). Herein, we did not studied the modifications in cAMP levels induced by hypercapnia, but Holmes and co-workers (2015) have showed that SQ2236, an inhibitor of transmembrane adenylyl cyclases, decreased the elevation evoked by hypercapnia by approximately 50%, a value that is quite similar with the 58% of reduction in CSN activity achieved by us with the non-selective blocker of A_2 receptors in hypercapnia.

In conclusion, the effect of adenosine in mediating the hypercapnic response in the CB is mediated by both A_{2A} and A_{2B} adenosine receptors.

Acknowledgements J.F.S. and B.F.M. are supported by PhD Grants from Portuguese Foundation for Science and Technology Reference PD/BD/105890/2014 and PD/BD/128336/2017, respectively.

References

Blain GM, Smith CA, Henderson KS, Dempsey JA (2010) Peripheral chemoreceptors determine the respiratory sensitivity of central chemoreceptors to CO_2. J Physiol 588:2455–2471

Conde SV, Monteiro EC (2004) Hypoxia induces adenosine release from the rat carotid body. J Neurochem 89:1148–1156

Conde SV, Obeso A, Vicario I, Rigual R, Rocher A, Gonzalez C (2006) Caffeine inhibition of rat carotid body chemoreceptors is mediated by A2A and A2B adenosine receptors. J Neurochem 98:616–628

Conde SV, Monteiro EC, Obeso A, Gonzalez C (2009) Adenosine in peripheral chemoreception: new insights into a historically overlooked molecule. Adv Exp Med Biol 648:159–174

Conde SV, Monteiro EC, Rigual R, Obeso A, Gonzalez C (2012) Hypoxic intensity: a determinant for the contribution of ATP and adenosine to the genesis of carotid body chemosensoryactivity. J Appl Physiol 112:2002–2010

Gonzalez C, Almaraz L, Obeso A, Rigual R (1994) Carotid body chemoreceptors: from natural stimuli to sensory discharges. Physiol Rev 74:829–898

Heeringa J, Berkenbosch A, de Goede J, Olievier CN (1979) Relative contribution of central and peripheral chemoreceptors to the ventilatory response to CO_2 during hyperoxia. Respir Physiol 37:365–379

Holmes AP, Nunes AR, Cann MJ, Kumar P (2015) Ecto-5′-nucleotidase, adenosine and transmembrane adenylyl cyclase signalling regulate basal carotid body chemoafferent outflow and establish the sensitivity to hypercapnia. Adv Exp Med Biol 860:279–289

Nattie E (1999) CO_2, brainstem chemoreceptors and breathing. Prog Neurobiol 59:299–331

Rigual R, Rico AJ, Prieto-Lloret J, de Felipe C, Gonzalez C, Donnelly DF (2002) Chemoreceptor activity is normal in mice lacking the NK1 receptor. Eur J Neurosci 16:2078–2084

Rodman JR, Curran AK, Henderson KS, Dempsey JA, Smith CA. (2001) Carotid body denervation in dogs: eupnea and the ventilatory response to hyperoxic hypercapnia. J Appl Physiol 91(1):328–35

Sebastião AM, Ribeiro JA (2009) Tuning and fine-tuning of synapses with adenosine. Curr Neuropharmacol 7(3):180–94

Smith CA, Blain GM, Henderson KS, Dempsey JA (2015) Peripheral chemoreceptors determine the respiratory sensitivity of central chemoreceptors to CO_2: role of carotid body CO_2. J Physiol 593:4225–4243

Acute Effects of Systemic Erythropoietin Injections on Carotid Body Chemosensory Activity Following Hypoxic and Hypercapnic Stimulation

12

David C. Andrade, Rodrigo Iturriaga, Florine Jeton, Julio Alcayaga, Nicolas Voituron, and Rodrigo Del Rio

Abstract

The carotid body (CB) chemoreceptors sense changes in arterial blood gases. Upon stimulation CB chemoreceptors cells release one or more transmitters to excite sensory nerve fibers of the carotid sinus nerve. While several neurotransmitters have been described to contribute to the CB chemosensory process less is known about modulatory molecules. Recent data suggest that erythropoietin (Epo) is involved in the control of ventilation, and it has been shown that Epo receptor is constitutively expressed in the CB chemoreceptors, suggesting a possible role for Epo in regulation of CB function. Therefore, in the present study we aimed to determine whether exogenous applications of Epo modulate the hypoxic and hypercapnic CB chemosensory responses. Carotid sinus nerve discharge was recorded in-situ from anesthetized adult male and female Sprague Dawley rats (350 g, n = 8) before and after systemic administration of Epo (2000 UI/kg). CB-chemosensitivity to hypoxia and hypercapnia was calculated by exposing the rat to F_iO_2 5–15% and F_iCO_2 10% gas mixtures, respectively. During baseline recordings at normoxia, we found no effects of Epo on CB activity both in male and

Senior authors Nicolas Voituron and Rodrigo Del Rio have equally contributed to this chapter.

D. C. Andrade
Laboratory of Cardiorespiratory Control, Department of Physiology, Faculty of Biological Sciences, Pontificia Universidad Católica de Chile, Santiago, Chile

R. Iturriaga
Laboratorio de Neurobiología, Facultad de Ciencias Biológicas, Pontificia Universidad Católica de Chile, Santiago, Chile

F. Jeton · N. Voituron
Laboratoire Hypoxie & Poumon, Universite Paris 13, Paris, France

J. Alcayaga
Laboratorio de Fisiología Celular, Universidad de Chile, Santiago, Chile

R. Del Rio (✉)
Laboratory of Cardiorespiratory Control, Department of Physiology, Faculty of Biological Sciences, Pontificia Universidad Católica de Chile, Santiago, Chile

Centro de Envejecimiento y Regeneración (CARE), Pontificia Universidad Católica de Chile, Santiago, Chile

Centro de Excelencia en Biomedicina de Magallanes (CEBIMA), Universidad de Magallanes, Punta Arenas, Chile
e-mail: rdelrio@bio.puc.cl

E. B. Gauda et al. (eds.), *Arterial Chemoreceptors*, Advances in Experimental Medicine and Biology 1071, https://doi.org/10.1007/978-3-319-91137-3_12

female rats. In addition, Epo had no effect on maximal CB response to hypoxia in both male and female rats. Epo injections enhanced the maximum CB chemosensory response to hypercapnia in female rats (before vs. after Epo, 72.5 ± 7.1 Hz vs. 108.3 ± 6.9 Hz, $p < 0.05$). In contrast, Epo had no effect on maximum CB chemosensory response to hypercapnia in male rats but significantly increased the response recovery times (time required to return to baseline discharge following hypercapnic stimulus) from 2.1 ± 0.1 s to 8.2 ± 2.3 s ($p < 0.05$). Taken together, our results suggest that Epo has some modulatory effect on the CB chemosensory response to hypercapnia.

Keywords

Carotid Body · Carotid sinus nerve · Erythropoietin · Hypoxia · Hypercapnia

12.1 Introduction

Regulation of breathing by arterial blood gases depend on peripheral and central chemoreceptors (Feldman et al. 2003; Guyenet 2014). The carotid Body (CB), the main peripheral chemoreceptor, is sensitive to changes in arterial blood gas composition (PO_2, PCO_2 and pH) (Gonzalez et al. 1992), initiating a ventilatory and sympathetic responses to maintain cardio-respiratory homeostasis (Gonzalez et al. 1995; Despas et al. 2006; Lopez-Barneo et al. 2016). Furthermore, recent studies indicate that peripheral chemoreceptors determine the sensitivity of the central chemoreceptors to CO_2 (Smith et al. 2015). Chemoreflex responses serve to maintain PCO_2 and pH within homeostatic parameters in the internal medium and to adjust the O_2 supply according to the metabolic demand (Gonzalez et al. 1992). Upon stimulation, the CB chemoreceptor cells release neurotransmitters that activate the sensory nerve fibers of the carotid sinus nerve (CSN) which relay this information to brainstem neurons that regulate breathing (Prabhakar and Peers 2014). In addition, the CB is sensitive to plasma levels of Erythropoietin (Epo) (Soliz et al. 2005). Previous studies suggest that CB chemotransduction is modulated by high Epo plasma levels (Soliz et al. 2007). EPO's role in synthesis of red blood cells is well known (Jelkmann 1992, 2011). However Epo is also produced by neurons and glial cells (Digicaylioglu et al. 1995). Furthermore, Epo-receptors (EpoR) are widely distributed in the areas known to be associated with respiratory control (Soliz et al. 2005) such as the Respiratory Rhythm Generator, the CBs and central chemoreceptor sensitive areas (i.e. nucleus of the solitarii tract and raphe). Interestingly, expression of Epo and its receptor has been observed in CB glomus cells (Lam et al. 2009). The presence of EpoR in these structures supports the notion that Epo participates in respiratory regulation and CB chemotransduction. Indeed, previous work indicates that Epo may regulate the hypoxic ventilatory response (HVR) in humans and mice in a sex-dependent manner (Jeton et al. 2017; Pichon et al. 2016; Soliz 2013; Soliz et al. 2012; Voituron et al. 2014). Thus, it is possible that Epo acting on EpoR located within the CB tissue may play a role in the regulation of the HVR. However, there is no direct evidence for the role played by Epo on CB chemosensory function.

Our goal was to assess whether exogenous applications of Epo could modulate the hypoxic and hypercapnic CB chemosensory responses. To assess this, carotid sinus nerve discharge was recorded *in-situ* from anesthetized adult male and female Sprague Dawley rats before and after systemic administration of Epo.

12.2 Methods

12.2.1 Ethical Approval

Experiments were performed in adult male (n = 4) and female (n = 4) Sprague Dawley rats (~350g). All experiments were approved by the Bioethical Committee of the P. Universidad Católica de Chile and were carried out under the guidelines of the American Physiological Society, the Guía para el Cuidado y Uso de los Animales de Laboratorio from CONICYT and the National Institutes of Health Guide for the Care and Use of Laboratory Animals.

12.2.2 Animal and Procedures

After deep anesthesia with sodium pentobarbital (40 mg/kg I.P.), rats (n = 8) were placed in supine position and the rectal temperature was maintained at 38.0 ± 0.5 °C with a regulated heating pad. Carotid sinus nerve discharge was recorded *in-situ* as previously described (Del Rio et al. 2010, 2011). Briefly, right carotid sinus nerve was isolated and placed on a pair of platinum electrodes and covered with warm mineral oil. Signal was then amplified (Grass P511, USA), filtered (30–500 Hz) and sent to an electronic spike-amplitude discriminator to select action potentials of given amplitude above the noise. Hyperoxic test (100% O_2) was used to set the threshold of CB sensory activity present in the electrophysiological recording. The selected action potentials were used to evaluate the CB chemosensory frequency of discharge (f_x expressed in Hz). Signals were acquired with an analog-digital system PowerLAB and analyzed with the Chart 7-Pro software (ADInstruments, Australia). CB-chemosensory sensitivity to hypoxia or hypercapnia was calculated by allowing the rat to spontaneously breathe F_iO_2 5–15% and F_iCO_2 3–7% gas mixtures, respectively, before and after 60 min of systemic administration of Epo (2000 UI/kg, 100 μl i.v. bolus).

12.2.3 Data Analysis and Statistics

The CB chemosensory frequency discharges (f_x) response-curves were fitted to the inspiratory PO_2 according to the following exponential function (Berkenbosch et al. 1991, 1997).

$$f_x = G\exp\left(-DPO_2\right) + A$$

In which, G is the overall gain, D is oxygen threshold and A is the value of CB chemosensory discharge measured during hyperoxic challenge (Dejour test). For the CB chemosensory response to hypercapnia, linear regression analysis was used. Results were expressed as mean ± standard errors of the mean (SEM). Differences were tested by Bonferroni test following two-ways ANOVA or by using Student T test. All analyses were performed with Graph Pad – Prism software (Graph Pad software, La Jolla, CA, USA). Differences were considered significant when $p < 0.05$.

12.3 Results

12.3.1 CB Chemosensory Activity in Male and Female Rats

Chemosensory activity in normoxia and during acute hypoxia did not differ between adult male and female rats (Table 12.1). There is a trend to have higher maximal responses to severe hypoxia in male rats compared to females. Indeed, the CB responses to F_iO_2 5% expressed as a percent change compared to the values at normoxia were 516.7 ± 31.4% in males and 404.0 ± 32.1% in females. However, this difference was not statistically different between both groups.

12.3.2 Effects of Epo Injection on CB Activity

During baseline recordings at normoxia, we found no effects of Epo on CB activity both in male and female rats (Fig. 12.1). In addition, acute Epo injection showed no effects in maximal CB response to hypoxia neither in CB oxygen sensitivity in male or female rats (Fig. 12.2).

Table 12.1 Carotid body chemosensory baseline activity in normoxia and in response to hypoxia

	♂	♀
F_iO_2 21% *fx* (Hz)	70.1 ± 14.1	80.9 ± 13.4
F_iO_2 10% *fx* (Hz)	310.6 ± 33.1	295.0 ± 29.0
F_iO_2 5% *fx* (Hz)	362.0 ± 22.0	327.0 ± 26.3

Data are showed as mean ± SEM. *fx*: Carotid body chemosensory activity; F_iO_2: Fraction of inspired oxygen

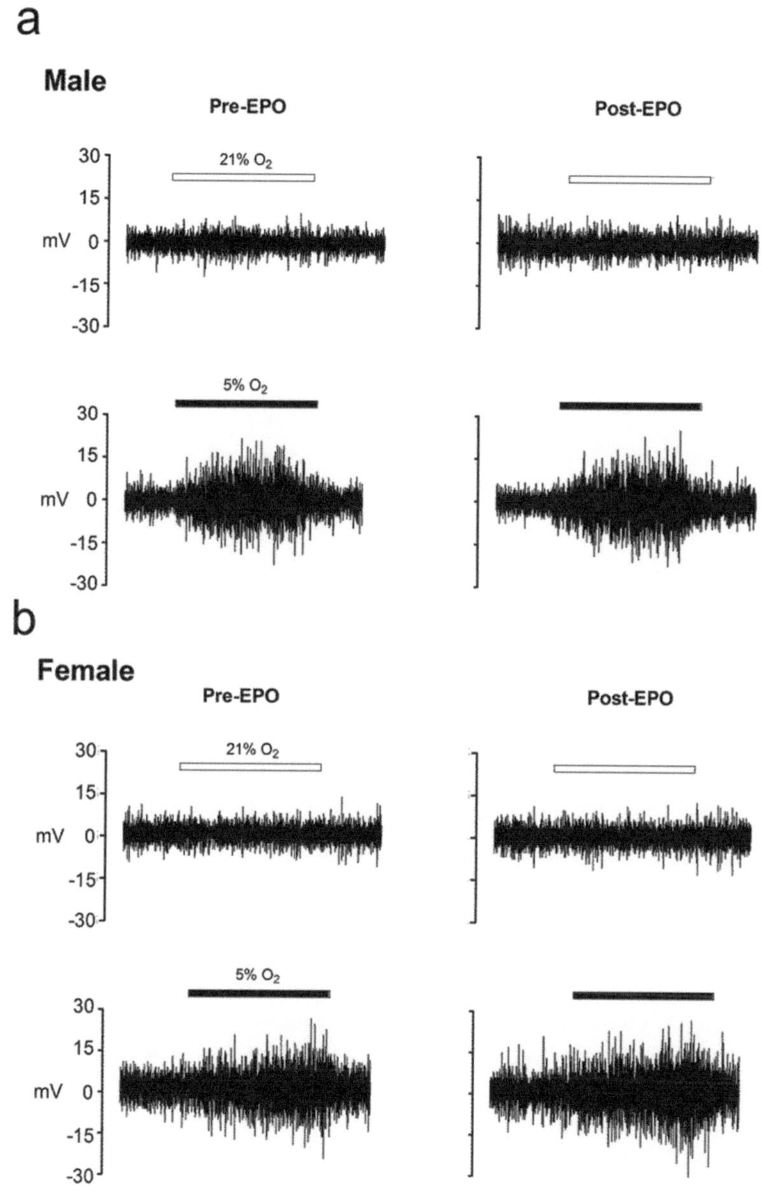

Fig. 12.1 Effects of Epo administration on carotid sinus nerve activity in male and female rats. (**a**), representative recordings of CB chemosensory activity *in situ* from male rats, during normoxic stimulation (Upper panel) before and after Epo i.v. injection. Note the absence of Epo effects on baseline CB activity under normoxia (Lower Panel), CB chemosensory activity during acute hypoxic (20s) stimulation before and after Epo administration. No effect of Epo on CB response to hypoxia was found. (**b**), representative CB recordings in female rats before and after Epo i.v. injection. Note the absence of Epo effects on baseline CB activity under normoxia

12.3.3 Effects of Epo on CB Chemosensory Response to Hypercapnia

The effect of Epo injections on CB activity is shown in Fig. 12.3. Interestingly, Epo administration increases the maximal CB chemosensory response to hypercapnia 10% CO_2 only in female rats (Fig. 12.4). Indeed, the female CB responses to hypercapnia were 72.5 ± 7.1 Hz vs. 108.3 ± 6.9 Hz ($P < 0.05$), before and after Epo, respectively. Contrarily, Epo has no effect on the maximal CB chemosensory response to hypercapnia in males but significantly increases the response recovery times (time required to return

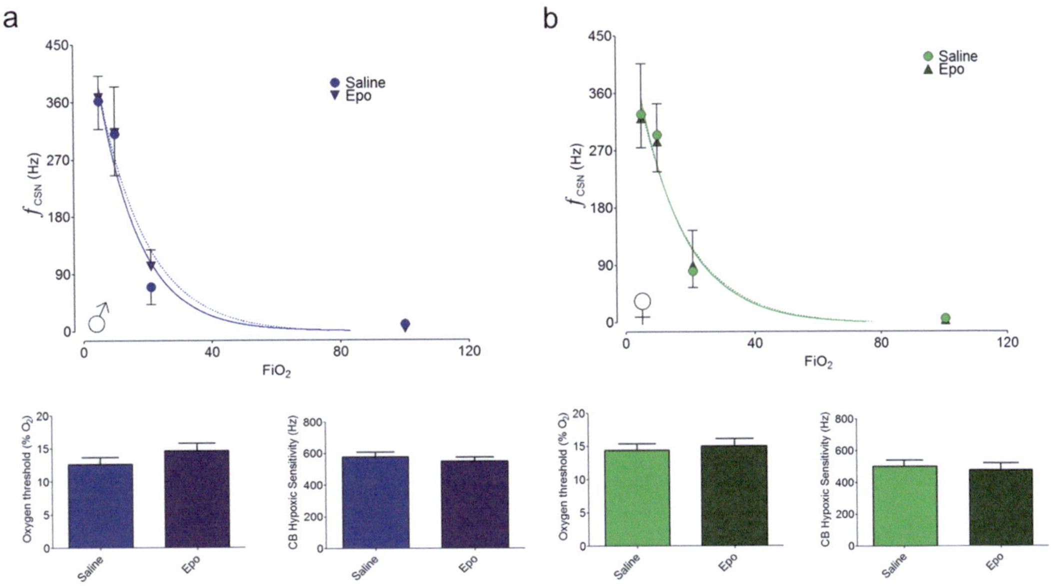

Fig. 12.2 Summary of the effects of Epo on carotid body chemosensory function in hypoxia. Upper panels, CB chemosensory response to hypoxia before and after Epo administration in (**a**) males and (**b**) females rats. $P > 0.05$, for CB chemosensory curves, Bonferroni test following two-ways ANOVA, $n = 4$ per group. Lower panels, Epo has no effect on oxygen threshold nor in CB hypoxic sensitivity in both males or females rats. $P > 0.05$, paired t test. $n = 4$ per group

to baseline discharge following hypercapnic stimulus) from 2.1 ± 0.1 s to 8.2 ± 2.3 s (Fig. 12.4).

12.4 Discussion

The main findings of our study were: (i) during normoxic condition Epo injection did not modify the CB chemosensory discharge in male and female rats; (ii) Epo injection did not change the CB chemosensory responses induced by hypoxia (10–5% O_2) neither in males nor in female rats; (iii) Epo injections enhanced the maximal CB chemosensory response to hypercapnia (10% CO_2) in female rats, and (iv) Epo injections increased the time needed to return to baseline discharge levels following hypercapnic stimulation in male rats. Together these results suggest that Epo modulate the CB chemosensory responses to hypercapnia in a sex-dependent manner.

Soliz et al. (2007) found that transgenic mouse line (Tg6), with high plasma levels of human Epo, did not show significant differences in resting minute ventilation compared to wild-type mice. Present results showed that Epo injections did not change CB chemosensory discharges in normoxia. Then, our results suggest that Epo is not related to the regulation of the tonic CB activity in normoxia. Therefore, agrees and extend previous studies showing no effect of Epo on breathing regulation under normoxic conditions.

It has been shown that Epo may regulate the ventilatory response to hypoxia in humans and mice (Jeton et al. 2017; Pichon et al. 2016; Soliz 2013; Soliz et al. 2012; Voituron et al. 2014). However, we did not observe significant differences before and after Epo injection on the CB chemosensory response to acute hypoxia. Furthermore, no sex differences were found on the CB chemosensory response to hypoxia. It is

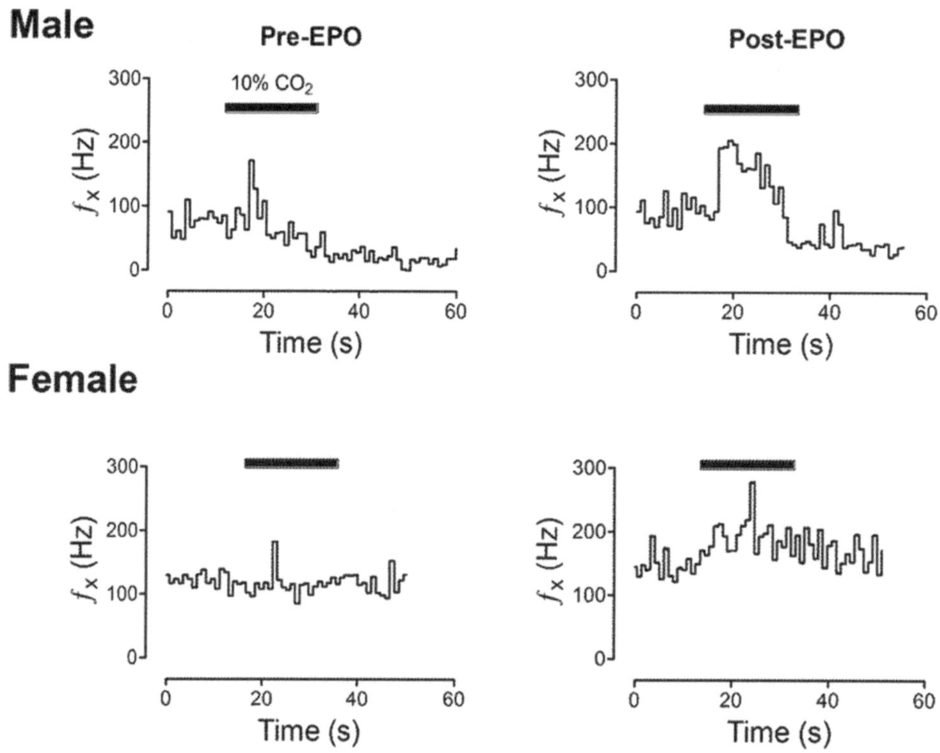

Fig. 12.3 Epo modulate the CB chemosensory responses to hypercapnia. Representative carotid sinus nerve recordings obtained in one male rat during acute (20s) hypercapnic stimulation (10%), before and after Epo i.v. injection (Upper panel) and one representative recording obtained in one female rat (Lower panel) during hypercapnia, before and after Epo administration

important to note that we only measure the effects of Epo on the CB chemoreceptor activity in response to hypoxia but not the chemoreflex-mediated hypoxic ventilatory response. Indeed, it has been proposed that Epo may regulate central brainstem areas related to chemoreflex control of the ventilation (i.e. NTS). Then, it is plausible that Epo may regulate the HVR through a central mechanism rather than a peripheral modulation of CB chemoreceptors.

In addition, we found that Epo modulate the CB response to hypercapnia in a sex-dependent manner. Indeed, in female rats Epo injections results in an increase CB chemosensory response to CO_2 without changing the duration of the response. In contrast, in male rats Epo only elicit increase in the duration of the CB chemosensory response to hypercapnia. Together, our data suggest that Epo modulate the CB nerve activity in response to hypercapnia acting on different EpoR expressed in male CBs compared to female CBs. Future studies are needed to address the differences, if any, on the expression profiles of EpoR on male and female CBs.

In summary our data suggest that Epo could modulate the CB chemosensory response to hypercapnia, but not to hypoxia in a sex-dependent manner. Additionally, Epo injections did not affects baseline CB chemosensory discharge in normoxia nor in male nor in female rats, suggesting that Epo did not contribute to tonic CB activity.

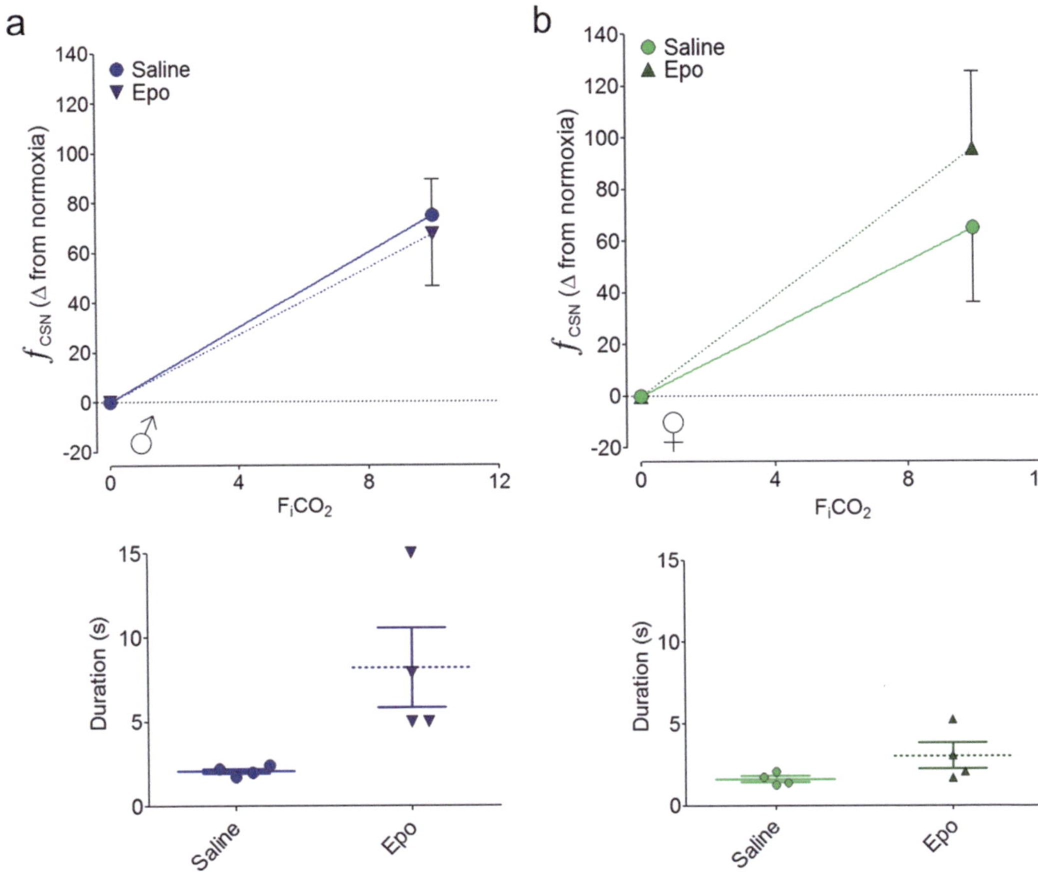

Fig. 12.4 Epo effects on CB chemosensory response to hypercapnia in adult male and female rats. *In situ* carotid sinus nerve recordings were obtained in spontaneously breathing animals during acute (20s) hypercapnic simulation. No effect of Epo was found on CB maximal response to hypercapnia in male rats (**a**, Upper panel). However, Epo administration results in a large increase in the duration of the response to hypercapnia (**a**, Lower panel). In contrast, Epo increases the CB response to hypercapnia (10%) in female rats (**b**) without affecting the duration of the CB response to hypercapnia (**b**, Lower panel). *, $P < 0.05$, Student t-test, $n = 4$ per group

Acknowledgements This work was supported by an ECOS-SUD program (N.V. action n°C16S03 and Programa de cooperación ECOS-CONICYT C16S03 to R.D.R.) and by Fondecyt grants #1140275 and #1180172 to R.D.R. and #1150040 to R.I. The "Relations Internationales" and the "Invited Professor" programs of the University Paris 13 also supported this study.

References

Berkenbosch A, Olievier CN, DeGoede J, Kruyt EW (1991) Effect on ventilation of papaverine administered to the brain stem of the anaesthetized cat. J Physiol 443:457–468

Berkenbosch A, Teppema LJ, Olievier CN, Dahan A (1997) Influences of morphine on the ventilatory response to isocapnic hypoxia. Anesthesiology 86:1342–1349

Del Rio R, Moya EA, Iturriaga R (2010) Carotid body and cardiorespiratory alterations in intermittent hypoxia: the oxidative link. Eur Respir J 36:143–150. https://doi.org/10.1183/09031936.00158109

Del Rio R, Munoz C, Arias P, Court FA, Moya EA, Iturriaga R (2011) Chronic intermittent hypoxia-induced vascular enlargement and VEGF upregulation in the rat carotid body is not prevented by antioxidant treatment. Am J Physiol Lung Cell Mol Physiol 301:L702–L711. https://doi.org/10.1152/ajplung.00128.2011

Despas F, Xhaët O, Senard JM, Verwaerde P, Jourdan G, Curnier D, Galinier M, Pathak A (2006) Chemoreflexes from physiology to practice. MT Cardio 2:321–327

Digicaylioglu M, Bichet S, Marti HH, Wenger RH, Rivas LA, Bauer C, Gassmann M (1995) Localization of specific erythropoietin binding sites in defined areas of the mouse brain. Proc Natl Acad Sci U S A 92:3717–3720

Feldman JL, Mitchell GS, Nattie EE (2003) Breathing: rhythmicity, plasticity, chemosensitivity. Annu Rev Neurosci 26:239–266. https://doi.org/10.1146/annurev.neuro.26.041002.131103

Gonzalez C, Almaraz L, Obeso A, Rigual R (1992) Oxygen and acid chemoreception in the carotid body chemoreceptors. Trends Neurosci 15:146–153

Gonzalez C, Vicario I, Almaraz L, Rigual R (1995) Oxygen sensing in the carotid body. Biol Signals 4:245–256

Guyenet PG (2014) Regulation of breathing and autonomic outflows by chemoreceptors. Comp Physiol 4:1511–1562. https://doi.org/10.1002/cphy.c140004

Jelkmann W (1992) Erythropoietin: structure, control of production, and function. Physiol Rev 72:449–489

Jelkmann W (2011) Regulation of erythropoietin production. J Physiol 589:1251–1258. https://doi.org/10.1113/jphysiol.2010.195057

Jeton F, Soliz J, Marchant D, Joseph V, Richalet JP, Pichon A, Voituron N (2017) Increased ventilation in female erythropoietin-deficient mouse line is not progesterone and estrous stage-dependent. Respir Physiol Neurobiol 245:98–104. https://doi.org/10.1016/j.resp.2017.07.002

Lam SY, Tipoe GL, Fung ML (2009) Upregulation of erythropoietin and its receptor expression in the rat carotid body during chronic and intermittent hypoxia. Adv Exp Med Biol 648:207–214. https://doi.org/10.1007/978-90-481-2259-2_24

Lopez-Barneo J, Gonzalez-Rodriguez P, Gao L, Fernandez-Aguera MC, Pardal R, Ortega-Saenz P (2016) Oxygen sensing by the carotid body: mechanisms and role in adaptation to hypoxia. Am J Physiol Cell Physiol 310:C629–C642. https://doi.org/10.1152/ajpcell.00265.2015

Pichon A et al (2016) Erythropoietin and the use of a transgenic model of erythropoietin-deficient mice. Hypoxia 4:29–39. https://doi.org/10.2147/HP.S83540

Prabhakar NR, Peers C (2014) Gasotransmitter regulation of ion channels: a key step in O2 sensing by the carotid body. Physiology 29:49–57. https://doi.org/10.1152/physiol.00034.2013

Smith CA, Blain GM, Henderson KS, Dempsey JA (2015) Peripheral chemoreceptors determine the respiratory sensitivity of central chemoreceptors to CO2: role of carotid body CO2. J Physiol 593:4225–4243. https://doi.org/10.1113/JP270114

Soliz J (2013) Erythropoietin and respiratory control at adulthood and during early postnatal life. Respir Physiol Neurobiol 185:87–93. https://doi.org/10.1016/j.resp.2012.07.018

Soliz J et al (2005) Erythropoietin regulates hypoxic ventilation in mice by interacting with brainstem and carotid bodies. J Physiol 568:559–571. https://doi.org/10.1113/jphysiol.2005.093328

Soliz J, Soulage C, Hermann DM, Gassmann M (2007) Acute and chronic exposure to hypoxia alters ventilatory pattern but not minute ventilation of mice overexpressing erythropoietin. Am J Physiol Regul Integr Comp Physiol 293:R1702–R1710. https://doi.org/10.1152/ajpregu.00350.2007

Soliz J, Khemiri H, Caravagna C, Seaborn T (2012) Erythropoietin and the sex-dimorphic chemoreflex pathway. Adv Exp Med Biol 758:55–62. https://doi.org/10.1007/978-94-007-4584-1_8

Voituron N et al (2014) Catalyzing role of erythropoietin on the nitric oxide central pathway during the ventilatory responses to hypoxia. Physiol Rep 2:e00223. https://doi.org/10.1002/phy2.223

13 Carotid Body Dysfunction in Diet-Induced Insulin Resistance Is Associated with Alterations in Its Morphology

Eliano Dos Santos, Joana F. Sacramento, Bernardete F. Melo, and Sílvia V. Conde

Abstract

The carotid body (CB) is organized in clusters of lobules containing type I cells and type II cells, in a ratio of approximately 4:1. The CB undergoes structural and functional changes during perinatal development, in response to a variety of environmental stimuli and in pathological conditions. Knowing that the CB acts as a metabolic sensor involved in the control of peripheral insulin sensitivity and that its overactivation contributes to the genesis of metabolic disturbances, herein we tested if diet-induced insulin resistance is associated with morphological alterations in the proportion of type I and type II cells in the CB. Diet induced insulin resistant model (HFHSu) was obtained by submitting *Wistar* rats to 14 weeks of 60% lipid-rich diet and 35% of sucrose in drinking water. The HFHSu group was compared with an aged-matched control group. Glucose tolerance and insulin sensitivity were measured in conscious animals before diet administration and 14 weeks after the diet protocol. The expression of tyrosine hydroxylase (TH) and nestin were assessed by immunohistochemistry to identify type I and type II cells, respectively. TH expression was also quantified by Western blot. As expected, 14 weeks of HFHSu diet induced a decrease in insulin sensitivity as well as in glucose tolerance. HFHsu diet increased the number of TH-positive type I cells by 192% and decreased nestin-postive type 2 cells by 74%. This increase in type II cells observed by immunohistochemistry correlates with an increase by 107% in TH expression quantified by Western blot. These results suggest that changes in CB morphology are associated with metabolic disturbances invoked by administration of a hypercaloric diet.

Keywords

Carotid body · Tyrosine hydroxylase · Hypercaloric diet · Insulin resistance

13.1 Introduction

The carotid body (CB) is organized in clusters of lobules containing type I cells, positive for tyrosine hydroxylase, and type II cells, positive for glial fibrillary acidic protein, in a ratio of approximately 4:1 (Nurse 2014). Type I cells, also known as glomus cells, produce many different neurotransmitters and peptide neuromodulators (Pardal et al. 2007; Porzionato et al. 2013). Type

E. Dos Santos · J. F. Sacramento · B. F. Melo
S. V. Conde (✉)
CEDOC, Centro de Estudos Doenças Crónicas, Nova Medical School, Faculdade de Ciências Médicas, Universidade Nova de Lisboa, Lisbon, Portugal
e-mail: silvia.conde@nms.unl.pt

E. B. Gauda et al. (eds.), *Arterial Chemoreceptors*, Advances in Experimental Medicine and Biology 1071, https://doi.org/10.1007/978-3-319-91137-3_13

I cells, also called sustentacular cells, envelop clusters of type I cells and have been until the nineties considered supportive cells. At the moment, type II cells have been proposed to be stem cell precursors for type I cells (Pardal et al. 2007) and/or amplifiers for type I cells that co-ordinate chemosensory transduction through interactions with the other cells of the carotid body (Tse et al. 2012; Porzionato et al. 2013). In fact, Nurse and Piskuric (2013) propose that paracrine signaling involving type II cells may play a role in chemotransduction, by acting as "ATP amplifiers". (Nurse and Piskuric 2013).

The CB undergoes structural and functional changes during perinatal development, and in response to a variety of environmental stimuli (Porzionato et al. 2013). In fact, numerous studies have documented both vascular dilation as well as hyperplasia and hypertrophy of glomus type I cells, in response to prolonged exposure to hypoxia (Hellstrom and Pequignot 1982; McGregor et al. 1984; Wang and Bisgard 2002; Chen et al. 2007; Conde et al. 2012). Furthermore, it was also described that with age, CBs enlarge and that this enlargement is accompanied by a decrease in the percentage of chemoreceptor tissue that can account for the decreased peripheral drive of ventilation (Conde et al. 2006).

In the last decades, the CB has been described as a major metabolic sensor (for a review see Conde et al. 2014). The CBs are involved in the control of peripheral insulin sensitivity, as the resection of the CB-sensitive nerve, the carotid sinus nerve, prevents and restores insulin sensitivity and glucose homeostasis in hypercaloric animal models of prediabetes and type II diabetes (Ribeiro et al. 2013; Sacramento et al. 2017, 2018). We have shown that insulin activates the CB, and proposed hyperinsulinemia to be one of the earliest events in a cascade that culminates in the CB overactivation leading to insulin resistance and glucose intolerance (Ribeiro et al. 2013). This overactivation of the CB in prediabetic animal models, was showed by an increase in the basal ventilation and in the ventilatory responses to hypoxia, by an increase in the release of dopamine and by the enlargement of the CBs (Ribeiro et al. 2013). In agreement with this overactivation of the CB, Cramer et al. (2014) showed that patients with type 2 diabetes exhibit CBs 20–25% larger than control volunteers.

Herein we have tested if diet-induced insulin resistance is associated with morphological alterations in the proportion of type I and type II cells in the CB.

13.2 Methods

Experiments were performed in 8 weeks old *Wistar* rats obtained from the vivarium of the Faculdade de Ciências Médicas, Universidade Nova de Lisboa, Lisboa, Portugal. After randomization, the animals were assigned to one of two groups, the high fat/high sucrose (HFHSu) diet group, a validated animal model of insulin resistance, achieved by submitting the animals to 14 weeks of 60% lipid-rich diet and 35% of sucrose in drinking water, and an age-matched control group, fed with a regular chow diet. Animals were kept under temperature and humidity control (21 ± 1 °C; 55 ± 10% humidity) with a 12 h light/12 h dark cycle and were given ad libitum access to food and water. Glucose tolerance and insulin sensitivity were measured in conscious animals before diet administration and 14 weeks after the diet protocol (Sacramento et al. 2017).

Rats were anaesthetized with sodium pentobarbital (Sigma, Madrid, Spain) (60 mg/kg i.p.) and the carotid bifurcation was collected and dissected along with the CB.

For immunohistochemistry analysis, CBs were fixed in 4% paraformaldehyde, embedded in OCT compound, sectioned (6 μm) into slides and frozen. The slides were then exposed to room temperature, rehydrated with distilled water and permeabilized with washing solution (PBS 1% + 0.1% Triton) before being submerged in blocking solution (PBS with goat serum and bovine serum albumin). After blocking, the slides were incubated overnight at 4 °C with a primary antibody against tyrosine hydroxylase (TH) (Abcam) and against Nestin (Abcam). After, the slides were washed and incubated with the secondary antibody - IgG anti-rabbit Alexa Fluor 594 (Abcam) washed again, incubated with DAPI (Santa Cruz) mounted with VECTASHIELD® Mounting Medium (Vector Laboratories), sealed with nail polish and kept in the dark at 4 °C. Images of the immunostained tissue were collected on the same day using the fluorescence microscope and analysed to quantify TH and Nestin positive cells with the Fiji app for ImageJ (https://imagej.nih.gov/ij/).

For the quantification of TH expression by Western Blot, the CBs were homogenized in Zurich (10 mM Tris-HCl, 1 mM EDTA, 150 mM NaCl, 1% Triton X-100, 1% sodium cholate, 1% SDS) with a cocktail of protease inhibitors. Briefly, after blocking, the membranes were incubated overnight at 4 °C with a mouse primary antibody against TH (Sigma-Aldrich) immunoreactivity (bands in the 60 kDa region). The membranes were washed with Tris-buffered saline with Tween (TBST) (0.02%) and incubated with a biotin-conjugated secondary antibody against mouse (Millipore) for 90 min at room temperature. The membranes were then washed with TBST (0.02%) and again incubated with ImmunoPure® Streptavidin, Horseradish Peroxidase Conjugated (Thermo Scientific), for 30 min at room temperature and developed with enhanced chemiluminescence reagents according to the manufacturer's instructions (ClarityTM Western ECL substrate, California, USA). Intensity of the signals was detected in a Chemidoc Molecular Imager (Chemidoc; BioRad, Madrid, Spain) and quantified using the Image Lab software (BioRad). The membranes were re-probed and tested with a mouse antibody for beta-actin (Sigma-Aldrich) immunoreactivity (bands in the 45 kDa region) to compare and normalize the expression of proteins with the amount of protein loaded.

Laboratory care was in accordance with the European Union Directive for Protection of Vertebrates Used for Experimental and Other Scientific Ends (2010/63/EU). Experimental protocols were approved by the Faculdade de Ciências Médicas Ethics Committee.

13.3 Results

As expected, 14 weeks of high-fat/high-sucrose (HFHsu) diet induced a decrease in insulin sensitivity as well as in glucose tolerance, and an increase in fasting glycemia (Table 13.1).

Moreover, HFHsu animals present an increase of 192% in the TH expression in Type I and a decrease of 74% in Nestin expression in type II cells (Fig. 13.1). This increase in type I cells observed by immunohistochemistry correlates with an increase by 107% in TH expression quantified by Western blot (Fig. 13.2).

13.4 Discussion

Herein we demonstrate that metabolic disturbances induced by administration of a hypercaloric diet induces morphological changes in the CB, demonstrated by an increase in TH-positive type I cells and a decrease in Nestin-positive type II cells.

The higher expression of type I cells as a consequence of a hypercaloric diet agrees with several studies that have documented structural and functional changes in the CB in response to a variety of other environmental stimuli (Porzionato et al. 2013; Hellstrom and Pequignot 1982; McGregor et al. 1984; Wang and Bisgard 2002; Chen et al. 2007). Moreover, CB hypertrophy occurs in several chronic diseases that occur in animal and humans (Kato et al. 2012; Felix et al. 2012; Sivridis et al. 2011; Bencini and Pulera 1991). Alterations in the amount of type I cells have also been described in ageing, but with opposite morphological changes. In fact, although the CBs enlarge with age there is a decrease in type I cells and accumulation of extracellular matrix in aged animals, which is consistent with the decreased magnitude of hypoxic responses and with the decrease in the release of catecholamines in response to hypoxia, but little effect on catecholamine content and basal release (Conde et al. 2006).

Herein, and as previously described we showed that rats become resistant to insulin as a

Table 13.1 Effect of high-fat/high-sucrose (HFHSu) diet on insulin sensitivity (assessed by an insulin tolerance test and expressed as the constant rate for glucose disappearance, KITT), fasting glycemia and glucose tolerance (assessed by an oral glucose tolerance test and expressed by the area under the curve, AUC)

	Control group		HFHSu group	
	Before diet (n = 6–8)	After 14 weeks of diet (n = 6–8)	Before diet (n = 6–8)	After 14 weeks of diet (n = 6–8)
K_{ITT} (%glucose/min)	4.65 ± 0.23	5.25 ± 0.17	4.25 ± 0.22	1.63 ± 0.17***
Glycemia (mg/dl)	84.83 ± 4.83	77.57 ± 2.60	92 ± 4.0	97.62 ± 2.63
AUC	21,918.66 ± 615.88	22,873.67 ± 489.70	20,466 ± 857.61	24,167.25 ± 640.29**

Data are means ± SEM of 6–8 animals. One and Two-Way ANOVA with Bonferroni multicomparison test: $^{**}p < 0.01$, $^{***}p < 0.001$ comparing HFHSu animals before and after 14 weeks of diet

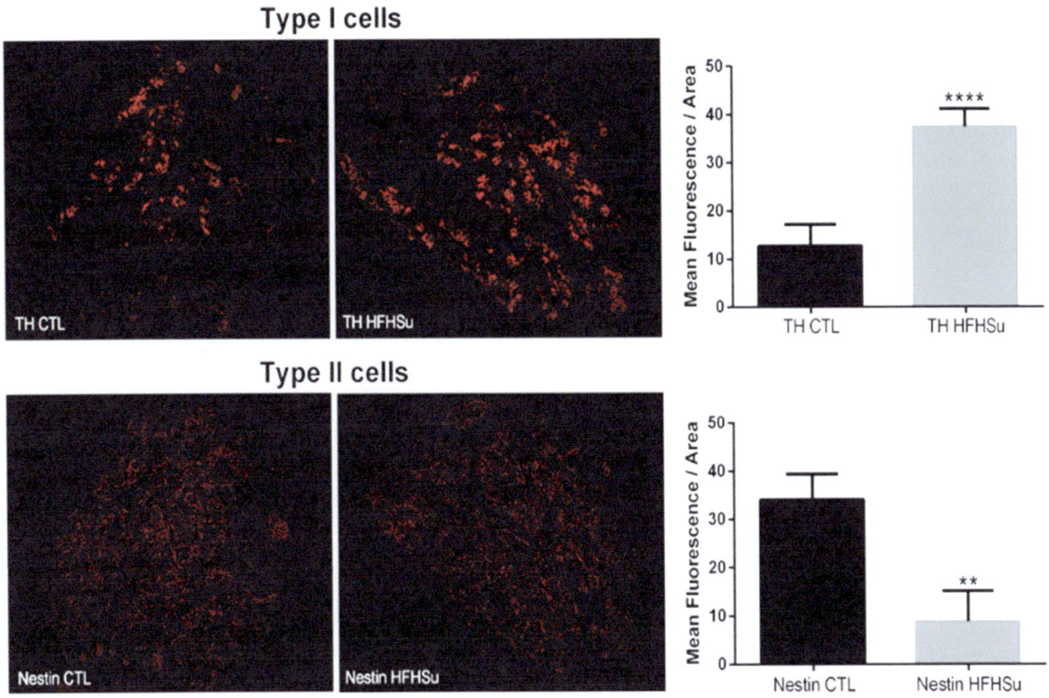

Fig. 13.1 Expression of tyrosine hydroxylase (TH) and nestin in CB cells assessed by immunohistochemistry, representing type I and type II cells, respectively. Bars represent means ± SEM; Unpaired t test: **$p < 0.005$ and ****$p < 0.0001$, TH expression in HFHSu vs CTL groups

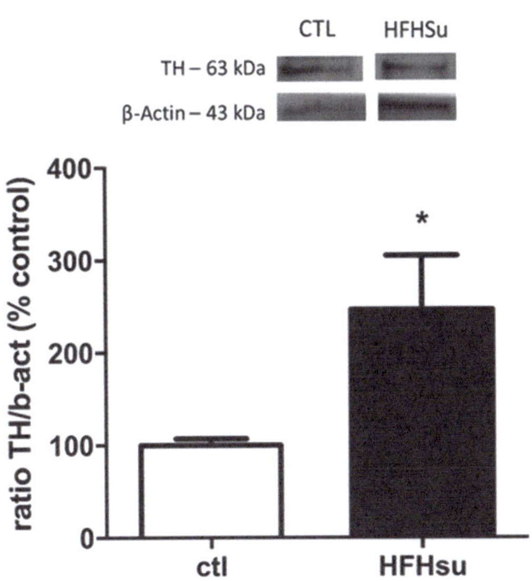

Fig. 13.2 Effect of HFHSu diet in TH expression in the rat CB, assessed by Western Blot. Data are means ± SEM. One-Way ANOVA with Bonferroni multicomparison test: *$p < 0.05$ comparing HFHSu with CTL values

consequence of administration of an HFHSu diet (Conde et al. 2017; Sacramento et al. 2018). We also showed in HFHSu animals an increase in the amount of type I cells, both by immunohistochemistry and Western-blot concomitantly with a decrease in nestin-positive cells. As CBs structurally express insulin receptors (Conde et al. 2014), these changes may reflect a compensatory response to an overactivation by hyperinsulinemia. In fact, Cramer et al. (2014) have previously shown that patients with diabetes mellitus type II, a disease largely associated with hypercaloric diets and insulin resistance, have 20–25% larger CBs (Cramer et al. 2014).

We postulate that CBs are able to develop a compensatory increase in number of type I cells in response to an overstimulation, by recruiting and inducing the differentiation of Nestin-positive type 2 cells. Also, sustained activation may lead to a stress state or the cells may reach their replication capacity inducing an apoptotic and/or a senescent state. Further morphology and physiology studies in these models must be performed.

Acknowledgments J.F.S. and B.F.M. are supported by PhD Grants from Portuguese Foundation for Science and Technology Reference PD/BD/105890/2014 and PD/BD/128336/2017, respectively.

References

Bencini C, Pulera N (1991) The carotid bodies in bronchial asthma. Histopathology 18:195e200

Chen J, He L, Liu X, Dinger B, Stensaas L, Fidone S (2007) Effect of the endothelin receptor antagonist bosentan on chronic hypoxia-induced morphological and physiological changes in rat carotid body. Am J Physiol Lung Cell Mol Physiol 292(5):L1257–L1262

Conde SV, Obeso A, Rigual R, Monteiro EC, Gonzalez C (2006) Function of the rat carotid body chemoreceptors in ageing. J Neurochem 99:711–723

Conde SV, Monteiro EC, Rigual R, Obeso A, Gonzalez C (2012) Hypoxic intensity: a determinant for the contribution of ATP and adenosine to the genesis of carotid body chemosensory activity. J Appl Physiol (1985) 112(12):2002–2010

Conde SV, Sacramento JF, Guarino MP, Gonzalez C, Obeso A, Diogo LN, Monteiro EC, Ribeiro MJ (2014) Carotid body, insulin, and metabolic diseases: unraveling the links. Front Physiol 5:418

Conde SV, Sacramento JF, Chew D, Melo BF, Donega M, Dopson W, Robinson A, Prieto-lloret J, Patel S, Holisnski B, Ramnarain N, Pikov V, Guarino MP (2017) Bilateral electrical modulation of carotid sinus nerve improves glucose homeostasis in rodent type 2 diabetes model. Diabetologia 60(Suppl 1):S1–S608

Cramer JA, Wiggins RH, Fudim M, Engelman ZJ, Sobotka PA, Shah LM (2014) Carotid body size on CTA: correlation with comorbidities. Clin Radiol 69(1):e33–e36

Felix AS, Rocha VN, Nascimento AL, de Carvalho JJ (2012) Carotid body remodelling in l-NAME-induced hypertension in the rat. J Comp Pathol 146(4):348–356

Hellstrom S, Pequignot JM (1982) Morphometric studies on intact and sympathectomised carotid bodies of long-term hypoxic rats. A light and electron microscopial study. In: Pallot DJ (ed) The peripheral arterial chemoreceptors. Croom Helm, London, pp 293–301

Kato K, Wakai J, Matsuda H, Kusakabe T, Yamamoto Y (2012) Increased total volume and dopamine β-hydroxylase immunoreactivity of carotid body in spontaneously hypertensive rats. Auton Neurosci 169(1):49–55

McGregor KH, Gil J, Lahiri S (1984) A morphometric study of the carotid body in chronically hypoxic rats. J Appl Physiol Respir Environ Exercise Physiol 57:1430–1438

Nurse CA (2014) Synaptic and paracrine mechanisms at carotid body arterial chemoreceptors. J Physiol 592(16):3419–3426

Nurse CA, Piskuric NA (2013) Signal processing at mammalian carotid body chemoreceptors. Semin Cell Dev Biol 24(1):22–30

Pardal R, Ortega-Sáenz P, Durán R, López-Barneo J (2007) Glia-like stem cells sustain physiologic neurogenesis in the adult mammalian carotid body. Cell 131(2):364–377

Porzionato A et al (2013) Inflammatory and immunomodulatory mechanisms in the carotid body. Respir Physiol Neurobiol 187(1):31–40

Ribeiro MJ, Sacramento JF, Gonzalez C, Guarino MP, Monteiro EC, Conde SV (2013) Carotid body denervation prevents the development of insulin resistance and hypertension induced by hypercaloric diets. Diabetes 62(8):2905–2916

Sacramento JF*, Chew DJ*, Melo BF, Donegá M, Dopson W, Guarino MP, Robinson A, Prieto-Lloret J, Patel S, Holinski BJ, Ramnarain N, Pikov V, Famm K, Conde SV (2018) Bioelectronic modulation of carotid sinus nerve activity as a treatment for type 2 diabetes. Diabetologia. 61(3):700–710

Sacramento JF, Ribeiro MJ, Rodrigues T, Olea E, Melo BF, Guarino MP, Fonseca-Pinto R, Ferreira CR, Coelho J, Obeso A, Seiça R, Matafome P, Conde SV (2017) Functional abolition of carotid body activity restores insulin action and glucose homeostasis in rats: key roles for visceral adipose tissue and the liver. Diabetologia 60(1):158–168

Sivridis E, Pavlidis P, Fiska A, Pitsiava D, Giatromanolaki A (2011) Myocardial hypertrophy induces carotid body hyperplasia. J Forensic Sci 56(Suppl 1):S90–S94

Tse A, Yan L, Lee AK, Tse FW (2012) Autocrine and paracrine actions of ATP in rat carotid body. Can J Physiol Pharmacol 90(6):705–711

Wang ZY, Bisgard GE (2002) Chronic hypoxia-induced morphological and neurochemical changes in the carotid body. Microsc Res Tech 59:168–177

14 Therapeutic Targeting of the Carotid Body for Treating Sleep Apnea in a Pre-clinical Mouse Model

Ying-Jie Peng, Xiuli Zhang, Jayasri Nanduri, and Nanduri R. Prabhakar

Abstract

Sleep apnea with periodic cessation of breathing during sleep is a highly prevalent respiratory disorder affecting an estimated 10% of adults. Patients with sleep apnea exhibit several co-morbidities including hypertension, stroke, disrupted sleep, and neurocognitive and metabolic complications. Emerging evidence suggests that a hyperactive carotid body (CB) chemo reflex is an important driver of apneas in sleep apnea patients. Gasotransmitters carbon monoxide (CO) and hydrogen sulfide (H_2S) play important roles in oxygen sensing by the CB. We tested the hypothesis that an augmented CB chemo reflex stemming from disrupted CO-H_2S signaling may lead to sleep apnea. This possibility was tested in mice deficient in hemeoxygenase-2 (HO-2), an enzyme involved in CO synthesis, which were shown to exhibit hyperactive CB activity due to high H_2S levels. We found that HO-$2^{-/-}$ mice exhibit a high incidence of apneas during sleep compared to wild type mice. Blocking the CB hyperactivity with L-propargylglycine, an inhibitor of cystathionine-γ-lyase (CSE), which catalyzes H_2S synthesis, prevented apneas in HO-$2^{-/-}$ mice. These findings suggest that targeting CB with inhibitors of CSE might be a novel therapeutic strategy for preventing sleep apnea.

Keywords

Hemeoxygenase-2 · Carbon monoxide · Cystathionine-γ-lyase · Hydrogen sulfide · Obstructive · Central sleep apnea

14.1 Introduction

Sleep disordered breathing (SDB) with obstructive (OSA) and central apnea (CSA) is a clinically prevalent respiratory disorder affecting an estimated 10% of the adult human population (Peppard et al. 2013). Patients with SDB exhibit a wide spectrum of pathologies including hypertension, stroke, neurocognitive and metabolic dysfunctions and disrupted sleep (St-Onge et al. 2016). Continuous positive airway pressure (CPAP) is the current treatment of choice for OSA (Combs et al. 2014; Ip et al. 2012; Veasey

Y.-J. Peng · X. Zhang · J. Nanduri
N. R. Prabhakar (✉)
Institute for Integrative Physiology and Center for Systems Biology of Oxygen Sensing, Biological Sciences Division, University of Chicago, Chicago, IL, USA
e-mail: nanduri@uchicago.edu

E. B. Gauda et al. (eds.), *Arterial Chemoreceptors*, Advances in Experimental Medicine and Biology 1071, https://doi.org/10.1007/978-3-319-91137-3_14

et al. 2006) and adaptive servo-ventilation is considered beneficial for CSA patients (Javaheri et al. 2016; Yamauchi et al. 2016). However, a substantial number of OSA patients do not respond to CPAP therapy (Antic et al. 2015; Gilmartin et al. 2005; Quan et al. 2012). Adaptive ventilation in heart failure patients exhibiting CSA showed 30% mortality with no demonstrable benefits for CSA (Cowie et al. 2015). These findings suggest that current treatment strategies are suboptimal for normalizing breathing in SDB patients. The estimated cost burden in unmanaged sleep apnea patients is exorbitantly high and ranges between ~$65 and 165 billion per year in the U.S. (Frost and Sullivan 2016; Harvard Medical School Division of Sleep Medicine 2010), highlighting a need for developing alternative therapeutic strategies for preventing SDB. However, development of additional therapeutic strategies critically depends on the availability of suitable animal model(s) exhibiting SDB. SDB patients often exhibit a mixed phenotype of OSA and CSA (Javaheri et al. 2009; Morgenthaler et al. 2006). Therefore, identifying animal model(s) exhibiting both OSA and CSA will facilitate the development of alternative therapeutic strategies.

Reflex arising from the carotid body (CB), the primary sensory organ for detecting hypoxia, is a major regulator of breathing (Kumar and Prabhakar 2012). Clinical studies have implicated abnormal CB chemo reflex in both genesis and downstream pathologies caused by sleep apnea (Dempsey et al. 2012; Kara et al. 2003; Mansukhani et al. 2015; Xie et al. 1995). Emerging evidence suggests that gasotransmitters play an important role in oxygen sensing by the CB (Makarenko et al. 2012; Peng et al. 2010, 2014; Prabhakar and Peers 2014; Yuan et al. 2015). Mice lacking hemeoxygenase-2 (HO-2), an enzyme responsible for the generation of endogenous carbon monoxide (CO), exhibit exaggerated CB response to hypoxia, due to increased cystathionine-γ-lyase (CSE)-derived H_2S synthesis (Yuan et al. 2015). These findings led us to hypothesize that HO-2 null mice exhibit high incidence of apnea, and pharmacologic blockade of H_2S synthesis prevents apnea. We tested this hypothesis by monitoring breathing under normoxia with whole body plethysmography continuously from 10 AM to 4 PM (which is the sleep time for rodents) in adult wild type and HO-$2^{-/-}$ mice of both genders.

14.2 Mice with Deficiency in HO-2 Exhibit Sleep Apnea

14.2.1 HO-$2^{-/-}$ Mice Exhibit High Incidence of Apnea and Hypopnea

Figure 14.1 shows examples of tracings of breathing in a wild type and a HO-$2^{-/-}$ mouse. Wild type mouse displayed relatively regular breathing during the 6 h period of recording. However, HO-$2^{-/-}$ mouse exhibited irregular breathing with apnea and hypopnea. We defined apnea as cessation of breathing for >2.5 breaths duration, excluding post sigh apnea, and hypopnea as

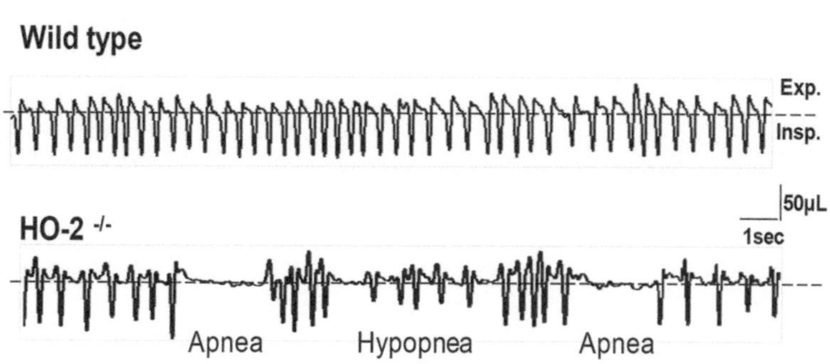

Fig. 14.1 Apnea and hypopnea in HO-2 null mice

Table 14.1 Incidence of apnea and hypopnea in wild type and HO-2$^{-/-}$ mice

Index (events/hr)	wild type (n = 40)	HO-2$^{-/-}$ (n = 70)
Apnea:		
<5	20	13
5~9	11	4
10~19	7	13
≥20	2	40
Hypopnea:		
<20	22	4
20~39	12	13
40~79	5	14
≥80	1	39

Table 14.2 Incidence of apnea and hypopnea during each state (events/stage hr) in wild type and HO-2$^{-/-}$ mice (n = 7 each)

	wild type	HO-2$^{-/-}$
Apnea:		
Wake	3 ± 2	5 ± 3
NREM	10 ± 4	29 ± 4*
REM	13 ± 7	60 ± 15**
Hypopnea:		
Wake	9 ± 3	17 ± 8
NREM	16 ± 7	55 ± 11*
REM	16 ± 5	70 ± 14**

$^*P < 0.05$; $^{**}P < 0.01$ (vs WT), data presented as mean ± SEM

breathing events with ≥30% reduction in tidal volume. As shown in Table 14.1, a majority of HO-2$^{-/-}$ mice showed greater than 20 events of apnea per hour (apnea index), ranging from 20 to 168 apneas/hour, whereas only 5% wild type mice exhibited such severe apneas. Likewise, 56% HO-2$^{-/-}$ mice showed greater than 80 events of hypopnea per hour (hypopnea index, Table 14.1), whereas only 2.5% wild type mice displayed similar hypopnea index. Chi-squared test showed that the distributions of apnea and hypopnea were significantly different between wild type and HO-2$^{-/-}$ mice ($P < 0.01$). These observations suggest that HO-2$^{-/-}$ mice exhibit high incidence of apnea and hypopnea.

14.2.2 HO-2$^{-/-}$ Mice Exhibit Sleep Apnea

Sleep-wake states significantly influence breathing. To examine the impact of sleep-wake states on the occurrence of apneas, we simultaneously monitored electroencephalogram (EEG), electromyogram of neck muscles (EMG) and breathing in un-sedated wild type and HO-2$^{-/-}$ mice. Sleep-wake states are determined based on analysis of combination of EEG spectrum and neck muscle tone. Wake state is characterized by EEG activity of mixed frequency with low amplitude and high muscle tone. Rapid eye movement (REM) sleep is identified by both the frequency of theta waves (6–9 Hz) and muscle atonia. Non-REM (NREM) sleep is characterized by high amplitude slow waves in the delta frequency range (1–4 Hz) and low muscle tone. Incidence of apnea and hypopnea during wake state were low and comparable between wild type and HO-2$^{-/-}$ mice (Table 14.2). NREM or REM sleep had little impact on apnea or hypopnea index in wild type mice. In striking contrast, apnea and hypopnea indices significantly increased during NREM sleep, and further increased during REM sleep in HO-2$^{-/-}$ mice (Table 14.2). These results demonstrate that HO-2$^{-/-}$ mice exhibit sleep apnea.

14.2.3 HO-2$^{-/-}$ Mice Exhibit Both Central and Obstructive Apnea

Apneas are classified as central and obstructive in nature. To characterize the apnea phenotype, we monitored inspiratory intercostal muscle electromyography activity with chronically implanted electrodes along with breathing in un-sedated HO-2$^{-/-}$ mice (n = 12). There were incidences of apnea with absence of both breathing movements and respiratory muscle activity, which were considered as central apneas. On other hand, some of the apneic events were associated with increased inspiratory muscle activity, which were considered as obstructive apneas. We found that a given HO-2$^{-/-}$ mouse exhibited both central and obstructive apnea during 6 h of recording. Average data showed that 59% of apnea events

Table 14.3 Incidence of obstructive versus central apnea in HO-$2^{-/-}$ mice

	Obstructive Apnea	Central Apnea
Apneas (% of Total)	59 ± 5	41 ± 5*

*$P < 0.05$, data presented as mean ± SEM

Table 14.4 Effect of hyperoxia and hypoxia on incidence of apnea in HO-$2^{-/-}$ mice

	21% O_2	90% O_2	15% O_2
Apnea index (% of 21% O_2)	100 ± 13	34 ± 5**	350 ± 62**

**$P < 0.01$ (vs 21% O_2), data presented as mean ± SEM

are of obstructive type, and 41% are of central type (Table 14.3).

14.3 Enhanced Carotid Body (CB) Chemo Reflex Contributes to Apnea in HO-2 Null Mice

We determined whether an exaggerated CB reflex contributes to high incidence of apnea in HO-$2^{-/-}$ mice. Hyperoxia (90% O_2), which is known to inhibit the CB activity, markedly reduced the number of apneas (Table 14.4). In contrast, hypoxia (15% O_2), a physiological activator of the CB, significantly increased the incidence of apnea (Table 14.4). Further experiments showed that hyperoxia reduced the incidence of both obstructive and central apneas. In contrast, hypoxic breathing increased the incidence of obstructive and central apnea by 3 and 2 fold, respectively. Although apneas could result from reduced chemosensitivity to CO_2 (Kumar et al. 2015), we found that stimulation of central CO_2 chemoreceptors with 2% CO_2 caused only a modest reduction (~32%) in the apnea index in HO-$2^{-/-}$ mice, which was due to reduced incidence of central apnea. These observations suggest that the exaggerated CB reflex is a major contributor to apneas in HO-$2^{-/-}$ mice.

14.4 Pharmacologic Blockade of H_2S Synthesis Prevents Apneas in HO-2 Mice

HO-$2^{-/-}$ mice exhibit increased CSE-derived H_2S production in the CB, and lowering H_2S production by genetic removal of CSE prevents the exaggerated CB response to hypoxia in HO-$2^{-/-}$ mice (Yuan et al. 2015). We found that apnea and hypopneas were absent in HO-2/CSE double knockout mice, suggesting that lowering H_2S production may prevent apnea. While genetic removal of CSE might not be a practical therapeutic strategy to treat sleep apnea clinically, we explored whether pharmacologic blockade of CSE-derived H_2S synthesis stabilizes breathing in HO-$2^{-/-}$ mice. Intra-peritoneal administration of L-propargylglycine (L-PAG), an inhibitor of CSE, reduced the incidence of apnea in a dose-dependent manner (1 ~ 150 mg/kg), with a median effective dose at 30 mg/kg (Table 14.5). L-PAG reduced both central and obstructive apneas. The effect of L-PAG at a dose of 30 mg. kg on apneas was seen within 2 h after administration and completely reversed after 24 h. More importantly, L-PAG (30 mg/kg) administration via oral route was equally effective in reducing the number of apneas and hypopneas. No sign of toxicity was observed with any of the doses of L-PAG tested.

14.5 Summary and Perspective

Our study establishes that HO-2 null mice exhibit a high incidence of apnea and hypopnea during sleep and might serve as a pre-clinical murine model of sleep apnea. Like patients with sleep apnea, HO-2 null mice display both central and obstructive apneas. The enhanced CB chemo reflex mediated by CSE-produced H_2S contributes to high incidence of apneas. Lowing H_2S production by pharmacologic blockade of CSE significantly prevent apneas in HO-2 null mice. Since sleep apnea is a multi-factorial respiratory disease, it would be important to determine whether CSE inhibitors prevent apneas in patients

Table 14.5 Effect of intraperitoneal administration of L-PAG on incidence of apnea in HO-$2^{-/-}$ mice

		L-PAG (mg/kg)					
	Vehicle	1	3	10	30	90	150
Apnea index (% of Vehicle)	100 ± 12	88 ± 9	70 ± 9	68 ± 6*	49 ± 5*	36 ± 5*	20 ± 6*

*$P < 0.05$; (vs Vehicle), data presented as mean ± SEM

with sleep apnea caused by other etiologies, such as heart failure, renal insufficiency and stroke.

Acknowledgements This work was supported by National Institutes of Health grants P01-HL-90554 and UH2-HL-123610.

References

Antic NA, Heeley E, Anderson CS, Luo Y, Wang J, Neal B, Grunstein R, Barbe F, Lorenzi-Filho G, Huang S, Redline S, Zhong N, McEvoy RD (2015) The sleep apnea cardio vascular endpoints (SAVE) trial: rationale, ethics, design, and progress. Sleep 38:1247–1257

Combs D, Shetty S, Parthasarathy S (2014) Advances in positive airway pressure treatment modalities for hypoventilation syndromes. Sleep Med Clin 9:315–325

Cowie MR, Woehrle H, Wegscheider K, Angermann C, d'Ortho MP, Erdmann E, Levy P, Simonds AK, Somers VK, Zannad F, Teschler H (2015) Adaptive servo-ventilation for central sleep apnea in systolic heart failure. N Engl J Med 373:1095–1105

Dempsey JA, Smith CA, Blain GM, Xie A, Gong Y, Teodorescu M (2012) Role of central/peripheral chemoreceptors and their interdependence in the pathophysiology of sleep apnea. Adv Exp Med Biol 758:343–349

Frost & Sullivan (2016) Hidden health crisis costing America billions: underdiagnosing and undertreating obstructive sleep apnea draining healthcare system. American Academy of Sleep Medicine, Darien

Gilmartin GS, Daly RW, Thomas RJ (2005) Recognition and management of complex sleep-disordered breathing. Curr Opin Pulm Med 11:485–493

Harvard Medical School Division of Sleep Medicine (2010) The price of fatigue: the surprising economic costs of unmanaged sleep apnea. Harvard Medical School Division of Sleep Medicine, Boston

Ip S, D'Ambrosio C, Patel K, Obadan N, Kitsios GD, Chung M, Balk EM (2012) Auto-titrating versus fixed continuous positive airway pressure for the treatment of obstructive sleep apnea: a systematic review with meta-analyses. Syst Rev 1:20

Javaheri S, Smith J, Chung E (2009) The prevalence and natural history of complex sleep apnea. J Clin Sleep Med 5:205–211

Javaheri S, Germany R, Greer JJ (2016) Novel therapies for the treatment of central sleep apnea. Sleep Med Clin 11:227–239

Kara T, Narkiewicz K, Somers VK (2003) Chemoreflexes–physiology and clinical implications. Acta Physiol Scand 177:377–384

Kumar P, Prabhakar NR (2012) Peripheral chemoreceptors: function and plasticity of the carotid body. Compr Physiol 2:141–219

Kumar NN, Velic A, Soliz J, Shi Y, Li K, Wang S, Weaver JL, Sen J, Abbott SB, Lazarenko RM, Ludwig MG, Perez-Reyes E, Mohebbi N, Bettoni C, Gassmann M, Suply T, Seuwen K, Guyenet PG, Wagner CA, Bayliss DA (2015) Regulation of breathing by CO_2 requires the proton-activated receptor GPR4 in retrotrapezoid nucleus neurons. Science 348:1255–1260

Makarenko VV, Nanduri J, Raghuraman G, Fox AP, Gadalla MM, Kumar GK, Snyder SH, Prabhakar NR (2012) Endogenous H2S is required for hypoxic sensing by carotid body glomus cells. Am J Physiol Cell Physiol 303:C916–C923

Mansukhani MP, Wang S, Somers VK (2015) Chemoreflex physiology and implications for sleep apnoea: insights from studies in humans. Exp Physiol 100:130–135

Morgenthaler TI, Kagramanov V, Hanak V, Decker PA (2006) Complex sleep apnea syndrome: is it a unique clinical syndrome? Sleep 29:1203–1209

Peng Y-J, Nanduri J, Raghuraman G, Souvannakitti D, Gadalla MM, Kumar GK, Snyder SH, Prabhakar NR (2010) H2S mediates O-2 sensing in the carotid body. Proc Natl Acad Sci U S A 107:10719–10724

Peng YJ, Makarenko VV, Nanduri J, Vasavda C, Raghuraman G, Yuan G, Gadalla MM, Kumar GK, Snyder SH, Prabhakar NR (2014) Inherent variations in CO-H2S-mediated carotid body O2 sensing mediate hypertension and pulmonary edema. Proc Natl Acad Sci U S A 111:1174–1179

Peppard PE, Young T, Barnet JH, Palta M, Hagen EW, Hla KM (2013) Increased prevalence of sleep-disordered breathing in adults. Am J Epidemiol 177:1006–1014

Prabhakar NR, Peers C (2014) Gasotransmitter regulation of ion channels: a key step in O2 sensing by the carotid body. Physiology (Bethesda) 29:49–57

Quan SF, Awad KM, Budhiraja R, Parthasarathy S (2012) The quest to improve CPAP adherence–PAP potpourri is not the answer. J Clin Sleep Med 8:49–50

St-Onge MP, Grandner MA, Brown D, Conroy MB, Jean-Louis G, Coons M, Bhatt DL, American Heart Association Obesity B. C., D.abetes, and Nutrition Committees of the Council on Lifestyle and Cardiometabolic Health, Young C. o. C. D. i. t., Cardiology C. o. C., Council a. S (2016) Sleep duration and quality: impact on lifestyle behaviors and cardiometabolic health: a scientific statement from the American Heart Association. Circulation 134:e367–e386

Veasey SC, Guilleminault C, Strohl KP, Sanders MH, Ballard RD, Magalang UJ (2006) Medical therapy for obstructive sleep apnea: a review by the Medical Therapy for Obstructive Sleep Apnea Task Force of the Standards of Practice Committee of the American Academy of Sleep Medicine. Sleep 29:1036–1044

Xie A, Rutherford R, Rankin F, Wong B, Bradley TD (1995) Hypocapnia and increased ventilatory responsiveness in patients with idiopathic central sleep apnea. Am J Respir Crit Care Med 152:1950–1955

Yamauchi M, Combs D, Parthasarathy S (2016) Adaptive servo-ventilation for central sleep apnea in heart failure. N Engl J Med 374:689

Yuan G, Vasavda C, Peng YJ, Makarenko VV, Raghuraman G, Nanduri J, Gadalla MM, Semenza GL, Kumar GK, Snyder SH, Prabhakar NR (2015) Protein kinase G-regulated production of H2S governs oxygen sensing. Sci Signal 8:ra37

Role of Estradiol Receptor Beta (ERβ) on Arterial Pressure, Respiratory Chemoreflex and Mitochondrial Function in Young and Aged Female Mice

15

Sofien Laouafa, Damien Roussel, François Marcouiller, Jorge Soliz, Aida Bairam, and Vincent Joseph

Abstract

We tested the hypothesis that ERβ is involved in respiratory control in female mice. We used young adult (5–6 months-old) and aged (17–18 months-old) ERβKO or wild-type controls (WT) female mice to assess arterial blood pressure (via a tail-cuff sensor) and indices of respiratory pattern (sighs and apneas – recorded by whole body plethysmography at rest). We also measured respiratory parameters at rest and in response to brief (<10 min) exposure to hypoxia (12% O_2) or hypercapnia (5% CO_2). Because ERβ is localized in mitochondria, and because estradiol and ERβ agonist increase mitochondrial O_2 consumption, we assessed the mitochondrial respiration (with a high-resolution oxygraph system) and the *in vitro* activity of the complex I of the electron transfer chain in samples of brain cortex in aged wild-type and ERβKO female mice. Compared to young WT mice, young ERβKO mice had elevated arterial blood pressure, but similar ventilatory responses to hypoxia and hypercapnia. In old ERβKO female mice compared to old WT mice, the arterial blood pressure was lower, the frequency of sighs was higher and the frequency of apneas was lower, and the hypoxic and hypercapnic ventilatory responses were reduced. In old ERβKO mice mitochondrial respiration and complex I activities in the brain cortex were lower than in WT mice. We conclude that ERβ has age-specific effects on vascular and respiratory functions in female mice.

Keywords

Estradiol receptor beta · Old mice · Arterial pressure · Chemoreflex · Mitochondria · Female mice

S. Laouafa
Centre de Recherche de l'Institut Universitaire de Cardiologie et de Pneumologie de Québec, Université Laval, Quebec City, QC, Canada

CNRS, UMR 5023, Université Claude Bernard Lyon 1, Villeurbanne, France

D. Roussel
CNRS, UMR 5023, Université Claude Bernard Lyon 1, Villeurbanne, France

F. Marcouiller · J. Soliz · A. Bairam · V. Joseph (✉)
Centre de Recherche de l'Institut Universitaire de Cardiologie et de Pneumologie de Québec, Université Laval, Quebec City, QC, Canada
e-mail: vincent.joseph@fmed.ulaval.ca

E. B. Gauda et al. (eds.), *Arterial Chemoreceptors*, Advances in Experimental Medicine and Biology 1071, https://doi.org/10.1007/978-3-319-91137-3_15

15.1 Introduction

Through their effects in the central nervous system, steroid hormones such as estradiol or progesterone play an essential role in the regulation of sexual behavior (Brinton et al. 2008; Micevych et al. 2015). These hormones are also involved in processes governing mood or cognition (Marrocco and McEwen 2016), and regulate homeostatic functions such as thermoregulation (Charkoudian et al. 2017) or respiratory control (Boukari et al. 2016, 2017; Behan and Wenninger 2008). Because the occurrence of sleep apnea is higher in men than in women and increases after menopause in women (Block et al. 1979; Bixler et al. 2001), it is of clinical relevance to better understand the effects of ovarian hormones on respiratory regulation. Earlier studies performed in cats suggested that the main effect of estradiol was limited to increase the expression of the progesterone receptor, thereby increasing the respiratory effect of progesterone (Bayliss et al. 1990). Thus, the effects of estradiol, and the role of the classical estradiol receptors (ERα and ERβ) in respiratory control remain poorly investigated. Nonetheless, we have reported that in newborn rats, long-term treatment with estradiol slightly reduced respiratory frequency and increased the frequency and length of apnea under resting conditions (Lefter et al. 2008). During hypoxia, estradiol treatment suppressed the normal decrease of body temperature, without altering the ventilatory response (Lefter et al. 2008). Furthermore, we showed recently that estradiol protects against the increased chemoreflex activity and respiratory instabilities induced by exposure to intermittent hypoxia (Laouafa et al. 2017b), conceivably due to protective effects of estradiol against the oxidative stress induced by intermittent hypoxia (Laouafa et al. 2017b), and likely involve effect of estradiol on mitochondrial respiration and production of reactive oxygen species in female rats exposed to intermittent hypoxia (Laouafa et al. 2017a). These results indicate that estradiol could have important roles on respiratory regulation that remains poorly understood. The effects of estradiol could be more important under specific pathophysiological conditions, and/or may be specific to sex- or age.

Classical ERs belong to the superfamily of steroid receptors, acting as transcription factors able to modulate the expression of target genes that have an "Estrogen Response Element" in their promoter region (Harris 2007). Two ERs have been described, ERα and ERβ, that are coded by different genes (Kuiper et al. 1996). We previously reported by immunohistochemistry that ERβ is expressed in the peripheral chemoreceptors of adult rats (Joseph et al. 2006), and ERβ mRNA and protein are densely expressed in central areas involved in respiratory control including the nucleus tractus solitarius, locus coeruleus, and the parabrachial nucleus (Shughrue et al. 1997; Shughrue and Merchenthaler 2001).

In the present study, we tested the hypothesis that ERβ is involved in respiratory control in female mice. We used young adult (5–6 months-old) and aged (17–18 months-old) ERβKO female mice to assess respiratory parameters at rest and in response to brief (<10 min) exposure to hypoxia (12% O_2) or hypercapnia (5% CO_2). Because previous studies showed elevated arterial pressure in young ERβKO female mice (Zhu et al. 2002), we also recorded arterial pressure in young and aged mice. Finally, because ERβ is localized to mitochondria, and because ERβ agonist and estradiol increase mitochondrial O_2 consumption (Irwin et al. 2012), we assessed the mitochondrial respiration and the *in vitro* activity of the complex I of the electron transfer chain in samples of brain cortex in aged wild-type and ERβKO female mice.

15.2 Material and Methods

We used female ERβKO mice (Krege et al. 1998) (B6.129P2-Esr2tm1Unc/J; n = 9) and their recommended wild-type control (WT: C57BL/6 J; n = 7) from The Jackson laboratory. The mice were kept in our animal care facility under standard conditions with food and water available ad-libitum. All experiments were approved by our local ethic committee (Ref: 2014–156) and strictly adhered to the rules of the Canadian

Council on Animal Care. We used the same mice at the age of 5–6 months (young) and 17–18 months (old) to perform measurement of arterial pressure and respiratory recordings. Old mice were sacrificed a few days after the last measurements and brain samples were used to assess mitochondrial respiration and in vitro activities of mitochondrial complex I.

15.2.1 Measurements of Arterial Pressure

Arterial blood pressure was measured as described previously in female rats (Laouafa et al. 2017b) by the tail cuff method and volume-pressure recording (CODA system – Kent Scientific, Torrington, CT, USA). Conscious mice were placed in a restrainer tube over a warmed blanket. After 30 min of habituation, several recordings were performed, separated at least by 5 min. We report the mean of the 3 lowest values for systolic, diastolic, and mean arterial pressures.

15.2.2 Respiratory and Metabolic Recordings

All recordings were performed as previously described for adult mice (Boukari et al. 2016; Marcouiller et al. 2014). Briefly, animals were placed in a whole-body plethysmograph chamber (Emka Technologies – Paris, France), flushed with fresh room air. For each animal, the flow rate was adjusted to avoid excessive built-up of CO_2 within the recording chamber. Flow rate fluctuated between 160 and 250 ml/min, CO_2 was measured in the outflowing air and was between 0.4 and 0.6%. Mice were left undisturbed for 4 h under room air to ensure a period of normoxic recordings representative of a resting state, during which mice spend much of their time asleep (Marcouiller et al. 2014; Bastianini et al. 2017). A subsampling pump was connected to the chamber for analysis of O_2%, CO_2%, and water pressure using dedicated gas analyzers (FC-10, CA-10 and RH-300 respectively, Sable Systems International, Las vegas, NV, USA). During recordings, values of O_2%, CO_2% and water pressure in the inflowing gas line were periodically measured and later used for calculation of O_2 consumption and CO_2 production using standard equations (Boukari et al. 2016). Rectal temperature was measured at the end of the normoxic recording period. Following the 4 h periods of recordings, the animals were exposed to brief periods of hypoxia (12% O_2–10 mins) and hypercapnia (5% CO_2–10 min) in randomized order, with 10–15 minutes between each test.

15.2.3 Analysis of Respiratory and Metabolic Rate Variables

All respiratory and metabolic variables were analyzed as performed previously in adult rats or mice (Laouafa et al. 2017b; Boukari et al. 2016; Marcouiller et al. 2014), to report respiratory frequency, tidal volume, minute ventilation, O_2 consumption rate, and CO_2 production under normoxic conditions and in response to hypoxia or hypercapnia. Because exposure to hypoxia was short (<10 min) and measurements are typically taken around the 5th minute of exposure, we estimated that rectal temperature and metabolic rate were similar in hypoxia and normoxia. We used standard criteria to report sigh, post-sigh apnea, and spontaneous apnea frequency under resting conditions (Laouafa et al. 2017b; Boukari et al. 2016; Marcouiller et al. 2014).

15.2.4 Mitochondrial Respiration Recordings on Permeabilized Brain Samples

In old mice, a few days after the respiratory measurements, the animals were sacrificed with an overdose of anesthetic (ketamine/xylasine), the brain cortex was rapidly dissected and immediately used to perform measurements of mitochondrial respiratory functions using an O2K Oroboros respirometer (Oroboros Instruments, Innsbruck, Austria).

After calibration of the Oroboros chambers, cortex samples were weighted (2–3 mg), and recordings of O_2 consumption performed at 37 °C in a respiration buffer (0.5 mM EGTA, 3 mM $MgCl_2$, 60 mM potassium lactobionate, 10 mM KH_2PO_4, 20 mM Hepes, 110 mM sucrose, 1 g/l BSA). Based on a previous study establishing the optimal conditions for measurements of mitochondrial respiration in permeabilized brain samples (Herbst and Holloway 2015), the samples are incubated in the recording chamber for 20 min with saponin (50 μg/ml). For each sample, we have differentiated mitochondrial respiration linked to NADH oxidation through the mitochondrial respiratory complex I (using pyruvate 5 mM, malate 0.2 mM, and glutamate 10 mM as substrates), and mitochondrial respiration linked to $FADH_2$ oxidation through the mitochondrial respiratory complex II (succinate 5 mM and rotenone 4 μM to block complex I activity). After equilibration with the substrates, ADP (500 μM) is added to the chambers to measure O_2 consumption under normal phosphorylating state (ATP synthesis – mitochondrial respiratory state 3). We then add cytochrome-c to assess the integrity of mitochondrial membranes, then oligomycin (2 μg/mg tissue) was added to block ATP synthesis and measure O_2 consumption due to leakage of protons in non-phosphorylating state (no ATP synthesis – state 4). Uncoupling is further exaggerated by adding graded bolus of carbonyl cyanide m-chlorophenylhydrazone (CCCP – 0.5 μM in 1 μl/bolus) until reaching stable O_2 consumption. We then blocked the activity of the complex 3 with antimycin A (2.5 μM) to measure non-mitochondrial O_2 consumption due to cytosolic oxidases. Finally, the maximum activity of the complex 4 (cytochrome c oxidase) is measured by using ascorbate (2 mM) and tetramethyl-*p*-phenylenediamine dihydrochloride (TMPD; 0.5 mM).

15.2.5 *In vitro* Activity of the Mitochondrial Complex I

Frozen cortex (20 mg/ml) were homogenized in 100 mM phosphate buffer KPI (100 mM K_2HPO_4, 100 mM KH_2PO_4, PH7,4) + EDTA 2 mM. Then, the tissues were centrifuged for 10 min at 1500 g, and the supernatant was collected and kept frozen at −80 °C. The concentration of proteins was determined by a standard colorimetric BCA assay kit (Thermo Scientific, catalogue # 23225) and complex I activity was normalized to protein concentration.

Complex I activity was assessed by measuring the disappearance rate of NADH by spectrophotometry. Briefly, the day of measurement, a solution containing 50 mM KPI buffer, 1 mM EDTA, 100 μM decylubiquinone, and 3.75 mg/ml BSA (10%) was prepared. A similar solution was prepared with rotenone (10 μM: inhibitor of complex I activity). 4 μl of each sample was added in two different wells. In the first well, we added 250 μl of solution containing rotenone, and in the second well we added 250 μl of cocktail without rotenone. We then added 3 mM of NADH into each well, and read the absorbance at 340 nm every 15 s for 3 min to monitor the rate of NADH disappearance. Since NADH can be consumed by complex I of the mitochondria and by cytochrome b5 reductase, the activity of the complex I was obtained by subtracting the slope of the well with rotenone (activity without complex I) from the slope of the well without rotenone (total activity).

15.2.6 Statistics

For arterial blood pressure, we used a student's t-test to assess if significant differences appeared between WT and ERβKO mice at each age. For respiratory variables at rest, we used a one-way ANOVA to assess if significant differences appear between groups (Young WT, young ERβKO, old WT, old ERβKO). When the ANOVA reported a significant effect ($P < 0.05$) we used a Fisher's LSD test to assess differences between WT and ERβKO mice with each age, and differences between old and young mice for each genotype. For the hypoxic or hypercapnic ventilatory response, we used 2 way-ANOVA for repeated measures, with genotype and gas exposure as grouping variables, followed by a post-hoc Fisher LSD test if significant effect of genotype or significant interaction between genotype and gas exposure appeared. All analyses were performed using GraphPad Prism 7. All values are reported as mean ± sem.

15.3 Results

15.3.1 Arterial Pressure in Young and Old WT and ERβKO Female Mice

As previously reported in ERβKO mice (Zhu et al. 2002), the systolic, mean, and diastolic arterial pressures were elevated in young ERβKO compared to wild-type (Fig.15.1 – upper panels). However, in older mice there was an opposite effect, with ERβKO mice showing reduced diastolic and mean arterial pressures than wild-type (Fig. 15.1 – lower panels).

15.3.2 Body Weight, Respiratory and Metabolic Variables at Rest in Young and Old WT and ERβKO Female Mice

Body weight was higher in ERβKO vs wild-type mice at both ages (Table 15.1), the mean of the difference between the WT and KO mice was 1.8 ± 0.8 g in young mice, and 14.2 ± 2.7 g in older mice, thus indicating a gradual increase in

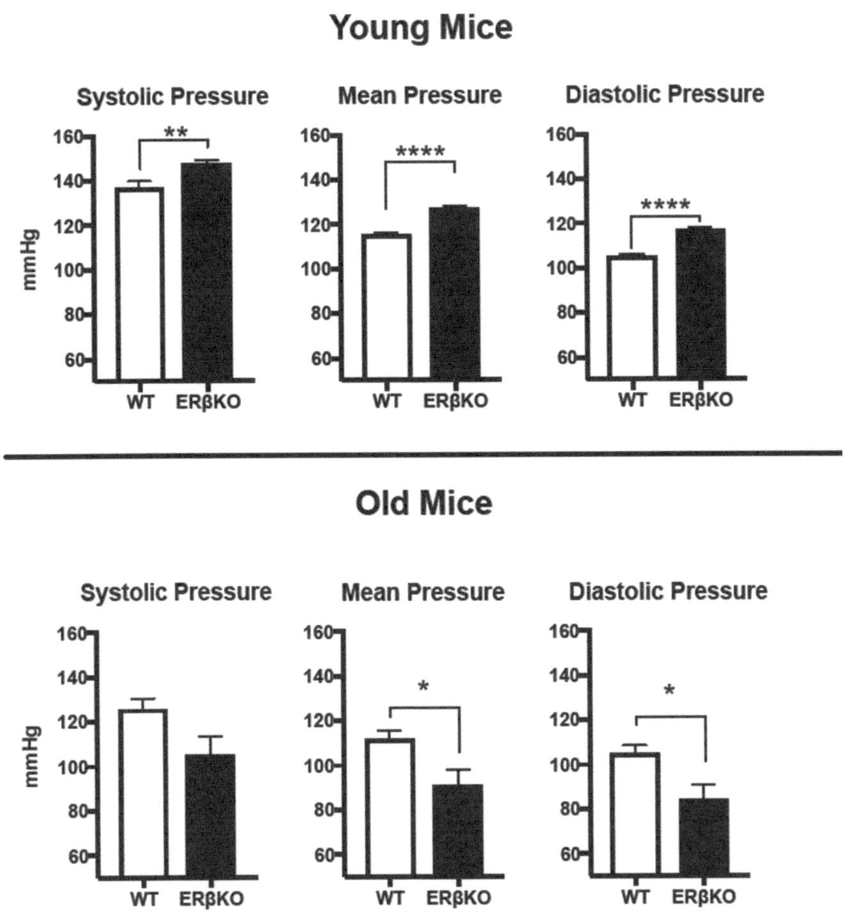

Fig. 15.1 Systolic, mean, and diastolic arterial pressure in young and old wild-type (WT) and ERβKO female mice. *, **, ****: $p < 0.05$, $p < 0.01$, and $p < 0.0001$ ERβKO vs WT

Table 15.1 Respiratory and metabolic variables recorded at rest in young (5–6 months) and old (17–18 months) wild-type (WT) and ERβKO female mice

		WT	βERKO
BW	Young Old	19.8 ± 0.3 28.4 ± 1.2°°°	21.5 ± 0.7* 42.5 ± 2.2***/°°°°
Tr	Young Old	36.0 ± 0.5 36.3 ± 0.3	37.0 ± 0.2 36.3 ± 0.3
fR	Young Old	163 ± 7 162 ± 9	149 ± 6 184 ± 9*/°°
V_T ml/100 g	Young Old	0.50 ± 0.10 0.65 ± 0.07	0.52 ± 0.08 0.58 ± 0.08
V_E ml/min/100 g	Young Old	83.5 ± 19.7 105 ± 14	80.8 ± 14.9 107 ± 20
VO_2 ml/min/100 g	Young Old	5.0 ± 0.6 4.3 ± 0.5	5.3 ± 0.7 4.1 ± 0.4
VCO_2 ml/min	Young Old	3.7 ± 0.4 2.9 ± 0.2	3.8 ± 0.4 2.7 ± 0.7°
VCO_2/VO_2	Young Old	0.772 ± 0.06 0.703 ± 0.059	0.733 ± 0.02 0.670 ± 0.028

Number of animals into each group: Young (7 WT and 8 KO), Old (6 WT and 9 KO)
BW body weight (grams), *Tr* rectal temperature (°C), *fR* respiratory frequency (breaths/min), V_T tidal volume (ml/100 g), V_E minute ventilation (=fR x V_T, in ml/min/100 g), VO_2 O_2 consumption rate (ml/min/100 g), VCO_2 CO_2 production rate (ml/min/100 g), VCO_2/VO_2 respiratory exchange ratio
*, ***$p < 0.05$, and $p < 0.001$ ERβKO vs WT (same age)
°°, °°°, °°°°$p < 0.01$, $p < 0.001$, and $p < 0.0001$ old vs young (same genotype)

body weight during aging in ERβKO mice. Previous studies reported an increased body weight in ERβKO mice, which was associated with increased fat mass when mice reached 10–12 months of age (Seidlova-Wuttke et al. 2012) but not in younger (4 months-old) animals (Ohlsson et al. 2000). This suggests an important role of ERβ to avoid excessive adipose tissue accumulation during aging in female mice, but this remains poorly documented (Davis et al. 2013).

All respiratory and metabolic parameters measured at rest were similar between WT and ERβKO mice. Compared to young ERβKO, in old KO mice respiratory frequency was higher, and CO_2 production rate was lower (Table 15.1). Interestingly, in old ERβKO mice the frequency of sigh was more elevated than in WT mice and in young KO mice (Fig. 15.2). The frequency of post-sigh apnea was reduced by aging, and in aged ERβKO mice the frequency of post-sigh and spontaneous apnea were drastically reduced compared to aged WT mice.

15.3.3 Hypoxic and Hypercapnic Ventilatory Responses in Young and Old WT and ERβKO Female Mice

During the exposures to hypoxia and hypercapnia in young mice, there were no differences between wild-type and ERβKO (Fig. 15.3). In aged mice, compared to wild-type controls ERβKO aged mice had a reduced ventilatory response to hypoxia due to a lower tidal volume (Fig. 15.4 – upper rows). In response to CO_2 exposure, minute ventilation was 39% lower in ERβKO compared to wild-type mice (Fig. 15.4 lower rows), again due to a lower elevation of tidal volume in response to the CO_2 exposure.

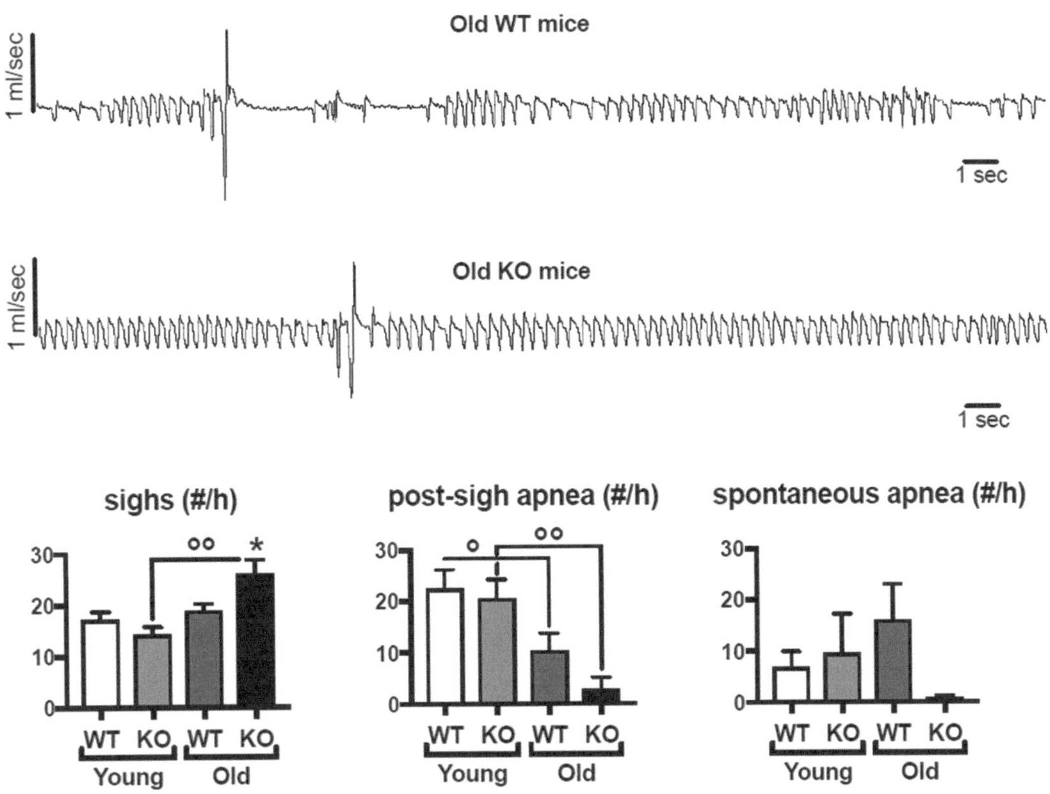

Fig. 15.2 Typical respiratory recordings from an old WT and an old ERβKO mice showing the occurrence of a sigh followed by apneas in the WT but not in the ERβKO mice. Lower panels: frequency of sigh, post-sigh apnea, and spontaneous apnea recorded at rest in young and old WT and ERβKO mice. *$p < 0.05$ ERβKO vs WT (same age). °, °°$p < 0.05$ and $p < 0.01$ old vs young (same genotype)

15.3.4 Mitochondrial Respiration and *In vitro* Activity of Electron Transport Chain Complexes in Young and Old WT and ERβKO Female Mice

Mitochondrial respiration was measured only in old WT and ERβKO mice using permeabilized cortical brain samples. When the respiratory activity was activated by substrates of complex I, the mitochondrial O_2 consumption was lower in ERβKO mice compared to WT control (Fig. 15.5). This was not observed with substrates of complex II, and there was no significant difference for the maximum activity of complex IV (cytochrome oxidase) between wild-type and ERβKO mice.

The activity of complex I measured *in vitro* by the disappearance rate of NADH was lower in ERβKO mice compared to WT (Fig. 15.5).

15.4 Discussion

The present studies show that deletion of ERβ in female mice leads to age-specific alterations of the cardio-vascular and respiratory systems. Interestingly, while elevated blood pressure was present in young ERβKO female mice, our results indicate that this effect does not persist in older ERβKO mice. In fact, in old ERβKO mice arterial blood pressure was lower than in old WT. In contrast, alterations in the respiratory pattern and chemoreflex functions appeared in old, but not in

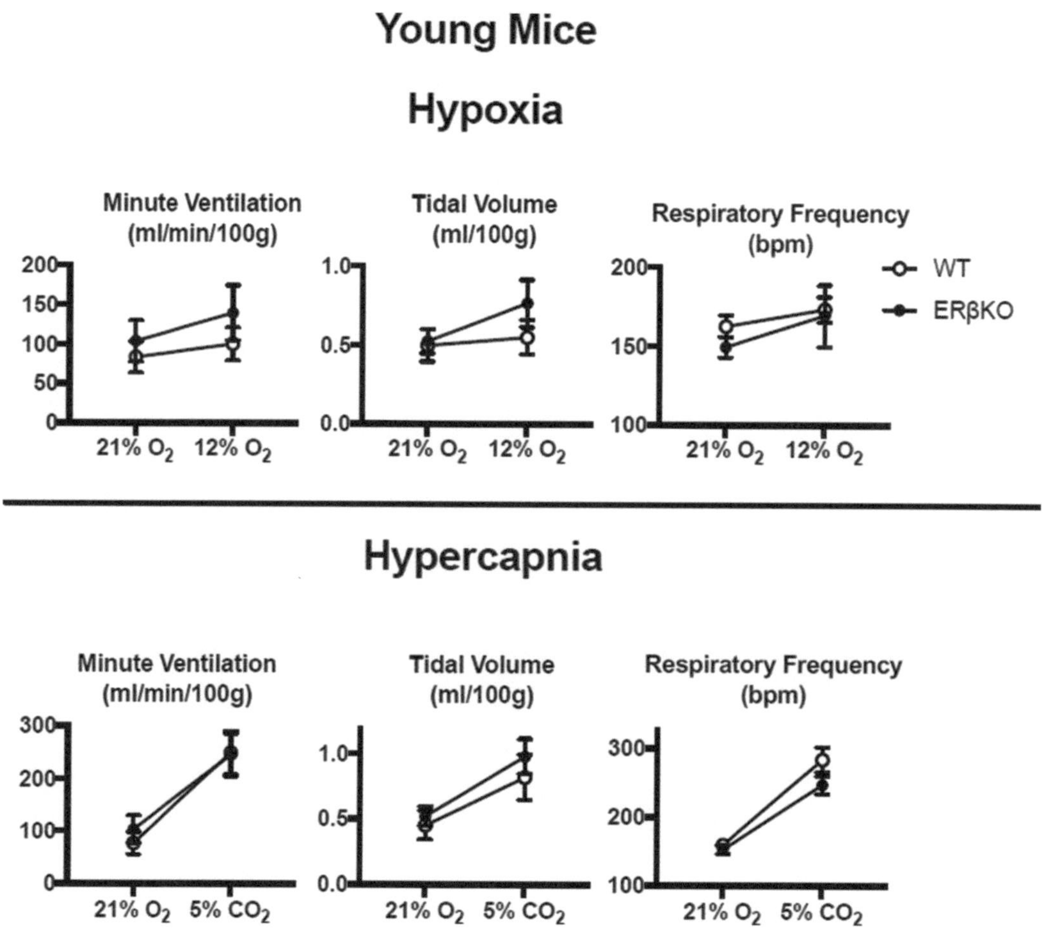

Fig. 15.3 Ventilatory responses to hypoxia (upper panels) and hypercapnia (lower panels) in young (5–6 months) wild-type (WT) and ERβKO female mice

young ERβKO female mice. In older mice, our results also show a reduced mitochondrial respiration in brain cortex through alteration of complex I activity.

15.4.1 Role of ERβ on the Cardiovascular System in Old Vs Young Female Mice

A recent systematic review based on the analysis of 88 studies (mostly on animal models or tissues) provides a comprehensive overview on the role of ERβ in the female cardiovascular system (Muka et al. 2016). ERβ is expressed in endothelial cells, and has an important role by a direct, vasodilatory action on blood vessels. ERβ regulates nitric oxide (NO) bio-availability and inflammatory markers in arterial walls, thus contributing to the regulation of arterial pressure (Muka et al. 2016). Our results showing elevated blood pressure in young female ERβKO mice are in line with previous studies, showing similar results (Zhu et al. 2002). We've been surprised to observe that arterial pressure is reduced in old ERβKO mice compared to their wild-type controls, and we are not aware of previous published report showing this effect. While it has been concluded that ERβ plays a vasodilatory role (Muka et al. 2016), some studies have reported that vasodilation mediated by an ERβ agonist does not require NO signaling, and that ERβ might

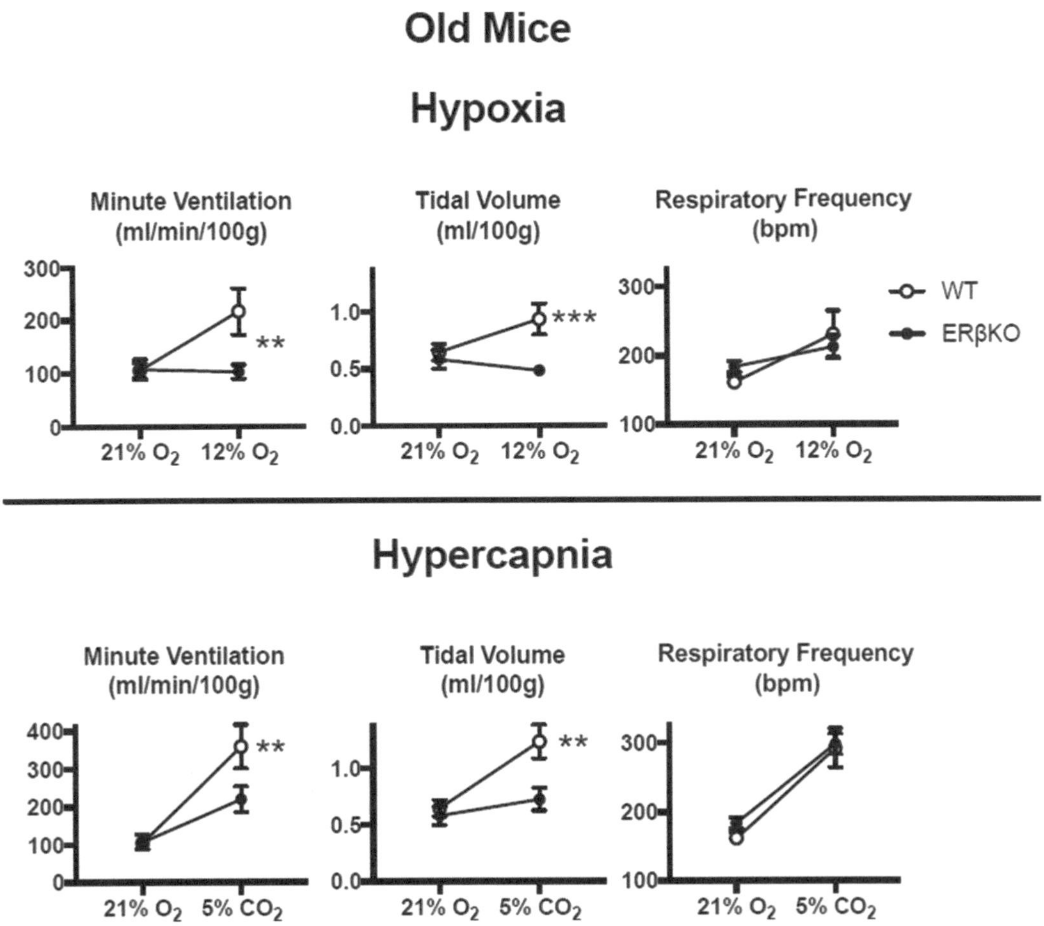

Fig. 15.4 Ventilatory responses to hypoxia (upper panels) and hypercapnia (lower panels) in old (17–18 months) wild-type (WT) and ERβKO female mice. **, ***$p < 0.01$, and $p < 0.001$ ERβKO vs WT

downregulate ERα- and NO-mediated vasodilation (Cruz et al. 2006). Furthermore, in a mouse model of accelerated senescence, age profoundly impacts the effect of estrogen-mediated regulation of NO and oxidative stress systems. In young females, estrogen increases the expression of endothelial NO synthase (eNOS) and decreases oxidative stress, but in aged mice estradiol treatment does not modulate eNOS expression. In isolated aorta of aged mice estradiol increases the expression of the NADPH oxidase subunit NOX1, and increases generation of superoxide anion (O_2^-, one of the major reactive oxygen species) by NADPH oxidase (Novensa et al. 2011). Interestingly, such reversal of estrogen action on the vascular system in aged mice was associated to increased expression level of ERβ and reduced expression level of ERα, leading to a high ratio of ERβ/ERα expression, (Novensa et al. 2011). It has been reported that ERβ inhibits ERα-dependent gene expression, and might thus opposes the vascular effects of ERα (Matthews and Gustafsson 2003). Since ERα is a potent modulator of vasodilation and NO synthesis (Traupe et al. 2007), we might therefore hypothesize that in aged ERβKO mice ERα signaling in blood vessels is increased, leading to vasodilation and reduced blood pressure. This scenario remains speculative, but we believe that it helps understanding our results.

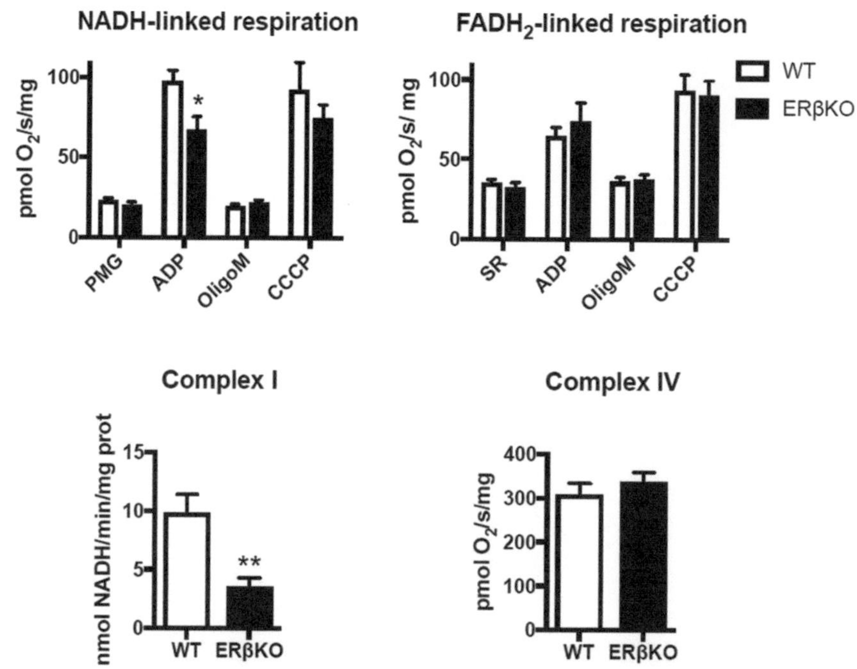

Fig. 15.5 Mitochondrial function in samples of permeabilized brain cortex in old (17–18 months) WT and ERβKO female mice. Upper panels: O_2 consumption rate (pmol O_2/second/mg tissue) with substrates of complex I (left – NADH-linked respiration), and substrates of complex II (right – $FADH_2$- linked respiration). PMG: Pyruvate, malate, glutamate (substrates of complex I), *SR* succinate, rotenone (substrate of complex II and blockade of complex I), *ADP* adenosine di-phosphate, *OligoM* Oligomycin A, *CCCP* carbonyl cyanide m-chlorophenylhydrazone. See text for further details. Lower panels: activity of complex I and IV of the electron transport chain (see text). *, **p < 0.05, and p < 0.01 ERβKO vs WT

15.4.2 Role of ERβ on the Respiratory Control System in Old Vs Young Female Mice

Contrasting with data showing higher blood pressure in young ERβKO mice, there was no evidence of changes of the respiratory control system in these animals. However, compared to their wild-type controls, old ERβKO had several alterations, including higher frequency of sighs recorded during sleep, but a much lower occurrence of post-sigh and spontaneous apneas, and reduced hypoxic and hypercapnic ventilatory responses. As mentioned in the introduction, ERβ mRNA and protein are both expressed in brainstem areas involved in respiratory control (such as the nucleus tractus solitarius, locus coeruleus, parabrachial nucleus), in hypothalamic nuclei (Shughrue and Merchenthaler 2001; Shughrue et al. 1997), and in peripheral chemoreceptors (Joseph et al. 2006). The fact that ERβ is present in the peripheral chemoreceptors and in the nucleus tractus solitarius, the main projection site of the peripheral chemoreceptors in the brainstem (Finley and Katz 1992), lend support to the hypothesis that ERβ modulates responses elicited by activation of the peripheral chemoreceptor, which is in line with the drastically attenuated hypoxic response in ERβKO female mice. Interestingly, the hypercapnic ventilatory response was not completely abolished but reduced by about 50–55% in ERβKO mice compared to wild-type controls. Furthermore, responses to hypoxia and hypercapnia were both reduced owing to a lower tidal volume, and the magnitude of the difference observed between WT and ERβKO for both hypoxic and hypercapnic ventilatory responses were similar. These findings suggest a common mechanism might be responsible for the altered hypoxic and hypercapnic ventilatory responses, possibly involving a reduction of the peripheral chemoreceptor drive in response to hypoxia and hypercapnia. Alternatively, a central inhibition and/or altered integrative processes common to respiratory responses to hypoxia and hypercapnia could also explain these results.

Other interesting aspects of our data are the differences of sigh and apnea frequency in normoxia between WT and ERβKO mice. As reported for alterations of the chemoreflex responses, these effects are observed only in old mice. Sighs are very common, and play an important, but often overlooked, role to regulate the respiratory control system, and could possibly have functions extending well beyond this system (Ramirez 2014). Because they are readily observable in brainstem slices, sighs are likely an intrinsic property of the respiratory control system, and in most cases, *in vitro* and *in vivo* sighs are followed by a cessation of breathing, referred to as "post-sigh apneas" (Ramirez 2014). Interestingly, while sigh frequency was increased in ERβKO mice compared to WT controls, the frequency of post-sigh apnea was drastically reduced in ERβKO mice. While we could speculate on the significance of these findings in light of our most recent understanding of sigh functions and underlying neurobiology (Ramirez 2014), we prefer to remain cautious, and simply report this intriguing phenotype observed in aged ERβKO female mice.

15.4.3 Role of ERβ on Mitochondrial Respiration in Old Female Mice

Finally, our data shows that ERβ deletion reduces mitochondrial respiratory activity and activity of electron transfer chain complex I in brain cortex samples. These results are intriguing because at the level of the peripheral chemoreceptors recent data support the hypothesis that the mitochondrial activity contributes to oxygen sensing (Fernandez-Aguera et al. 2015; Chang 2017; McElroy and Chandel 2017). Therefore, if similar effects of ERβ deletion are present in the peripheral chemoreceptors of ERβKO mice, this may likely contribute to the reduced hypoxic ventilatory response in older mice. These data are in line with the effect of estradiol on mitochondrial function, which are mediated by ERα and ERβ (Yang et al. 2004; Irwin et al. 2012; Razmara et al. 2008). The modulations of mitochondrial function by E_2 and ERs are thought to contribute to cardio-vascular and neurological protective effects in women before menopause, and to explain the longer life expectancy of women compared to men (Nilsen and Brinton 2004; Arnold et al. 2012; Irwin et al. 2012; Razmara et al. 2008).

15.5 Conclusion

We conclude that KO mice are a reliable model to assess the role of specific estradiol receptor on respiratory regulation. We started our studies with young ERβKO mice, and because there was no evidence for altered respiratory control system, we decided to keep these animals until they reached an older age. Our data reveal that in 17–18 months-old mice, the deletion of ERβ leads to several alterations of the respiratory control system, including reductions of chemoreflex functions. Because this is associated with a lower mitochondrial respiratory activity in the central nervous system, our results show that ERβ acts at different levels of the oxygen transport cascade (from pulmonary ventilation to mitochondrial O_2 consumption). This may be relevant in pathological or physiological conditions in which O_2 homeostasis is compromised (sleep apnea, altitude, lung diseases), and research efforts to further understand the role of estradiol and ERs in these conditions should be further developed.

References

Arnold S, Victor MB, Beyer C (2012) Estrogen and the regulation of mitochondrial structure and function in the brain. J Steroid Biochem Mol Biol 131(1–2):2–9. https://doi.org/10.1016/j.jsbmb.2012.01.012

Bastianini S, Alvente S, Berteotti C, Lo Martire V, Silvani A, Swoap SJ, Valli A, Zoccoli G, Cohen G (2017) Accurate discrimination of the wake-sleep states of mice using non-invasive whole-body plethysmography. Sci Rep 7:41698. https://doi.org/10.1038/srep41698

Bayliss DA, Cidlowski JA, Millhorn DE (1990) The stimulation of respiration by progesterone in ovariectomized cat is mediated by an estrogen-dependent hypothalamic mechanism requiring gene expression. Endocrinology 126:519–527

Behan M, Wenninger JM (2008) Sex steroidal hormones and respiratory control. Respir Physiol Neurobiol 164(1–2):213–221. https://doi.org/10.1016/j.resp.2008.06.006

Bixler EO, Vgontzas AN, Lin HM, Ten Have T, Rein J, Vela-Bueno A, Kales A (2001) Prevalence of sleep-disordered breathing in women: effects of gender. Am J Respir Crit Care Med 163(3 Pt 1):608–613

Block A, Boysen P, Wynne J, Hunt L (1979) Sleep apnea, hypopnea and oxygen desaturation in normal subjects. A strong male predominance. N Engl J Med 300:513–517

Boukari R, Rossignol O, Baldy C, Marcouiller F, Bairam A, Joseph V (2016) Membrane progesterone receptor-beta, but not -alpha, in dorsal brain stem establishes sex-specific chemoreflex responses and reduces apnea frequency in adult mice. J Appl Physiol 121(3):781–791. https://doi.org/10.1152/japplphysiol.00397.2016

Boukari R, Laouafa S, Ribon-Demars A, Bairam A, Joseph V (2017) Ovarian steroids act as respiratory stimulant and antioxidant against the causes and consequences of sleep-apnea in women. Respir Physiol Neurobiol 239:46–54. https://doi.org/10.1016/j.resp.2017.01.013

Brinton RD, Thompson RF, Foy MR, Baudry M, Wang J, Finch CE, Morgan TE, Pike CJ, Mack WJ, Stanczyk FZ, Nilsen J (2008) Progesterone receptors: form and function in brain. Front Neuroendocrinol 29(2):313–339. https://doi.org/10.1016/j.yfrne.2008.02.001

Chang AJ (2017) Acute oxygen sensing by the carotid body: from mitochondria to plasma membrane. J Appl Physiol 1985:jap 00398 02017. doi:https://doi.org/10.1152/japplphysiol.00398.2017

Charkoudian N, Hart ECJ, Barnes JN, Joyner MJ (2017) Autonomic control of body temperature and blood pressure: influences of female sex hormones. Clin Auton Res 27(3):149–155. https://doi.org/10.1007/s10286-017-0420-z

Cruz MN, Douglas G, Gustafsson JA, Poston L, Kublickiene K (2006) Dilatory responses to estrogenic compounds in small femoral arteries of male and female estrogen receptor-beta knockout mice. Am J Physiol Heart Circ Physiol 290(2):H823–H829. https://doi.org/10.1152/ajpheart.00815.2005

Davis KE, Neinast MD, Sun K, Skiles WM, Bills JD, Zehr JA, Zeve D, Hahner LD, Cox DW, Gent LM, Xu Y, Wang ZV, Khan SA, Clegg DJ (2013) The sexually dimorphic role of adipose and adipocyte estrogen receptors in modulating adipose tissue expansion, inflammation, and fibrosis. Mol Metab 2(3):227–242. https://doi.org/10.1016/j.molmet.2013.05.006

Fernandez-Aguera MC, Gao L, Gonzalez-Rodriguez P, Pintado CO, Arias-Mayenco I, Garcia-Flores P, Garcia-Perganeda A, Pascual A, Ortega-Saenz P, Lopez-Barneo J (2015) Oxygen sensing by arterial chemoreceptors depends on mitochondrial complex I signaling. Cell Metab 22(5):825–837. https://doi.org/10.1016/j.cmet.2015.09.004

Finley JC, Katz DM (1992) The central organization of carotid body afferent projections to the brainstem of the rat. Brain Res 572(1–2):108–116

Harris HA (2007) Estrogen receptor-beta: recent lessons from in vivo studies. Mol Endocrinol 21(1):1–13. https://doi.org/10.1210/me.2005-0459

Herbst EA, Holloway GP (2015) Permeabilization of brain tissue in situ enables multiregion analysis of mitochondrial function in a single mouse brain. J Physiol 593(4):787–801. https://doi.org/10.1113/jphysiol.2014.285379

Irwin RW, Yao J, To J, Hamilton RT, Cadenas E, Brinton RD (2012) Selective oestrogen receptor modulators differentially potentiate brain mitochondrial function. J Neuroendocrinol 24(1):236–248. https://doi.org/10.1111/j.1365-2826.2011.02251.x

Joseph V, Doan VD, Morency CE, Lajeunesse Y, Bairam A (2006) Expression of sex-steroid receptors and steroidogenic enzymes in the carotid body of adult and newborn male rats. Brain Res 1073–1074:71–82. https://doi.org/10.1016/j.brainres.2005.12.075

Krege JH, Hodgin JB, Couse JF, Enmark E, Warner M, Mahler JF, Sar M, Korach KS, Gustafsson JA, Smithies O (1998) Generation and reproductive phenotypes of mice lacking estrogen receptor beta. Proc Natl Acad Sci USA 95(26):15677–15682

Kuiper GG, Enmark E, Pelto-Huikko M, Nilsson S, Gustafsson JA (1996) Cloning of a novel receptor expressed in rat prostate and ovary. Proc Natl Acad Sci USA 93(12):5925–5930

Laouafa S, Bairam A, Soliz J, Roussel D, Joseph V (2017a) Estradiol receptor agonists α and β protect against brain mitochondrial dysfunction in a model of sleep apnea. FASEB J 31(1 Supplement):696.6–696.6

Laouafa S, Ribon-Demars A, Marcouiller F, Roussel D, Bairam A, Pialoux V, Joseph V (2017b) Estradiol protects against cardiorespiratory dysfunctions and oxidative stress in intermittent hypoxia. Sleep 40(8). https://doi.org/10.1093/sleep/zsx104

Lefter R, Doan VD, Joseph V (2008) Contrasting effects of estradiol and progesterone on respiratory pattern and hypoxic ventilatory response in newborn male rats. Respir Physiol Neurobiol 164(3):312–318. https://doi.org/10.1016/j.resp.2008.07.026

Marcouiller F, Boukari R, Laouafa S, Lavoie R, Joseph V (2014) The nuclear progesterone receptor reduces post-sigh apneas during sleep and increases the ventilatory response to hypercapnia in adult female mice. PLoS One 9(6):e100421. https://doi.org/10.1371/journal.pone.0100421

Marrocco J, McEwen BS (2016) Sex in the brain: hormones and sex differences. Dialogues Clin Neurosci 18(4):373–383

Matthews J, Gustafsson JA (2003) Estrogen signaling: a subtle balance between ER alpha and ER beta. Mol Interv 3(5):281–292. https://doi.org/10.1124/mi.3.5.281

McElroy GS, Chandel NS (2017) Mitochondria control acute and chronic responses to hypoxia. Exp Cell Res 356(2):217–222. https://doi.org/10.1016/j.yexcr.2017.03.034

Micevych PE, Wong AM, Mittelman-Smith MA (2015) Estradiol membrane-initiated signaling and female reproduction. Compr Physiol 5(3):1211–1222. https://doi.org/10.1002/cphy.c140056

Muka T, Vargas KG, Jaspers L, Wen KX, Dhana K, Vitezova A, Nano J, Brahimaj A, Colpani V, Bano A, Kraja B, Zaciragic A, Bramer WM, van Dijk GM, Kavousi M, Franco OH (2016) Estrogen receptor beta actions in the female cardiovascular system: a systematic review of animal and human studies. Maturitas 86:28–43. https://doi.org/10.1016/j.maturitas.2016.01.009

Nilsen J, Brinton RD (2004) Mitochondria as therapeutic targets of estrogen action in the central nervous system. Curr Drug Targets CNS Neurol Disord 3(4):297–313

Novensa L, Novella S, Medina P, Segarra G, Castillo N, Heras M, Hermenegildo C, Dantas AP (2011) Aging negatively affects estrogens-mediated effects on nitric oxide bioavailability by shifting ERalpha/ERbeta balance in female mice. PLoS One 6(9):e25335. https://doi.org/10.1371/journal.pone.0025335

Ohlsson C, Hellberg N, Parini P, Vidal O, Bohlooly YM, Rudling M, Lindberg MK, Warner M, Angelin B, Gustafsson JA (2000) Obesity and disturbed lipoprotein profile in estrogen receptor-alpha-deficient male mice. Biochem Biophys Res Commun 278(3):640–645. https://doi.org/10.1006/bbrc.2000.3827

Ramirez JM (2014) The integrative role of the sigh in psychology, physiology, pathology, and neurobiology. Prog Brain Res 209:91–129. https://doi.org/10.1016/B978-0-444-63274-6.00006-0

Razmara A, Sunday L, Stirone C, Wang XB, Krause DN, Duckles SP, Procaccio V (2008) Mitochondrial effects of estrogen are mediated by estrogen receptor alpha in brain endothelial cells. J Pharmacol Exp Ther 325(3):782–790. https://doi.org/10.1124/jpet.107.134072

Seidlova-Wuttke D, Nguyen BT, Wuttke W (2012) Long-term effects of ovariectomy on osteoporosis and obesity in estrogen-receptor-beta-deleted mice. Comp Med 62(1):8–13

Shughrue PJ, Merchenthaler I (2001) Distribution of estrogen receptor beta immunoreactivity in the rat central nervous system. J Comp Neurol 436(1):64–81

Shughrue PJ, Lane MV, Merchenthaler I (1997) Comparative distribution of estrogen receptor-alpha and -beta mRNA in the rat central nervous system. J Comp Neurol 388(4):507–525

Traupe T, Stettler CD, Li H, Haas E, Bhattacharya I, Minotti R, Barton M (2007) Distinct roles of estrogen receptors alpha and beta mediating acute vasodilation of epicardial coronary arteries. Hypertension 49(6):1364–1370. https://doi.org/10.1161/HYPERTENSIONAHA.106.081554

Yang SH, Liu R, Perez EJ, Wen Y, Stevens SM Jr, Valencia T, Brun-Zinkernagel AM, Prokai L, Will Y, Dykens J, Koulen P, Simpkins JW (2004) Mitochondrial localization of estrogen receptor beta. Proc Natl Acad Sci USA 101(12):4130–4135. https://doi.org/10.1073/pnas.0306948101

Zhu Y, Bian Z, Lu P, Karas RH, Bao L, Cox D, Hodgin J, Shaul PW, Thoren P, Smithies O, Gustafsson JA, Mendelsohn ME (2002) Abnormal vascular function and hypertension in mice deficient in estrogen receptor beta. Science 295(5554):505–508. https://doi.org/10.1126/science.1065250

16 Chronic Heart Failure Abolishes Circadian Rhythms in Resting and Chemoreflex Breathing

Robert Lewis, Bryan T. Hackfort, and Harold D. Schultz

Abstract

Physiological systems often display 24 h rhythms that vary with the light/dark cycle. Disruption of circadian physiological rhythms have been linked to the progression of various cardiovascular diseases, and advances in the understanding of these rhythms have led to novel interventions and improved clinical outcomes. Although respiratory function has been known to vary between the light and dark periods, circadian rhythms in breathing have been understudied in clinical conditions. In the current study, we have begun to assess light/dark variations in respiration in chronic heart failure (CHF), a condition associated with abnormal resting and chemoreflex breathing as well as exercise intolerance. CHF was induced using coronary artery ligation and verified using echocardiography. Sham animals underwent a thoracotomy without coronary artery ligation. Tidal volume, respiratory frequency, and minute ventilation were all determined by whole body plethysmography under resting conditions and in response to chemoreflex challenges during the light and dark periods. Light/dark differences in voluntary exercise were assessed using a running wheel. The sham control group showed light/dark differences in resting and chemoreflex breathing, as well as arterial pressure, and these effects were eliminated in the CHF group. Both groups completed more rotations on the running wheel during the dark period compared to during the light period. The data suggest that CHF disrupts cardiovascular and respiratory circadian rhythms.

Keywords

Circadian rhythm · Chronic heart failure · Chronobiology · Plethysmography

16.1 Introduction

Many cellular and physiological processes display cyclic patterns with the same periodicity as the 24 h. light/dark cycle (Partch et al. 2014), and these circadian rhythms appear to be disrupted in a variety of clinical conditions (Durgan and Young 2010; Ayala et al. 2013). Among the circadian rhythms in physiology, rhythmic changes in cardiovascular function have perhaps been most studied. Research into circadian rhythms in cardiovascular physiology and pathophysiology has led to recent clinical efforts to utilize knowledge of these rhythms in treating cardiovascular diseases (Tsimakouridze et al. 2015), and these

R. Lewis (✉)
A.T. Still University School of Osteopathic Medicine, Mesa, AZ, USA
e-mail: robertlewis@atsu.edu

B. T. Hackfort · H. D. Schultz
Department of Cellular and Integrative Physiology at the University of Nebraska Medical Center, Omaha, NE, USA

E. B. Gauda et al. (eds.), *Arterial Chemoreceptors*, Advances in Experimental Medicine and Biology 1071, https://doi.org/10.1007/978-3-319-91137-3_16

efforts have been shown to reduce adverse cardiac remodeling, buffer the morning surge in blood pressure, and reduce the number of adverse cardiovascular events (Tsimakouridze et al. 2015). While advances in chronobiology have improved outcomes for patients with cardiovascular diseases, less is known about circadian rhythms in breathing.

Chronic heart failure (CHF) is a condition characterized by disordered breathing, intermittent hypoxia, exercise intolerance, and chronically elevated sympathetic tone (Morrissey et al. 2011). Patients with reduced ejection fraction CHF do not show circadian rhythms in blood pressure (Komori et al. 2016). Typically, in healthy humans, blood pressure decreases during the night by about 10–20% of the mean daytime values (Floras et al. 1978). CHF patients are more likely than the general population to show abnormal light/dark blood pressure rhythms, and the particular pattern of blood pressure rhythm appears to vary with the type of CHF (Komori et al. 2016). CHF with preserved ejection fraction shows a "riser" pattern in which blood pressure is greater during the sleeping period, and CHF patients with reduced ejection fraction fail to show light/dark variations in blood pressure (Komori et al. 2016). While circadian cardiovascular rhythms have been examined in CHF, no one has yet to assess circadian rhythms in breathing.

Respiratory function has been shown to vary with the light/dark cycle (Mortola 2004). While light/dark variations in breathing have consistently been observed using continuous monitoring of respiration (Mortola and Seifert 2002; Seifert and Mortola 2002), respiratory measurements taken at two discrete time points have produced conflicting results. Light/dark variations in resting breathing were detected in neonate rats (Saiki and Mortola 1995) but not in adult rats (Peever and Stephenson 1997) when observed at two time points. Peever and Stephenson (1997) assessed ventilation in Sprague-Dawley rats at two timepoints and found that chemoreflex breathing to hypercapnia increased during the night, but resting breathing showed no change between the light and dark periods. Notably, the respiratory rates for these animals at rest were considerably fast (96 ± 6 bpm during the day and 103 ± 8 during the night), which may have masked differences between the light and dark periods. In our experience, respiratory rates of adult Sprague-Dawley rats are much faster after first placing the animal in the plethysmography chamber, and respiratory rates decline to around 50–70 bpm during the day. It is unclear why the respiratory rates in the Peever and Stephenson study remained elevated during the day. In the current study, we assessed resting and chemoreflex breathing using whole body plethysmography at two time points, separated by about 12 h, in a sham operated control group and a group with CHF induced by coronary artery ligation. Subsequently, light/dark variations in spontaneous exercise were also assessed using a running wheel.

16.2 Methods

16.2.1 Animals

Adult male Sprague Dawley rats (300–450 g) were used in this study. Animals were held under a 12:12 light/dark cycle with lights on at 6 am. All procedures were approved by the Institutional Animal Care and Use Committee at the University of Nebraska Medical Center.

16.2.2 Whole Body Plethysmography

Tidal volume (V_T), respiratory frequency (fR), and minute ventilation (V_E) were all determined by whole body plethysmography. V_T was measured using a differential pressure transducer (Kent Scientific) and amplifier (MP45, Validyne Engineering Corporation) prior to recording through a Power Lab System (AD Instruments) and Lab Chart v8.1.5. Rats were given at least 45 min to acclimate to the chamber, and then exposed to a series of gas challenges (10% O_2 with 2% CO_2, 5% CO_2, 90% O_2, 5% CO_2 with 90% O_2) with 3–5 min of recovery time in between in each challenge. Chemoreflex breathing was assessed during the day (8 am–2 pm) and during the night (8 pm–2 am). V_T was normalized to body weight (kg). Chemoreflex sensitivity was quantified by subtracting V_E during a given challenge (10%O_2, 5% CO_2, and 90%O_2 with 5%CO_2) from baseline V_E.

16.2.3 Coronary Artery Ligation

Heart failure was induced via coronary artery ligation (n = 5) as previously described (Del Rio et al. 2013). Sham animals underwent thoracotomy without coronary artery ligation (n = 5). Six weeks following surgery, heart failure was verified using echocardiography (Vevo 770; Visualsonics). Two-dimensional short-axis view of the left ventricle was used to assess end-diastolic and end-systolic diameter as well as end-diastolic and end-systolic volume of the left ventricle. Animals with an ejection fraction <45% were considered to be in heart failure.

16.2.4 Scurry Wheel

Voluntary exercise was assessed using Scurry Wheel (Lafayette Instruments, Model 80859S) which consists of a running wheel (inner diameter 35.56 cm; width 10.92 cm) inside a cage (40.64 x 50.80 x 20.96 cm). The scurry wheel was connected to an eight channel Powerlab (AD Instruments), and wheel rotations were recorded in LabChart v8.1. To reduce the effects of the novel environment on behavior, each animal was given 24 h to acclimate to the cage. Animals had ad libitum access to food and water during the scurry wheel experiment. One heart failure animal died before completing the scurry wheel, reducing the n for the CHF group to four.

16.2.5 Blood Pressure

To assess light/dark variations in blood pressure, a separate group of animals were implanted with blood pressure telemetry (Datasciences International, either PA-C40 or HD-S10) in the left femoral artery. Surgical implantation occurred under 2–5% isoflurane balanced with oxygen, and animals were given at least 2 weeks to recover. Following recovery from telemetry implantation, these animals underwent either coronary artery ligation (n = 5) or a sham surgery (n = 4), as described above. At least 2 weeks after the thoracotomy, these animals were placed in the cage with the running wheel, as described above, and blood pressure was recorded along with running wheel rotations. Segments of the blood pressure recordings in which the animal remained at rest for an extended period of time (>25 min) were selected to assess light/dark variations in blood pressure. Two animals that were infarcted had ejection fractions greater than 45%, so they were excluded from the study.

16.2.6 Statistics

All data are presented as mean ± SEM, unless otherwise noted. One-way or two-way ANOVAs with Fisher's LSD used for pairwise comparisons were used to analyze the data. $p < 0.05$ was considered statistically significant.

16.3 Results

16.3.1 Echocardiographic Data

Bodyweights between the sham (366 ± 12) and CHF (339 ± 7) groups did not differ ($p = 0.08$). Differences were seen in cardiac structure and function (Table 16.1). The CHF group had greater end-diastolic ($p < 0.01$) and end-systolic ($p < 0.01$) volumes, greater end-diastolic ($p = 0.02$) and end-systolic ($p < 0.01$) diameters, and lower ejection fraction ($p < 0.0001$) and fractional shortening ($p < 0.0001$).

16.3.2 Circadian Rhythms in Resting and Chemoreflex Breathing

Resting and chemoreflex breathing was assessed using whole-body plethysmography at two discreet time points, separated by approximately

Table 16.1 Cardiac function

Condition	BW (g)	LVEDD (mm)	LVESD (mm)	LVEDV (μL)	LVESV (μL)	SV (μL)	EF (%)	FS (%)
Sham	441 ± 27.8	0.83 ± 0.06	0.56 ± 0.05	132 ± 25	44 ± 8	88 ± 17	66.8 ± 3.55	31.8 ± 3.24
CHF	484 ± 16.9	1.21 ± 0.07*	1.03 ± 0.07*	353 ± 51*	228 ± 39*	123 ± 12*	36.4 ± 2.65*	15.5 ± 1.26*

*Statistically different from sham; sham, n = 5; chf n = 5

12 h. The sham-operated control group displayed light/dark variations in V_E when breathing room air under baseline ($p = 0.040$) and post-baseline ($p = 0.038$) conditions (Fig. 16.1a). The sham group also showed light/dark differences in V_E during chemoreflex breathing, specifically hypoxia ($p < 0.001$), hypercapnia ($p < 0.001$), and hyperoxia with hypercapnia ($p < 0.001$) (Fig. 16.1a). In contrast, chronic heart failure eliminated circadian rhythms in resting breathing under the baseline ($p = 0.57$) and post-baseline ($p = 0.32$) conditions as well as during chemoreflex breathing (hypoxia $p = 0.87$, hypercapnia $p = 0.07$, and hyperoxia with hypercapnia $p = 0.47$) (Fig. 16.1a). There was also a main effect of heart failure status ($p < 0.001$) with the CHF group displaying greater V_E than the sham control group (Fig. 16.1a).

Analysis of fR data showed a main effect of heart failure status ($p < 0.001$) with the CHF showing greater respiratory rates and a main effect of time ($p = 0.02$) with respiratory rates increasing during the dark (Fig. 16.1b). The sham group showed light/dark differences in fR under resting conditions (baseline: $p = 0.02$; post-baseline: $p = 0.04$) and in response to hypoxia ($p = 0.01$), but not in response to hypercapnia ($p = 0.54$), hyperoxia ($p = 0.22$), or hyperoxia with hypercapnia ($p = 0.15$) (Fig. 16.1b). Respiratory rate did not differ between light and dark periods in the CHF group ($p = 0.16$) (Fig. 16.1b).

CHF animals had greater V_T than did sham animals ($p < 0.05$), but no main effect of time ($p = 0.06$) or interaction ($p = 0.48$) was seen (Fig. 16.1c).

16.3.3 Circadian Rhythms in Chemoreflex Sensitivity

Chemoreflex sensitivity was also assessed in sham and CHF animals during the light and dark periods. The data showed a statistically significant time x condition interaction ($p = 0.005$), with the sham group displaying greater sensitivity during the dark compared to during the light period ($p = 0.01$) while the CHF group showed no difference in chemoreflex sensitivity between the day and the night ($p = 0.12$) (Fig. 16.2).

16.3.4 Circadian Rhythms in Voluntary Exercise

Voluntary locomotor activity was assessed using a running wheel, following a 24 h. period in which each animal acclimated to the novel cage containing the running wheel (Fig. 16.3). The total number of rotations was determined for the light and for the dark periods. Wheel running showed a main effect of time ($p < 0.001$) with no main effect of heart failure status ($p = 0.69$) or interaction ($p = 0.73$). The sham group ran 128 ± 14 rotations during the light period and 582 ± 241 rotations during the dark period ($p < 0.05$). The CHF group also ran more during the dark. However, there was no difference between the sham and CHF groups.

16.3.5 Circadian Rhythms in Blood Pressure

Circadian rhythms in blood pressure and voluntary exercise were assessed in rats with chronically implanted telemetry. Mean arterial pressure

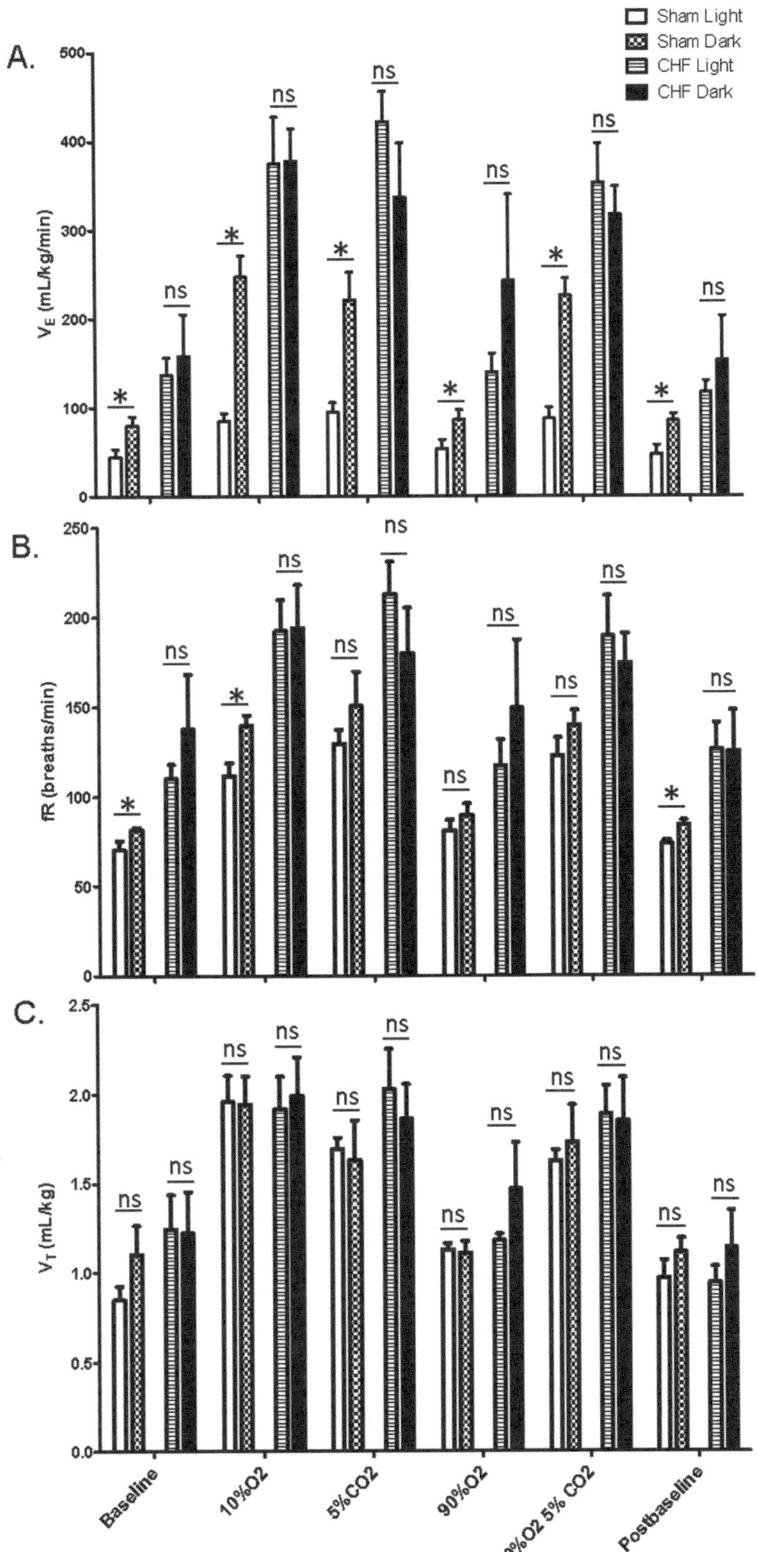

Fig. 16.1 Light/Dark variations in resting and chemoreflex breathing. (**a**) Minute ventilation (VE), (**b**) respiratory frequency (fR), and (**c**) tidal volume (VT) were measured during the light and dark periods. Data are presented as mean ± SEM

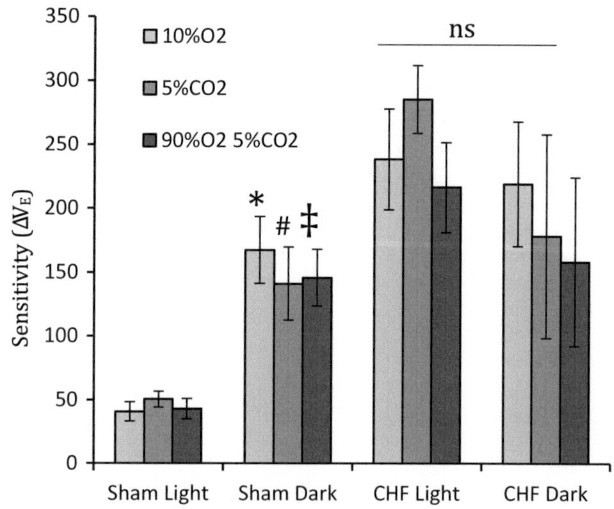

Fig. 16.2 Chemoreflex sensitivity. Light/dark variations in chemoreflex sensitivity by subtracting the V_E during baseline breathing for each group from each chemoreflex challenge.* statistically different from sham light 10% O_2; # statistically different from sham light 5% CO_2; ‡ statistically different from sham light 90%O_2 5%CO_2; *ns* not significant

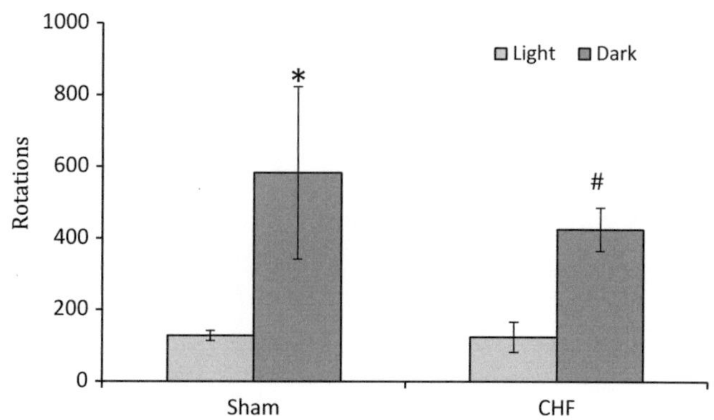

Fig. 16.3 Voluntary exercise. The number of running wheel rotations was recorded for the sham (n = 5) and CHF (n = 4) groups during the light and dark periods. *statistically significant from sham light; # statistically significant from CHF light

(MAP) differed between the light and dark periods in the sham (light, 102 ± 2.13 mmHg; dark, 111 ± 3.49 mmHg; p = 0.03) but not in the CHF group (light, 97 ± 3.80 mmHg; dark, 85 ± 1.36 mmHg; p > 0.05). Similarly, systolic pressure was greater during the dark compared to the light in the sham group (light, 128 ± 1.14 mmHg; dark, 136 ± 1.62 mmHg; p = 0.03) but not the CHF group (light, 119 ± 8.96 mmHg; dark, 123 ± 9.42 mmHg; p > 0.05). Diastolic pressure did not differ between the light and dark periods in either the sham (light, 95 ± 4.86 mmHg; dark, 93 ± 4.6 mmHg; p = 0.75) or CHF group (light, 85 ± 1.36 mmHg; dark, 88 ± 2.93 mmHg; p = 0.82). The sham and CHF groups did not differ from each other in MAP (p = 0.14), systolic pressure (p = 0.21), or diastolic pressure (p = 0.22).

16.4 Discussion

Respiratory function has been shown to have a circadian periodicity and vary with the light/dark cycle (Mortola 2004). A previous study that acquired respiratory measurements at discrete time points did not find light/dark differences in resting breathing in adult Sprague-Dawley rats (Peever and Stephenson 1997); however, we assessed breathing during the light and dark periods, and we detected light/dark variations in V_E in both resting and chemoreflex breathing as well as light/dark differences in chemoreflex sensitivity in the control group. The elevated respiratory rates (~90 bpm) of the former study may have masked light/dark variations in resting breathing. In contrast, the control group of the current study showed respiratory rates of 70 ± 3.87 bpm during the light period and 81 ± 3.73 bpm during the dark. The current study demonstrates that measurements of respiratory function taken at discrete time points can be used to detect light/dark variations in breathing in healthy adult rats. Theses findings are consistent with previous studies that found greater HVR (Seifert and Mortola 2002) and HCVR (Peever and Stephenson 1997) during the dark in rats.

In addition to finding circadian rhythms in resting and chemoreflex breathing in healthy adult rats, we also assessed these rhythms in CHF rats with reduced ejection. To our knowledge, this is the first study to assess circadian rhythms in respiration in CHF. Light/dark variations in respiration and arterial pressure appear to be abolished in CHF with reduced ejection fraction. Our study corroborates previous work on blood pressure rhythms in human heart failure patients (Komori et al. 2016) and mice with myocardial infarctions (Mousa et al. 2014). The current study expands previous findings in showing that CHF also eliminates respiratory circadian rhythms. Although CHF rats did not show a circadian rhythm in breathing, chemoreflex sensitivity, or blood pressure, the CHF group showed light/dark variation in voluntary exercise. These results suggest that CHF impairs circadian rhythmicity of autonomic function with respect to neural control of breathing and cardiovascular function before significantly impacting circadian variations in behavioral locomotor activity.

Also, surprisingly, the CHF group exercised the same amount as the sham control group. The reason for this finding is unclear. Exercise intolerance in CHF appears to be mediated largely by skeletal muscle myopathy (Piepoli and Crisafulli 2014). Thus, it is conceivable that while ejection fraction was decreased and respiratory function was abnormal, skeletal muscle myopathy was not sufficiently advanced to impair locomotor activity.

The mechanism underlying the respiratory circadian rhythm in the control group and the lack of light/dark variations in the CHF group have not been determined but 24 h variations in carotid body chemoreceptor sensitivity is a promising target. A series of ex vivo experiments in the rat carotid body suggest that carotid body sensitivity varies between the light and dark periods (Chen et al. 2005; Tjong et al. 2006). The firing rate of the rat carotid body increases significantly in response to hypoxia in the presence of melatonin (Chen et al. 2005; Tjong et al. 2006), a hormone that is released during the night in both humans (Grivas and Savvidou 2007) and rats (Lynch et al. 1984). Altered carotid body function may explain the lack of light/dark variations in breathing in CHF. While the responsiveness of the carotid body in the presence of melatonin may promote light/dark variations in healthy controls, carotid body function is exaggerated at rest and in response to hypoxia and hypercapnia in CHF (Andrade et al. 2015). The heightened activity of chemosensory cells of the carotid body induced by CHF may mask light/dark differences in resting and chemoreflex breathing. Future studies should assess the role of melatonin in modulating resting and chemoreflex breathing during the light and dark periods in healthy controls and explore how this process may be altered in CHF.

Acknowledgements The authors would like to acknowledge Mary Ann Zink and Kaye Talbitzer for their technical assistance in completing this project.

This work was supported by NIH PO1-HL062222-17.

References

Andrade DC, Lucero C, Toledo C, Madrid C, Marcus NJ, Schultz HD, Del Rio R (2015) Relevance of the carotid body Chemoreflex in the progression of heart failure. Biomed Res Int 2015:467597. https://doi.org/10.1155/2015/467597

Ayala DE, Moya A, Crespo JJ, Castineira C, Dominguez-Sardina M, Gomara S, Sineiro E, Mojon A, Fontao MJ, Hermida RC, Hygia Project I (2013) Circadian pattern of ambulatory blood pressure in hypertensive patients with and without type 2 diabetes. Chronobiol Int 30(1–2):99–115. https://doi.org/10.3109/07420528.2012.701489

Chen Y, Tjong YW, Ip SF, Tipoe GL, Fung ML (2005) Melatonin enhances the hypoxic response of rat carotid body chemoreceptor. J Pineal Res 38(3):157–163. https://doi.org/10.1111/j.1600-079X.2004.00187.x

Del Rio R, Marcus NJ, Schultz HD (2013) Inhibition of hydrogen sulfide restores normal breathing stability and improves autonomic control during experimental heart failure. J Appl Physiol 114(9):1141–1150. https://doi.org/10.1152/japplphysiol.01503.2012

Durgan DJ, Young ME (2010) The cardiomyocyte circadian clock: emerging roles in health and disease. Circ Res 106(4):647–658. https://doi.org/10.1161/circresaha.109.209957

Floras JS, Jones JV, Johnston JA, Brooks DE, Hassan MO, Sleight P (1978) Arousal and the circadian rhythm of blood pressure. Clin Sci Mol Med 55(Suppl 4):395s–397s

Grivas TB, Savvidou OD (2007) Melatonin the "light of night" in human biology and adolescent idiopathic scoliosis. Scoliosis 2:6. https://doi.org/10.1186/1748-7161-2-6

Komori T, Eguchi K, Saito T, Hoshide S, Kario K (2016) Riser pattern: another determinant of heart failure with preserved ejection fraction. J Clin Hypertens (Greenwich) 18(10):994–999. https://doi.org/10.1111/jch.12818

Lynch HJ, Deng MH, Wurtman RJ (1984) Light intensities required to suppress nocturnal melatonin secretion in albino and pigmented rats. Life Sci 35(8):841–847

Morrissey RP, Czer L, Shah PK (2011) Chronic heart failure. Am J Cardiovasc Drugs 11(3):153–171. https://doi.org/10.2165/11592090-000000000-00000

Mortola JP (2004) Breathing around the clock: an overview of the circadian pattern of respiration. Eur J Appl Physiol 91(2–3):119–129. https://doi.org/10.1007/s00421-003-0978-0

Mortola JP, Seifert EL (2002) Circadian patterns of breathing. Respir Physiol Neurobiol 131(1–2):91–100

Mousa TM, Schiller AM, Zucker IH (2014) Disruption of cardiovascular circadian rhythms in mice post myocardial infarction: relationship with central angiotensin II receptor expression. Physiol Rep 2(11). https://doi.org/10.14814/phy2.12210

Partch CL, Green CB, Takahashi JS (2014) Molecular architecture of the mammalian circadian clock. Trends in Cell Biol 24(2):90–99. https://doi.org/10.1016/j.tcb.2013.07.002

Peever JH, Stephenson R (1997) Day-night differences in the respiratory response to hypercapnia in awake adult rats. Respir Physiol 109(3):241–248

Piepoli MF, Crisafulli A (2014) Pathophysiology of human heart failure: importance of skeletal muscle myopathy and reflexes. Exp Physiol 99(4):609–615. https://doi.org/10.1113/expphysiol.2013.074310

Saiki C, Mortola JP (1995) Hypoxia abolishes the morning-night differences of metabolism and ventilation in 6-day-old rats. Can J Physiol Pharmacol 73(1):159–164

Seifert EL, Mortola JP (2002) Circadian pattern of ventilation during prolonged hypoxia in conscious rats. Respir Physiol Neurobiol 133(1–2):23–34

Tjong YW, Chen Y, Liong EC, Tipoe GL, Fung ML (2006) Chronic hypoxia modulates the function and expression of melatonin receptors in the rat carotid body. J Pineal Res 40(2):125–134. https://doi.org/10.1111/j.1600-079X.2005.00286.x

Tsimakouridze EV, Alibhai FJ, Martino TA (2015) Therapeutic applications of circadian rhythms for the cardiovascular system. Front Pharmacol 6:77. https://doi.org/10.3389/fphar.2015.00077

17 High Fat Feeding in Rats Alters Respiratory Parameters by a Mechanism That Is Unlikely to Be Mediated by Carotid Body Type I Cells

Ryan J. Rakoczy, Richard L. Pye, Tariq H. Fayyad, Joseph M. Santin, Barbara L. Barr, and Christopher N. Wyatt

Abstract

The carotid bodies (CB) respond to changes in blood gases with neurotransmitter release, thereby increasing carotid sinus nerve firing frequency and ultimately correcting the pattern of breathing. It has previously been demonstrated that acute application of the adipokine leptin augments the hypoxic sensory response of the intact *in-vitro* CB (Pye RL, Roy A, Wilson RJ, Wyatt CN. FASEB J 30(1 Supplement):983.1, 2016) and isolated CB type I cell (Pye RL, Dunn EJ, Ricker EM, Jurcsisn JG, Barr BL, Wyatt CN. Arterial chemoreceptors in physiology and pathophysiology. Advances in experimental medicine and biology. Springer, Cham, 2015). This study's aim was to examine, *in-vivo*, if elevated leptin modulated CB function and breathing.

Rats were fed high fat or control chow for 16-weeks. High fat fed (HFF) animals gained significantly more weight compared to control fed (CF) animals and had significantly higher serum leptin levels compared to CF. Utilizing whole-body plethysmography, HFF animals demonstrated significantly depressed breathing compared to CF at rest and during hypoxia. However, amplitudes in the change in breathing from rest to hypoxia were not significantly different between groups. CB type I cells were isolated and intracellular calcium levels recorded. Averaged and peak cellular hypoxic responses were not significantly different.

Despite a small but significant rise in leptin, differences in breathing caused by high fat feeding are unlikely caused by an effect of leptin on CB type I cells. However, the possibility remains that leptin may have *in-vivo* postsynaptic effects on the carotid sinus nerve; this remains to be investigated.

Keywords

Carotid body · Leptin

R. J. Rakoczy · R. L. Pye · T. H. Fayyad · B. L. Barr
C. N. Wyatt (✉)
Department of Neuroscience, Cell Biology, & Physiology, Wright State University, Dayton, OH, USA
e-mail: Christopher.Wyatt@wright.edu

J. M. Santin
Division of Biological Sciences, University of Missouri, Columbia, MO, USA

E. B. Gauda et al. (eds.), *Arterial Chemoreceptors*, Advances in Experimental Medicine and Biology 1071, https://doi.org/10.1007/978-3-319-91137-3_17

17.1 Introduction

Obese individuals suffer from many pathologies, one of particular interest being obesity hypoventilatory syndrome (OHS) (Berger et al. 2009). The adipokine leptin, found in excess in the circulating serum of the obese, has been shown to have profound effects on increasing breathing in wild-type animals and those unable to produce functional leptin proteins (ob^{-}/ob^{-}). For example, focal injections of leptin into central breathing centers (i.e., nucleus tractus solitarius) causes dose-dependent increases in minute ventilation in wild-type rats (Inyushkina et al. 2009). Ob^{-}/ob^{-} mice, normally suffering from hypoventilation, also show significant increases in minute ventilation after exogenous, peripheral infusion of obese levels of leptin (O'Donnell et al. 1999). However, despite high circulating levels of this respiratory stimulant in obese individuals, hypoventilation persists (Campo et al. 2007).

Capable of crossing the blood brain barrier and interacting with central regions to trigger satiety (i.e., hypothalamus) and increase ventilation (i.e., NTS), leptin also has receptors in the periphery through which it may act to modulate breathing. It was recently demonstrated that functional leptin receptors, both long (Ob-Rb) and short isoforms (Ob-Ra – Ob-Rf) exist on the type I cell of the rat and human carotid body (Porzionato et al. 2011). Pye et al. (2016) observed that obese quantities of exogenous leptin (200 ng/mL) administered in acute doses to intact, arterially perfused carotid bodies, increased their baseline firing rates and sensory responses to hypoxia. This experimental evidence demonstrated leptin's potential ability to modulate breathing via its actions on the carotid body.

It was the aim of the following study to build upon the work completed by Pye et al. in 2015 and 2016. In the present study, we aimed to increase circulating serum leptin levels, *in-vivo*, through a high fat feeding paradigm. High fat feeding has been demonstrated to lead to excessive weight gain and a subsequent rise in serum leptin levels (Van Heek et al. 1997; Knight et al. 2010). We hypothesized that chronic overstimulation of leptin receptors (i.e., due to hyperleptinemia) on carotid body type I cells would cause their downregulation and/or insensitivity to leptin, potentially leading to a blunting of the carotid body mediated hypoxic response. Whole-body plethysmography was used to test breathing parameters and measure the acute hypoxic ventilatory response (HVR). Intracellular calcium measurement, in isolated type I cells, was used to indicate whether any depression in breathing was due to leptin's actions on these oxygen-sensing organs. We hypothesized that HFF animals would have blunted hypoxic responses due to a decrease in the hypoxic sensory responses of the type I cells of the carotid body.

17.2 Methods

Thirty-six adult, male, Sprague-Dawley rats (Envigo) were randomly assigned to either a control chow (CF; n = 18; TD.06414, Harlan) or an ingredient-matched high-fat chow (HFF; n = 18; TD.08806, Harlan) for 16 weeks *ad libitum*; animals weight was measured weekly.

17.2.1 Ethics

All procedures and experiments were conducted in accordance with local, state, and federal guidelines relating to animal welfare and treatment. Wright State University's Institutional Animal Care and Use Committee approved experimental protocols described herein. Authors declare no conflict of interest.

17.2.2 Metabolism

Resting metabolism was assessed after 16 weeks of feeding using an indirect calorimetric method as previously described by Frappel et al. (1992). A respiratory exchange ratio [RER: VCO_2 (mL of CO_2 min^{-1} kg^{-1})/VO_2 (mL of O_2 min^{-1} kg^{-1})] was then calculated for each animal.

17.2.3 Whole-Body Plethysmography

Breathing parameters (tidal volume, frequency) were measured using whole-body plethysmography (DSI Finepointe) as previously described by Nirogi et al. (2012). Briefly, acclimated animals had baseline breathing recordings collected in room air before exposure to an acute hypoxic stimulus (10% O_2).

17.2.4 Carotid Body Dissection & Type I Cell Isolation

After 16-weeks of feeding, unfasted, adult Sprague-Dawley rats were deeply anesthetized (isoflurane: 4.5% in O_2) before euthanasia via decapitation. Carotid bodies were then dissected and type I cells isolated as previously described (Burlon et al. 2009).

17.2.5 Blood Collection & Leptin Assay

Immediately post-decapitation, whole blood was collected passively by draining of the euthanized animal's trunk. Blood was centrifuged and serum aliquoted and frozen −15 °C (<4 months) before being thawed, diluted both 5- and 20-fold, and used for quantification of leptin via ELISA (KRC2281; Novex). Samples were diluted and tested (Syngery H1; BioTek) in duplicates while following manufacturers protocol, equating to four tests per sample of serum.

17.2.6 Carotid Body Type I Cell Ca^{2+}-Imaging

A ratiometric calcium-sensitive fluorescent dye, FURA-2AM (Invitrogen) was used to visualize and quantify intracellular calcium in isolated carotid body type I cells as previously described (Pye et al. 2015).

17.2.7 Statistics

Two-tailed Student's unpaired t-tests were used to analyze data with significance set at $p < 0.05$. Data are presented as means ± the standard error of the mean. One star: $p < .05$; two stars: $p < .01$; three stars: $p < .001$; four stars: $p < .0001$.

17.3 Results

17.3.1 Effects of Diet on Physiology

Thirty-six male Sprague-Dawley rats, separated into two equal groups (n = 18), were fed either a high fat or control chow for 16 weeks. After 16 weeks, HFF animals weighed significantly more than CF (512.6 ± 14.70 vs. 444.1 ± 7.090 g, $p < .001$, Fig. 17.1a). Respiratory exchange ratios (RER) at rest were calculated using an indirect

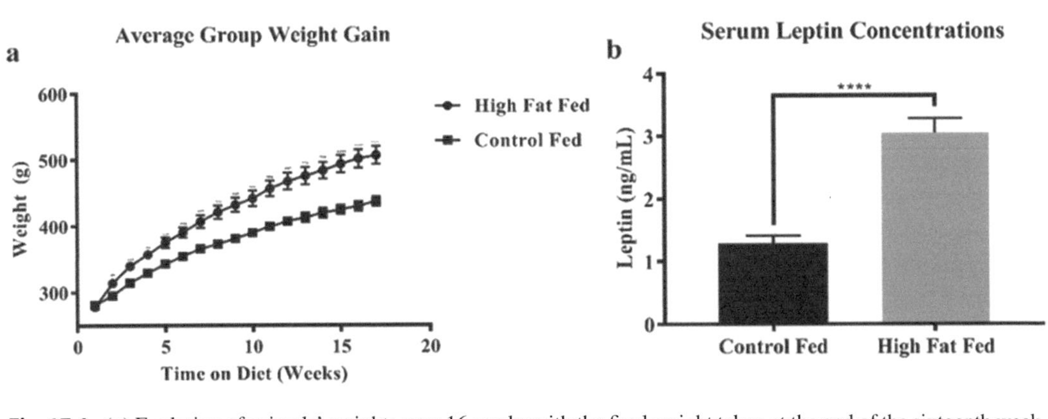

Fig. 17.1 (**a**) Evolution of animals' weights over 16 weeks with the final weight taken at the end of the sixteenth week. (**b**) Serum leptin concentrations were significantly higher in the HFF group compared to the CF group (n = 18/n = 18)

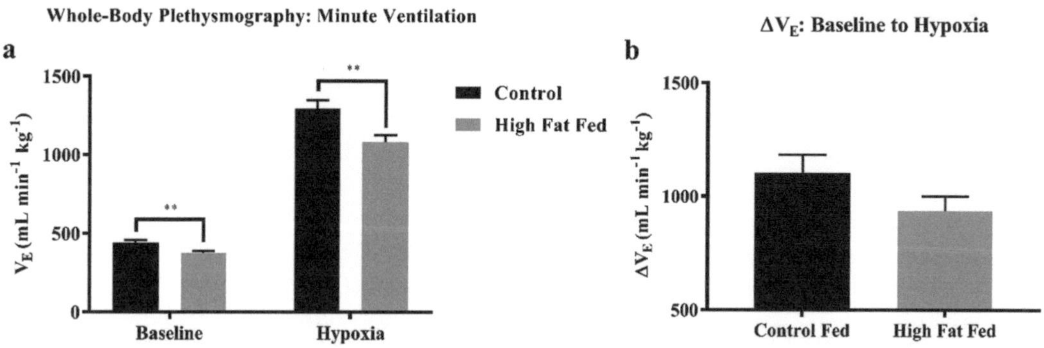

Fig. 17.2 (**a**) Minute ventilation (V_E) during rest or exposure to a 10% O_2 hypoxic stimulus. (**b**) Assessment of oxygen-sensitivity by analyzing the amplitude of the change in breathing (ΔV_E) from baseline to hypoxia

calorimetric method. After 16 weeks on the diet HFF animals' metabolic fuel shifted toward the use of fatty acids (0.77 ± 0.018 vs. 0.88 ± 0.037 RER, $p < .01$); a function of lowered CO_2-production in HFF animals (10.02 ± 0.1609 vs. 12.15 ± 0.418 mL CO_2 min^{-1} kg^{-1}, $p < .0001$). Oxygen consumption between groups was not significantly different.

After the 16th week on the diet regimen, animals were sacrificed, blood collected and processed, before being used for quantification of endogenous leptin levels via ELISA. Standards ran in duplicate were significantly correlated (r^2: 0.997, $p < .05$). Sixteen weeks of HFF resulted in significantly higher leptin levels in HFF animals compared to CF (3.05 ± 0.24 vs. 1.29 ± 0.12 ng/mL, $p < 0.0001$, Fig. 17.1b).

17.3.2 Effects of Diet on Breathing

Whole-body plethysmography was used to measure breathing parameters in unrestrained, conscious animals during rest and acute exposure to hypoxia. HFF animals demonstrated significantly depressed baseline breathing compared to CF (373.9 ± 13.46 vs. 442.8 ± 15.29 mL min^{-1} kg^{-1}, $p < .01$). HFF animals also exhibited attenuated breathing during exposure to hypoxia (1084.1 ± 46.25 vs. 1298.2 ± 52.61 mL min^{-1} kg^{-1}, $p < .01$, Fig. 17.2a). Attenuated minute ventilation was a function of lowered tidal volume, not frequency, among HFF animals compared to CF during rest (6.412 ± 0.212 vs. 8.163 ± 0.263 mL breath^{-1} kg^{-1}, $p < .0001$) and hypoxia (8.587 ± 0.287 vs. 10.63 ± 0.354 mL breath^{-1} kg^{-1}, $p < .0001$).

Assessment of oxygen sensitivity at the level of the whole animal was accomplished by subtracting the V_E values during hypoxia from those taken during baseline. This simple transformation allows for comparison of the absolute change in breathing between groups. Oxygen sensitivity at the level of the whole animal remained unchanged between CF and HFF groups, data shown in Fig. 17.2b.

17.3.3 Effects of Diet on CB Type I Cell O_2-Sensing

A calcium-sensitive dye, FURA-2, was used to measure intracellular calcium levels in isolated CB type I cells from HFF and CF animals. Analyzing both the peak cellular responses (Fig. 17.3c) and the averaged cellular responses (Fig. 17.3d) during hypoxic exposure shows nearly identical responses from cells of HFF (n = 16) and CF (n = 16) animals. Cellular data is time-matched to the whole-animal breathing data; both analyses concur in suggesting O_2-sensitivity is not changed by a high fat diet at the level of the whole-animal and CB type I cell.

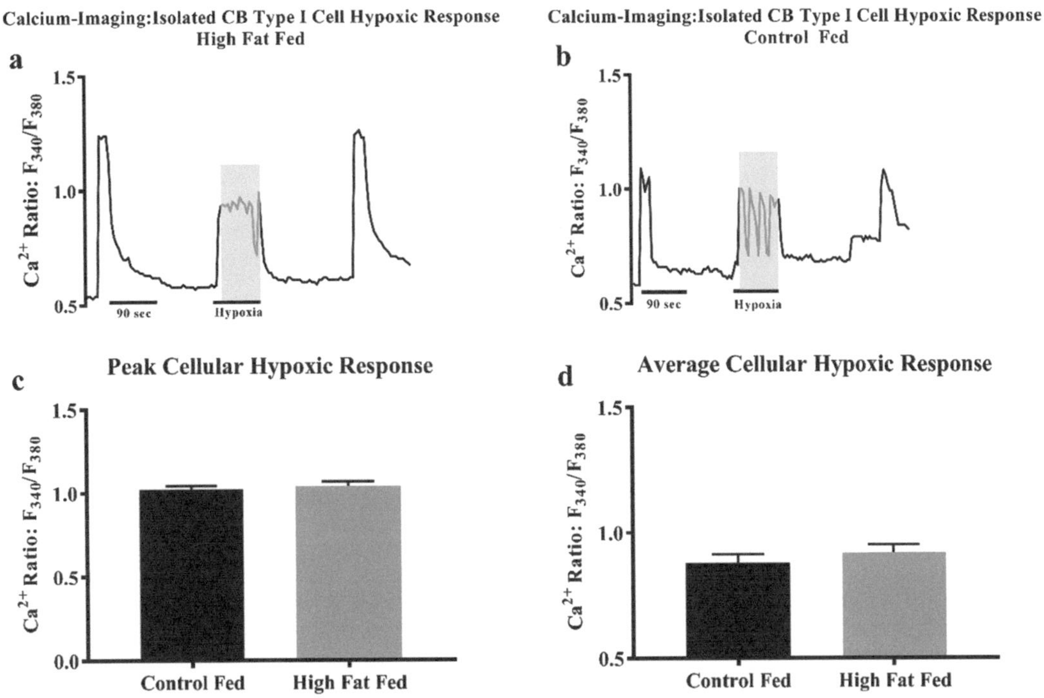

Fig. 17.3 Example calcium-imaging recordings are shown in (**a**) and (**b**). High fat feeding over 16-weeks did not change oxygen-sensitivity of isolated CB type I cells (n = 16/n = 16) when analyzed as either peak cellular responses (**c**) or averaged cellular responses (**d**) (70-second data bin: gray box in **a** and **b**) during hypoxia

17.4 Conclusion

These experiments demonstrated that a high fat diet in rats led to significant physiological changes including weight gain, hyperleptinemia, lowered carbohydrate metabolism (decreased RER), and depression of breathing during baseline and hypoxic exposure. Although respiratory depression is evident, the mechanism underpinning this attenuation does not appear to be mediated by the CB type I cell. Similar hypoxic responses (ΔV_E) between groups (see Fig. 17.2b), during the CB-mediated phase of the biphasic hypoxic response (≤90 s; Cummings and Wilson 2005), provide further evidence for lack of involvement of the carotid bodies in the development of respiratory depression.

Future experiments would aim to further increase the animal weight on HFF, and thus leptin levels, through addition of simple sugars to drinking water, lengthening the time on the diet, and/or supplementing the chow with ingredient-matched treats. Leptin levels in the experiments by Pye et al.'s (2015, 2016) were nearly 65-fold larger than measured in our HFF animals and likely explain why differences were not seen at the CB type I cellular level.

Acknowledgements We would like to thank Dr. Lynn Hartzler, Lori and Dale Goss, and the Wright State University LAR staff and veterinarians for their support with this project. This work was funded in part by NIH grant RO1HL091836.

References

Berger KI, Goldring RM, Rapoport DM (2009) Obesity hypoventilation syndrome. Semin Respir Crit Care Med 30(3):253–261

Burlon DC, Jordan HL, Wyatt CN (2009) Presynaptic regulation of isolated neonatal rat carotid body type I cells by histamine. Respir Physiol Neurobiol 168(3):218–223

Campo A, Frühbeck G, Zulueta JJ, Iriarte J, Seijo LM, Alcaide AB, Salvador J (2007) Hyperleptinaemia, respiratory drive and hypercapnic response in obese patients. Eur Respir J 30(2):223–231

Cummings KJ, Wilson RJ (2005) Time-dependent modulation of carotid body afferent activity during and after intermittent hypoxia. Am J Physiol Regul Integr Comp Physiol 288(6):1571–1580

Frappell PB, Dotta A, Mortola JP (1992) Metabolism during normoxia, hyperoxia, and recovery in newborn rats. Can J Physiol Pharmacol 70(3):408–411

Inyushkin AN, Inyushkina EM, Merkulova NA (2009) Respiratory responses to microinjections of leptin into the solitary tract nucleus. Neurosci Behav Physiol 39:231–240

Knight ZA, Hannan KS, Greenberg ML, Friedman JM (2010) Hyperleptinemia is required for the development of leptin resistance. PLoS One 5(6):1–8

Nirogi R, Shanmuganathan D, Jayarajan P, Abraham R, Kancharla B (2012) Comparison of whole body and head out plethysmography using respiratory stimulant and depressant in conscious rats. J Pharmacol Toxicol Methods 65(1):37–43

O'Donnell CP, Schaub CD, Haines AS, Berkowitz DE, Tankersley CG, Schwartz AR, Smith PL (1999) Leptin prevents respiratory depression in obesity. Am J Respir Crit Care Med 159(5):1477–1484

Porzionato A, Rucinski M, Macchi V, Stecco C, Castagliuolo I, Malendowicz LK, De Caro R (2011) Expression of leptin and leptin receptor isoforms in the rat and human carotid body. Brain Res 1385:56–67

Pye RL, Dunn EJ, Ricker EM, Jurcsisn JG, Barr BL, Wyatt CN (2015) Acutely administered leptin increases $[Ca^{2+}]_i$ and BK_{Ca} currents but does not alter chemosensory behavior in rat carotid body Type I cells. In: Peers C, Kumar P, Wyatt C, Gauda E, Nurse C, Prabhakar N (eds) Arterial chemoreceptors in physiology and pathophysiology. Advances in experimental medicine and biology, vol 860. Springer, Cham, pp 61–67

Pye RL, Roy A, Wilson RJ, Wyatt CN (2016) The effect of obese levels of leptin on peripheral chemoreception. FASEB J 30(1 Supplement):983.1

Van Heek M, Compton DS, France CF, Tedesco RP, Fawzi AB, Graziano MP, Davis HR Jr (1997) Diet-induced obese mice develop peripheral, but not central, resistance to leptin. J Clin Investig 99(3):385–390

18 Leptin in the Commissural Nucleus Tractus Solitarii Increases the Glucose Responses to Carotid Chemoreceptors Activation by Cyanide

Mónica Lemus, Cynthia Mojarro, Sergio Montero, Valery Melnikov, Mario Ramírez-Flores, and Elena Roces de Álvarez-Buylla

Abstract

Leptin is a protein hormone that plays a key role in the regulation of energy balance and glucose homeostasis. Leptin and all leptin receptor isoforms are present in the carotid bodies, but its precise function in glucose regulation and metabolism is not yet known. The aim of this study was to determine whether exogenous leptin, microinjected into the commissural nucleus tractus solitarii (cNTS), preceding sodium cyanide (NaCN) injection into the circulatory isolated carotid sinus (ICS), *in vivo,* modifies hyperglycemic reflex (HR) and brain glucose retention (BGR). In anesthetized Wistar rats (sodium pentobarbital, i.p. 3.3 mg/100 g/saline, Pfizer, Mex), arterial and venous blood samples were collected from silastic catheters implanted in the abdominal aorta and jugular sinus. Exogenous leptin (50 ng/20 nL of aCSF) or leptin vehicle (20 nL of aCSF) microinjected (stereotaxically) into the cNTS 4 min before NaCN (5 μg/100 g/50 μL saline into ICS) (experimental 1 [E1] and control 1[C1] groups, respectively) significantly increased HR and BGR compared with their basal values, but the increase was bigger in the E1 group. When leptin or aCSF were injected into the cNTS before saline (E2 and C2 groups, respectively) glucose responses did not vary when compared with their basal levels. Leptin and its receptors in the cNTS cells probably contribute to their sensitization during hypoxia.

Keywords

Brain glucose retention · Carotid body chemoreceptors · cNTS- hyperglycemic reflex · Leptin

Contribution Statement: Sergio Montero and Elena Roces de Álvarez-Buylla have equally contributed to this chapter.

M. Lemus · C. Mojarro
E. R. de Álvarez-Buylla (✉)
Centro Universitario de Investigaciones Biomédicas, Universidad de Colima, Colima, Mexico
e-mail: rab@ucol.mx

S. Montero (✉)
Centro Universitario de Investigaciones Biomédicas, Universidad de Colima, Colima, Mexico

Facultad de Medicina, Universidad de Colima, Colima, Mexico
e-mail: checo@ucol.mx

V. Melnikov · M. Ramírez-Flores
Facultad de Medicina, Universidad de Colima, Colima, Mexico

E. B. Gauda et al. (eds.), *Arterial Chemoreceptors*, Advances in Experimental Medicine and Biology 1071, https://doi.org/10.1007/978-3-319-91137-3_18

18.1 Introduction

It has been hypothesized that chronic exposure of different organs (adipose tissue, liver, pancreas and skeletal muscle) to high levels of leptin causes insulin resistance and hyperglycemia (Denroche et al. 2012), however the precise mechanism involved in the above effect is still not clear (Wjidan et al. 2015). Since the central nervous system (CNS) plays a critical role in mediating the majority of leptin actions on energy homeostasis, we and others believe that the glucose-increasing effects of this hormone are mediated by the CNS (Morton and Schwartz 2011). In the CNS, leptin inhibits glucose production and increases tissue glucose utilization in diabetic animals (Elmquist et al. 1999), besides, leptin could be an effective treatment in lipoatrophic diabetic mice (Gavrilova et al. 2000). Recent studies have shown that Ob-Rb leptin receptors expressed, predominantly, in the commissural nucleus tractus solitarii (cNTS) (Ciriello and Moreau 2013), could be essential in the hyperglycemic reflex (HR) and brain glucose retention (BGR) establishment induced by the local application of sodium cyanide (NaCN) to the carotid body chemoreceptors (CBch) (Alvarez-Buylla and de Alvarez-Buylla 1988; Montero et al. 2000). We showed that exogenous leptin, microinjected into the cNTS before a NaCN infusion in the circulatory isolated carotid sinus (ICS) *in vivo* increased HR with BGR, and further establish that the cNTS has an ability to participate in glucose homeostasis due to leptin sensitization during hypoxia.

18.2 Materials and Methods

18.2.1 Ethics statement

All procedures were carried out in accordance with the United States National Institutes of Health Guide for the Care and Use of Laboratory Animals, and the Bioethics and Biosecurity Committee of the University of Colima (No. 2013–02).

18.2.2 Animals and General Surgery

Experiments were conducted on Wistar rats (280–300 g), housed individually under a 12:12-h light-dark cycle at 22–23 °C. The rats were fed a laboratory rodent diet with water *ad libitum*, but food was removed 12 h before surgery. Silastic catheters filled with heparin (1000 IU/mL, Inhepar, Pisa, Mexico) were inserted into the carotid artery, abdominal aorta, and into the jugular sinus without interrupting their circulation as previously described (Alvarez-Buylla and de Alvarez-Buylla 1988). Briefly, the rats were anesthetized with sodium pentobarbital (3.3 mg/100 g/saline i.p, Pfizer, México), and anesthesia was maintained by continuous i.p. infusion of sodium pentobarbital (0.063 mg/100 g/min/saline). Buprenorphine analgesia (0.03 mg/kg, Temgesic, Schering-Plough, Mexico) was provided before the surgical procedures. The correct placement of the catheters was verified at the end of the experiments. The rats were artificially ventilated; respiratory rate and tidal volume were adjusted to maintain the pO_2, pCO_2 and pH homeostatic values.

18.2.3 Drugs

Buprenorphine (Schering Plough, Temgesic, México, 0.005 mg/100 g i.m.) (Cowan et al. 1977). Cerebrospinal fluid, freshly prepared (aCSF, NaCl 145 mM, KCl 2.7 mM, $MgCl_2$ 1.0 mM, $CaCl_2$ 1.2 mM, ascorbic acid $C_6H_8O_6$ 0.2 mM, NaH_2PO_4 2.0 mM, in 100 mL tri-distilled water with a pH of 7.3–7.4) (20 nL) (McNay and Sherwin 2004). Leptin (Sigma, México) (50 ng/20 nL of aCSF). Saline (Pisa, México) (0.9% 50 nL) (Alvarez-Buylla and de Alvarez-Buylla 1988). Sodium cyanide (NaCN) (Fluka Biochemika, Sigma, Mexico – 5 μg/100 g/50 nL saline) in the ICS (Alvarez-Buylla and de Alvarez-Buylla 1988). Sodium pentobarbital (Pfizer, Méx.) (3 mg/100 g saline) (Alvarez-Buylla and de Alvarez-Buylla1988).

18.2.4 Experimental Protocol

Thirty two rats were used. The animals were divided at random in 4 groups: (a) experimental 1, infusion of leptin in the cNTS and NaCN in the ICS (n = 8); (b) experimental 2, infusion of leptin in the cNTS and saline in the ICS (n = 8); c) control 1, infusion of the vehicle of leptin (aCSF) in the cNTS and NaCN in the ICS (n = 8); d) control 2, infusion of aCSF in the cNTS and saline in the ICS (n = 8).

18.2.5 Measurements

Glucose concentration in blood was determined with a digital glucometer (ACCU -CHEK Performa, Roche, Mexico) in μmol/mL. Arterial blood (5 μL) was collected via the catheter inserted into the abdominal aorta. Venous blood (5 μL) was collected from the jugular sinus, via de jugular catheter. In each rat, two arterial and two venous basal samples were taken, with a 5 min interval; the values were then averaged to obtain a basal level of glucose concentration. After the injection of NaCN or saline into the ICS (t = 0), four experimental samples were then collected at t = 5 min, t = 10 min, t = 20 min and t = 30 min. The blood volume taken from the cannulated vessels amounted ~ 0.3% of the total circulating volume. In order to compensate the volume of blood, after each pair of samples taken, 10 μL of saline was injected i.v. Carotid blood flow was monitored with an ultrasonic pulsed doppler flowmeter (Bioengineering, Univ. Iowa), placed around the right common carotid artery without altering blood circulation, and was recorded in mL/g/min (Alvarez-Buylla et al. 2003; Lemus et al. 2011). Brain glucose retention was calculated in μmoL/g/min multiplying the arterio-venous glucose differences by the carotid blood flow. Arterio-venous glucose difference was determined as mentioned before (Alvarez-Buylla and de Alvarez-Buylla1988) (Fig. 18.1).

18.2.6 Microinjections into the cNTS

After placing the rats' head in the stereotaxic (Stoelting stereotaxic frame, Wood Dale, IL USA), and adjusting the incisor bar to get a flat skull position (Paxinos and Watson 1986), occipital craniotomy was performed with a 1/32-inch burr to expose the brain surface after cutting the dura membrane. Access to cNTS was done with a glass micropipette (40–50 μm tip diameter) connected to a Hamilton microsyringe (0.1–0.5 μL) with silastic tubing (Alvarez-Buylla and de Alvarez-Buylla 1988). The pipette was lowered into the cNTS, using lambda as midpoint region (the lambda coordinates were: AP = −5.1 mm, L = −0.1 mm, V = 8.1 mm). The injected volume was measured by direct observation of the fluid meniscus in the micropipette. After fitting the catheters, the rats were stabilized for 30 min in order to minimize surgery effects on the experimental results. At the end of each experiment, the injection site was marked with methylene blue (50 nL) injected by the same pipette (Lemus et al. 2011). At the end of the experiments, the anesthetized rats were decapitated and the brains kept at −70 °C (Ultra freezer Thermo Scientific Revco, Walthan, MA. USA). The brainstems were cut in coronal sections (40 μm) in a cryomicrotome (Leica CM-1800) after 0.5% cresyl violet staining (Lemus et al. 2011). Dehydrated sections were cover slipped with Entellan (Merck, Darmstadt, Germany). The slides were analyzed under an Axio Imager bright-field microscope (Axio Imager, Carl Zeiss, Munich Germany) equipped with a digital camera using AxioVision (version 4.8, 2009) software (Carl Zeiss, Munich, Germany).

18.2.7 Carotid Body Chemoreceptor Stimulation

Was performed by slowly injecting 5 μg/kg of NaCN in 0.25 mL saline through a 27-gauge needle and a thin catheter (PE-10; Clay Adams)

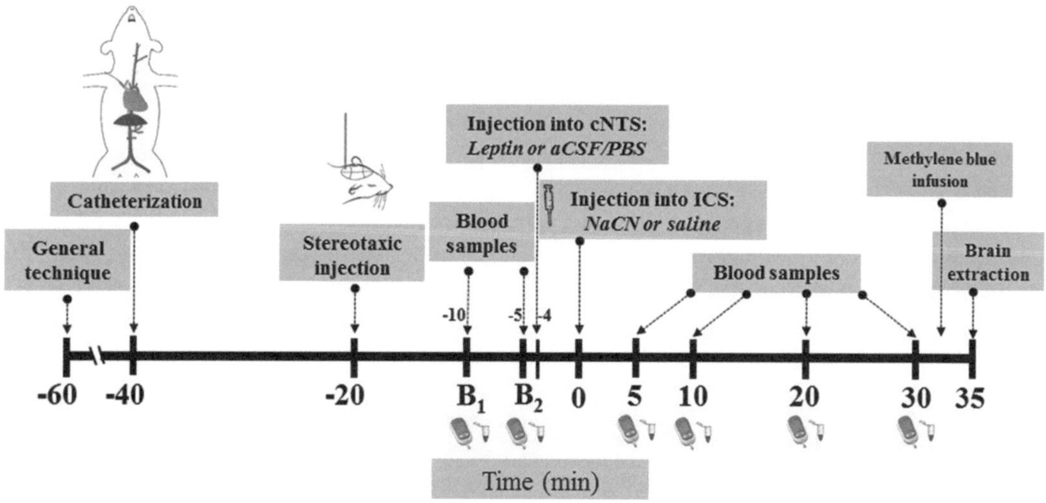

Fig. 18.1 Experimental design. *aCSF* artificial cerebrospinal fluid, *cNTS* commissural nucleus tractus solitarii, *ICS* isolated carotid sinus, *NaCN* sodium cyanide

into the left ICS, to avoid baroreceptor stimulation. The technique to isolate de left CS from general circulation is described in detail in a previous paper (Alvarez-Buylla and de Alvarez-Buylla 1988). The brief interruption of carotid circulation during NaCN stimulation does not produce cerebral ischemia (Ueki et al. 1988). With this technique, only the left ICS was exposed to NaCN, which was cleared out into a washing catheter within 15–16 s.

18.2.8 Data Analysis

The data are means ± S.E.M. Basal arterial blood glucose values in our anesthetized animals at t = 0 min were stable and ranged between 0.42 and 0.47 μmol/mL. The comparisons between groups were made with analysis of variance (ANOVA) and t-test. Paired Student t-test compares basal *vs* experimental data in the same animal.

18.3 Results

When leptin was infused into the cNTS 4 min before NaCN injection or saline in the ICS (E1 and E2 experimental groups, respectively, n = 8), a significant increase in HR ($p < 0.05$) and BGR were observed at all the times studied only in E1 group, when compared to rats in control 1 group that only received aCSF, the vehicle to leptin (n = 8); the HR rose from 5.2 ± 0.05 to 9.6 ± 0.13 μmol/mL ($p < 0.05$), reaching a BGR rise up to 1.8 ± 0.09 μmol/g/min) ($p < 0.05$). By the contrary, the rats in experimental 2 group (n = 8) that received saline in the ICS after leptin in the cNTS, did not change the HR nor the BGR (Fig. 18.2).

In C1 and C2 control groups, that received aCSF instead of leptin in the cNTS, arterial glucose concentration only rose when NaCN was injected into the ICS, and the levels increased from 5.2 ± 0.05 to 7.8 ± 0.30 μmol/mL at 20 min post injection ($p < 0.05$) (Fig. 18.2). With respect to BGR, only control 1 group, with NaCN into the ICS, showed significant increases in all the times studied with a maximum rise at 10 min post injection, from 0.32 ± 0.08 to 1.3 ± 0.09 μmol/g/min ($p < 0.05$) (Fig. 18.2).

When saline infusion was made in the ICS after aCSF in the cNTS (control 2 group, n = 8), no changes were observed, neither in the HR nor in the BGR (Fig. 18.2). Importantly, leptin vehicle (aCSF) in the cNTS induced a small increase in BGR between t = 15 and t = 30 min, and this increase was significantly smaller than

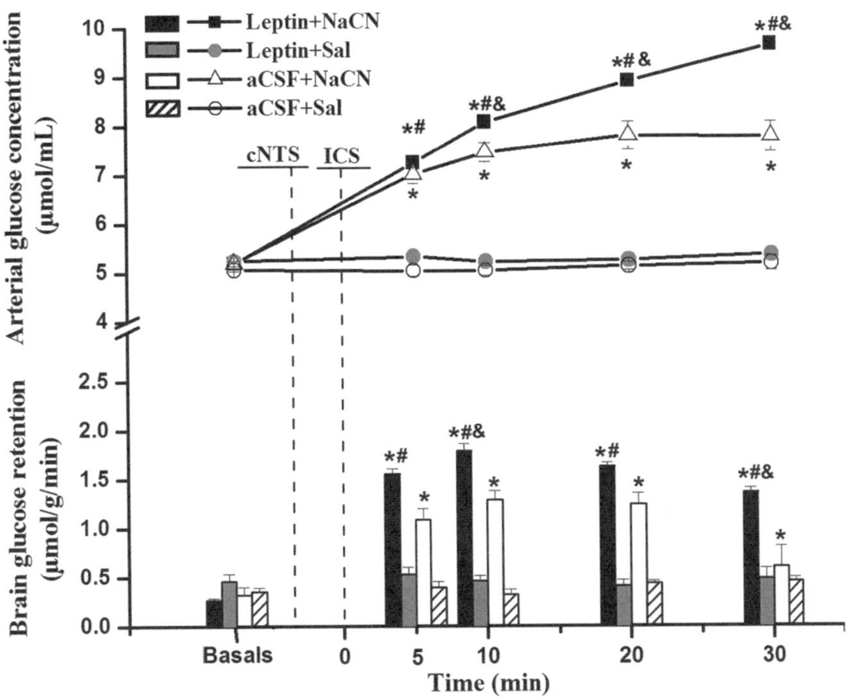

Fig. 18.2 Hyperglycemic reflex and brain glucose retention after leptin infusion or aCSF in the cNTS, applied 4 min before the carotid body chemoreceptors stimulation with cyanide or control experiments with saline. Commissural nucleus tratus solitarius (cNTS); isolated carotid sinus (ICS). *Significantly different from its basal value; #significantly different from experimental 2 group (leptin + Sal); & significantly different from control 1 group (aCSF + NaCN). Values are means ± SEM, $p < 0.05$

the one obtained with NaCN. It is interesting to note that HR and BGR only rose when the CBR received the NaCN stimulus.

18.4 Discussion

This study provided the first evidence of cNTS leptin potentiating effect on glucose responses, induced by the activation of the carotid body chemoreceptor reflex (CBch) by NaCN. Leptin infusion into the cNTS before CBch by anoxic stimulation (NaCN) in the ICS, significantly increased HR and BGR when compared with control rats, in which CBch were stimulated without a central leptin injection. In the same way, when mean BGR levels were compared with their own basal, an increase was also observed between both groups, beginning at 5 min after the NaCN injection into the ICS. These results indicated a potentiating effect of leptin on glucose regulation. Discrete microinjections of leptin into the caudal NTS pressor sites elicites an increase in cardiovascular variables due to both, parasympathetic inhibition and sympathetic excitation (Ciriello and Caverson 2014). Although we did not measure autonomic activity in our study, the changes observed in HR and BGR could be due to an increased sympathetic activity following CBch activation by NaCN (Alvarez-Buylla et al. 1997). The effects of glycemic potentiation after leptin, analyzed here, are probably due to a decrease in insulin secretion (García-Jiménez et al. 2015) followed by an increase in blood glucose levels (Denroche et al. 2012). Intracerebroventricular administration of leptin promotes hepatic gluconeogenesis (Liu et al. 1998) decreasing insulin levels, through the sympathetic nervous system activation (Park et al. 2010). Therefore, leptin injected into the cNTS seems to potentiate the hyperglycemic response after anoxic stimulation of CBch through the excitation of sympathetic nervous system with insulin secretion inhibition and glucagon secretion activation (Mizuno et al. 1998). Our results suggested that this adipocyte-derived hormone could be an important factor linking obesity, metabolic syndrome, type-2 diabetes, and cardiovascular disease (García-Jiménez et al. 2015).

The lack of leptin effects on glycemic variables in the absence of CBch anoxic stimulation observed in this study could be due to a lack of phosphorylation of STAT 3 or cFos protein expression in POMC neurons in NTS (Huo et al. 2006). Intermittent hypoxia did not alter STAT3 levels within the carotid body, suggesting that STAT3 activation in the carotid bodies of animals exposed to hypoxia is the result of Ob-Rb receptor activation (Messenger et al. 2013). The subsequent reduction in Ob-Rb levels after hypoxia is likely due to the increased exposure of this receptor to leptin, which is known to induce a down-regulation of Ob-Rb (Uotani et al. 1999). In conclusion our results showed that leptin in the cNTS increases both, the HR and the BGR, in response to anoxic stimulus applied to the CBch. This leptin effect was not observed in rats without the anoxic stimulus in the CBch. It will be interesting to investigate further if Ob-Rb antagonist administration in the cNTS reverts leptin effects on HR and BGR observed after CBch anoxic stimulation.

Acknowledgements Our work was supported by Consejo Nacional de Ciencia y Tecnología, México (CB20121-177047).

Conflict of Interest Statement The authors declare no conflict of interest.

References

Alvarez-Buylla R, de Alvarez-Buylla ER (1988) Carotid sinus receptors participate in glucose homeostasis. Respir Physiol 72:347–360

Alvarez-Buylla R, Alvarez-Buylla E, Mendoza H, Montero SA, Alvarez-Buylla A (1997) Pituitary and adrenals are required for hypoglycemic reflex initiated by stimulation of CBR with cyanide. Am J Physiol (Reg Integr Comp Physiol) 272:R92–R399

Alvarez-Buylla R, Huberman A, Montero S, Lemus M, Valles V, de Alvarez-Buylla ER (2003) Induction of brain glucose uptake by a factor secreted into cerebrospinal fluid. Brain Res 994:24–133

Ciriello J, Caverson M (2014) Carotid chemoreceptor afferent projections to leptin receptor containing neurons in nucleus of the solitary tract. Peptides 58:30–35

Ciriello J, Moreau JM (2013) Systemic administration of leptin potentiates the response of neurons in the nucleus of the solitary tract to chemoreceptor activation in the rat. Neurosci 229:887–899

Cowan A, Lewis JW, Macfarlane IR (1977) Agonist and antagonist properties of buprenorphine, a nociceptive agent. Br J Pharmacol 60:537–545

Denroche H, Huynh F, Kieffer T (2012) The role of leptin in glucose homeostasis. J Diabetes Invest 3:115–129

Elmquist JK, Elias CF, Saper CB (1999) From lesions to leptin: hypothalamic control of food intake and body weight. Neuron 22:221–232

García-Jiménez S, Bernal G, Martínez SMF, Monroy NA, Jaimes TC, Meneses AA et al (2015) Serum leptin is associated with metabolic syndrome in obese Mexican subjects. J Clin Lab Anal 29:5–9

Gavrilova O, Marcus-Samuels B, Graham D, Kim JK, Shulman GI, Castle AL et al (2000) Surgical implantation of adipose tissue reverses diabetes in lipoatrophic mice. J Clin Invest 105:271–278

Huo L, Grill HJ, Bjorbaek C (2006) Divergent regulation of propiomelanocortin neurons by leptin in the nucleus of the solitary tract and the arcuate hypothalamic nucleus. Diabetes 55:567–573

Lemus M, Montero S, Leal C, Portilla de Buen E, Luquin S, García J et al (2011) Nitric oxide infused in the solitary tract nucleus blocks brain glucose retention induced by carotid chemoreceptor stimulation. Nitric Oxide 25:385–395

Liu L, Karkanis GB, Hawkins M, Barzilai N, Wang L, Rossetti L (1998) Intracerebroventricular leptin regultes hepatic but not peripheral glucose fluxes. J Biol Chem 273:31160–31167

McNay EC, Sherwin RS (2004) From artificial cerebrospinal fluid (aCSF) to artificial extracellular fluid (aECF): microdialysis perfusate composition effects on in vivo brain ECF glucose measurements. J Neurosci Methods 132:35–43

Messenger SA, Moreau JM, Ciriello J (2013) Effect of chronic intermittent hypoxia on leptin and leptin receptor protein expression in the carotid body. Brain Res 1513:51–60

Mizuno A, Murakami T, Otani S, Kuwajima M, Shima K (1998) Leptin affects pancreatic endocrine functions through the sympathetic nervous system. Endocrinol 139:3863–3870

Montero S, Yarkov A, Álvarez-Buylla R (2000) Carotid chemoreceptors participation in brain glucose regulation. Adv Exp Med Biol 475:749–760

Morton GJ, Schwartz MW (2011) Leptin and the central nervous system control of glucose metabolism. Physiol Rev 91:389–411

Park S, Ahn IS, Kim da S (2010) Central infusion of leptin improves insulin resistance and suppresses b-cell function, but notb-cell mass, primarily through the sympathetic nervous system in a type 2 diabetic rat model. Life Sci 86:854–862

Paxinos G, Watson C (1986) The rat brain in stereotaxic coordinates, 2nd edn. Academic, NewYork

Ueki M, Linn F, Hossmann K-A (1988) Functional activation of cerebral blood flow and metabolism before and after global ischemia of rat brain. J Cereb Blood Flow Metab 8:486–494

Uotani S, Bjørbaek C, Tornøe J, Flier JS (1999) Functional properties of leptin receptor isoforms: internalization and degradation of leptin and ligand-induced receptor downreg-ulation. Diabetes 48:279–286

Wjidan K, Ibrahim E, Caszo B, Gnanou J, Singh H (2015) Dysregulation of glucose homeostasis following chronic exogenous administration of leptin in healthy Sprague-Dawley rats. JCDR 9:OF06–OF09

19 Chronic Intermittent Hypoxia in Premature Infants: The Link Between Low Fat Stores, Adiponectin Receptor Signaling and Lung Injury

Na-Young Kang, Julijana Ivanovska, Liran Tamir-Hostovsky, Jaques Belik, and Estelle B. Gauda

Abstract

Premature infants have chronic intermittent hypoxia (CIH) that increases morbidity, and the youngest and the smallest premature infants are at the greatest risk. The combination of lung injury from inflammation/oxidative stress causing low functional residual capacity combined with frequent short apneas leads to CIH. Adiponectin (APN) is an adipose-derived adipokine that protects the lung from inflammation and oxidative stress. Premature and small for gestational age (SGA) infants have minimal body fat and low levels of circulating APN. To begin to understand the potential role of APN in lung protection during lung development, we characterized the developmental profile of APN and APN receptors (AdipoR1 and AdipoR2) protein and mRNA expression in the newborn rat lung at fetal day (FD) 19, and postnatal days (PD) 1, 4, 7, 10, 14, 21, and 28. Protein levels in lung homogenates were measured by western blot analyses; relative mRNA expression was detected by quantitative PCR (qPCR); and serum high molecular weight (HMW) APN was measured using enzyme-linked immunosorbent assay (ELISA). Results: APN protein and mRNA levels were lowest at FD19 and PD1, increased 2.2-fold at PD4, decreased at PD10, and then increased again at PD21. AdipoR1 protein and mRNA levels peaked at PD1, followed by a threefold drop by PD4, and remained low until PD21. AdipoR2 protein and mRNA levels also peaked at PD1, but remained high at PD4, followed by a 1.7-fold drop by PD10 that remained low by PD21. Serum APN levels detected by ELISA did not differ from PD4 to PD28. To date, this is the first report characterizing APN and APN receptor protein and mRNA expression in the rat lung during development. The developmental stage of the newborn rat lung models that of the premature human infant; both are in the saccular stage of lung development. In the newborn rat lung, alveolarization begins at PD4, peaks at PD10, and ends at PD21. Importantly, we found that AdipoR1 receptor protein and mRNA expression is lowest during lung alveolarization (PD4 to PD21). Thus, we speculate that low levels of AdipoR1 during lung alveolarization contributes to the increased susceptibility to developing acute

N.-Y. Kang · J. Ivanovska · L. Tamir-Hostovsky
J. Belik
The Hospital for Sick Children, Division of Neonatology, University of Toronto, Toronto, ON, Canada

E. B. Gauda (✉)
Department of Pediatrics, The Hospital for Sick Children, University of Toronto, Toronto, ON, Canada
e-mail: estelle.gauda@sickkids.ca

E. B. Gauda et al. (eds.), *Arterial Chemoreceptors*, Advances in Experimental Medicine and Biology 1071, https://doi.org/10.1007/978-3-319-91137-3_19

lung edema and chronic lung injury such as bronchopulmonary dysplasia (BPD) in premature human infants.

Keywords

Adiponectin · Alveolarization · Bronchopulmonary dysplasia · Developmental profile · Lung development

19.1 Introduction

CIH that occurs in premature infants is the combination of several biological and physiological factors that come together to create the perfect storm (Di Fiore et al. 2013; Gauda and McMaster 2017). These factors include (1) the influence of the carotid body on baseline breathing and (2) low functional residual capacity decreasing oxygen stores. The carotid body contributes 40% to baseline breathing in premature infants, and thus their breathing is characterized by short apneas and periodic breathing. Because of lung injury resulting in low functional residual capacity, rapid oxygen desaturations occurs with even short apneas. Adiponectin (APN) is a pleiotropic adipokine, produced mainly by adipocytes. It is best known for its role in metabolism, promoting insulin-sensitivity and improved glucose uptake. Relevant to our work, APN also has potent anti-inflammatory and antioxidant effects (Sliman et al. 2013). APN mediates its effects via APN receptors, AdipoR1 and AdipoR2. AdipoR1 and AdipoR2 are ubiquitously expressed throughout the body including the lung, with the highest level of AdipoR1 expression in skeletal muscle, while AdipoR2 is predominately expressed in the liver. Premature and small for gestational age (SGA) infants have minimal body fat, and therefore, low levels of circulating APN. Moreover, premature infants have immature lungs and are susceptible to acute lung injury from exposure to inflammation and oxidative stress. We speculate that the low levels of APN may cause or at least contribute to their increased risk for early lung edema and later bronchopulmonary dysplasia (BPD). In adult models, APN protects the lung from inflammatory and oxidative injury (Sliman et al. 2013; Xu et al. 2013). However, the biological role of APN in the lung during early development, and how the lack of endogenous APN contributes to an increased susceptibility to acute and chronic lung injury, has yet to be explored. Thus, we studied the developmental profile of APN and APN receptor (AdipoR1 and AdipoR2) protein and mRNA expression in the newborn rat, in which lung development during the first 2 weeks after birth parallels that of premature human infants born between 23 and 25 weeks of gestation.

19.2 Methods

19.2.1 Animals

Experiments involving animals were conducted in accordance with the Canadian Council on Animal Care Guidelines, and the use and care of animals were in accordance with the Declaration of Helsinki Conventions. Animal studies were approved by the Animal Care Committee of The Hospital for Sick Children. Sprague-Dawley rat pups from fetal day (FD) 19 to postnatal day (PD) 21 were used. At different time-points (FD19, PD1, PD4, PD7, PD10, PD14, PD21, and PD28) 8 male pups were euthanized. Lung and heart tissues were then collected and flash frozen in liquid nitrogen. Blood was also taken, and serum was separated and frozen. A total of 12 litters (n = 8 male pups per litter) were studied for each time point.

19.2.2 Western Blot Analyses

Tissues previously stored at –80 °C were homogenized in RIPA cell lysis buffer (0.1 M $NaPO_4$, 1% (wt/vol) sodium deoxycholate, 1% (vol/vol) Nonidet P-40, 20% SDS, 0.877% (wt/vol) NaCl, 0.074% (wt/vol) EDTA, pH 7.2) containing protease and phosphatase inhibitors. The homogenates were incubated for 20 min on ice and then

Table 19.1 Rat (Rattus Norvegicus) primer sequences for quantitative PCR (qPCR)

Gene (RefSeq accession No.)	Forward 5′–3′	Reverse 5′–3′
Rat adiponectin (NM_144744.3)	GAGACGCAGGTGTTCTTGGT	GGAACATTGGGGACAGTGAC
Rat ADIPOR1 (NM_207587.1)	GTTGTACCCACCATGCACTTT	GGTTGGACACCCCATAGAAGT
Rat ADIPOR2 (NM_001037979.1)	GCCATTATCGTCTCTCAGTGG	GTCACATTTGCCAGGAAAGAA

sonicated for 20 s, vortexed and incubated for 10 min on ice. Samples were then centrifuged at 12,000 RPM for 10 min. Protein concentrations were quantified by Bradford spectrophotometry assay, using known concentrations of bovine serum albumin (BSA) as a standard. Lung tissue lysates containing 50 μg of total protein were boiled for 5 min in SDS sample buffer (0.125 M Tris·HCl, 4% SDS, 20% glycerol, and 10% 2-mercaptoethanol, 0.004% bromphenol blue, pH 6.8). Equal amounts of protein were then loaded onto a gel, subjected to SDS polyacrylamide gel electrophoresis for 1 h at 150 V, and transferred onto polyvinylidene difluoride membranes. The membranes were blocked in TBST (20 mM Tris base, 150 mM NaCl, pH 8.0, with 0.1% Tween 20) for 1 h and then incubated with anti-adiponectin antibody (26 kDa, Novus Biologicals), anti-AdipoR1 antibody (43 kDa, Novus Biologicals), or anti-AdipoR2 antibody (44 kDa, Novus Biologicals) overnight at 4°C. To compensate for differences in protein loading, each membrane was incubated with the anti-beta-actin (β-actin) antibody (42 kDa, Santa Cruz Biotechnologies) for 2 h at room temperature. The membrane was washed and incubated with anti-rabbit IgG secondary antibody conjugated to horseradish peroxidase for 1 h at room temperature. After several washes with TBST, protein bands were visualized using an enhanced chemiluminescence method. Images were digitally captured using a MicroChemi chemiluminescent system, and band intensities were quantified using ImageJ software.

19.2.3 qPCR

RNA was extracted and reverse transcribed from tissue samples stored in RNAlater®. qPCR was performed on a Stratagene MX3000P qPCR system using SYBR Green qPCR Master Mix (Qiagen). Cycling conditions were 95°C for 10 min followed by 40 cycles of 95°C for 15 s and 60°C for 60 s. A standard curve for each primer set was performed using rat lung cDNA standard (Agilent), to ensure that reaction efficiency was equivalent to primers for the housekeeping gene, β-actin. The expression of β-actin was unaffected by the experimental conditions. Primer sequences are listed in Table 19.1. Following each standard curve reaction, molecular weight standards (O-GeneRuler; Fermentas, Burlington, Ontario, Canada) and PCR products underwent DNA gel electrophoresis to confirm the presence of a single product of predicted length. A dissociation (melt) curve and a no-template control were run for each set of samples to exclude nonspecific product formation and reaction contamination. Samples were run in duplicate, and expression of the gene of interest was normalized to β-actin. Change in expression relative to the calibrator samples was calculated by the 2Δct method using Stratagene MxPro software (version 4.01).

19.2.4 Elisa

Rat HMW APN ELISA kits from Amsbio (Cambridge, MA, USA) were used to detect the serum concentrations of HMW APN. Frozen serum samples were thawed and all reagents were allowed to reach room temperature. Samples and HMW APN standards were applied to an antibody-precoated microplate, and then the conjugate was added. After incubating and washing, samples were incubated with substrate and the reaction was stopped using stop solution. Optical density was determined using a microplate reader at 450 nm.

19.2.5 Statistical Analysis

All values are expressed as means ± SE (n = 7–8 per group). Data were subjected to one-way analysis of variance (ANOVA) followed by pairwise multiple comparisons using the Holm-Sidak method with $p < 0.05$ indicating significance.

19.3 Results

19.3.1 Adiponectin Protein Developmental Profile

As shown in Table 19.2, total lung APN protein levels were lowest at FD19 and PD1 which then increased fourfold by PD7, followed by a twofold decrease from PD7 to PD10. Thereafter, APN protein levels increased from PD10 by PD21 and then decreased slightly at PD28.

19.3.2 AdipoR1 Protein Developmental Profile

As shown in Table 19.3, total lung AdipoR1 protein levels peaked at PD1, decreased threefold by PD4, and remained low until PD14. AdipoR1 protein levels then increased at PD14 to levels comparable to levels at PD1. There was a subsequent fall in AdipoR1 at PD21 and levels remained low thereafter.

Table 19.2 Western blot analysis for total lung expression of APN (26 kDa) normalized to β-actin (43 kDa) in the newborn rat at fetal day (FD) 19, and postnatal days (PD) 1, 4, 7, 10, 14, 21 and 28

Age	Mean ± SE
FD19	0.43 ± 0.03
PD1	0.57 ± 0.06
PD4	1.27 ± 0.09*
PD7	1.61 ± 0.07
PD10	0.82 ± 0.09**
PD14	0.94 ± 0.06
PD21	1.25 ± 0.11***
PD28	0.98 ± 0.07

Note. Values represent means ± SE for n = 7–8 samples per group analyzed by one-way ANOVA. *$p < 0.05$ compared to all groups except PD21 and PD28. **$p < 0.05$ compared to FD19, PD4, PD7, and PD21. ***$p < 0.05$ compared to FD19, PD1, PD7, PD10, and PD14

Table 19.3 Western blot analysis for total lung expression of AdipoR1 (43 kDa) normalized to β-actin (43 kDa) in the newborn rat at fetal day (FD) 19, and postnatal days (PD) 1, 4, 7, 10, 14, 21 and 28

Age	Mean ± SE
FD19	0.74 ± 0.10
PD1	1.33 ± 0.31
PD4	0.45 ± 0.07*
PD7	0.73 ± 0.04
PD10	0.57 ± 0.12*
PD14	1.30 ± 0.12
PD21	0.69 ± 0.09*
PD28	0.89 ± 0.05

Note. Values represent means ± SE for n = 7–8 samples per group analyzed by one-way ANOVA. *$p < 0.05$ compared to PD1 and PD14

Table 19.4 Western blot analysis for total lung expression of AdipoR2 (44 kDa) normalized to β-actin (43 kDa) in the newborn rat at fetal day (FD) 19, and postnatal days (PD) 1, 4, 7, 10, 14, 21 and 28

Age.	Mean ± SE
FD19	0.90 ± 0.08
PD1	1.23 ± 0.08
PD4	1.02 ± 0.10*
PD7	0.77 ± 0.04
PD10	0.60 ± 0.06**
PD14	0.96 ± 0.15
PD21	0.61 ± 0.07**
PD28	1.22 ± 0.05

Note. Values represent means ± SE for n = 7–8 samples per group analyzed by one-way ANOVA. *$p < 0.05$ compared to PD10 and PD21. **$p < 0.05$ compared to PD1, PD4, PD14, and PD28

19.3.3 AdipoR2 Protein Developmental Profile

As shown in Table 19.4, total lung AdipoR2 protein levels increased by 1.8-fold from FD19 to PD1 and peaked at PD1. There was a threefold drop from PD4 to PD10, and then an increase at PD14 to levels comparable to levels at PD1. AdipoR2 protein levels subsequently dropped at PD21, and then increased again at PD28 to levels comparable to levels at PD1.

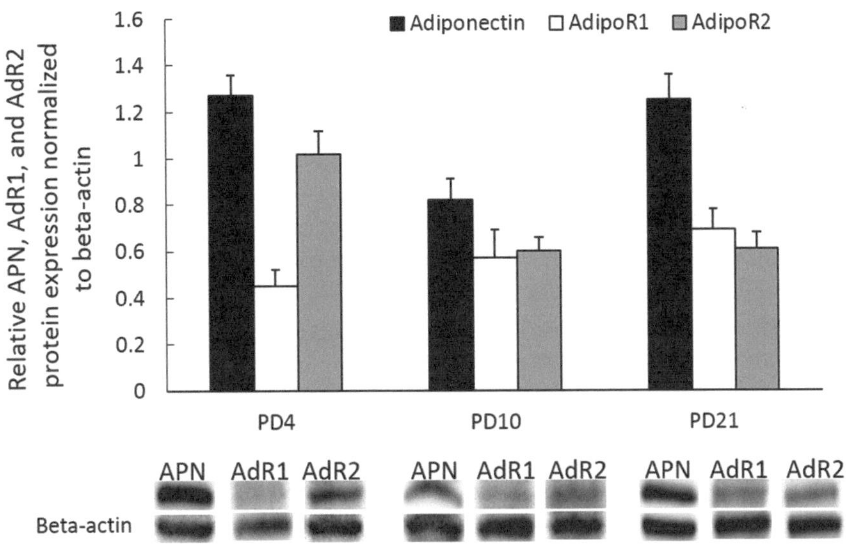

Fig. 19.1 Western blot analysis for total lung expression of adiponectin (26 kDa), AdipoR1 (43 kDa), and AdipoR2 (44 kDa) normalized to β-actin in the newborn rat during alveolarization (postnatal days (PD) 4, 10 and 21). APN is adiponectin, AdR1 and AdR2 are adiponectin receptor 1 and 2, respectively. See Tables 19.1, 19.2 and 19.3 for a more complete expression profile

19.3.4 Adiponectin, AdipoR1, and AdipoR2 Western Blot Analyses During Alveolarization

As shown in Fig. 19.1, APN protein levels are highest at PD4, decreased at PD10, and then at PD21 increased to levels comparable to PD4 levels. AdipoR1 protein levels are lowest at PD4 and increased slightly at PD10 and again at PD21. AdipoR2 protein levels are highest at PD4, subsequently decreased at PD10, and remained low at PD21.

19.3.5 Adiponectin, AdipoR1, and AdipoR2 mRNA Developmental Profile

The pattern of mRNA relative expression for lung APN, AdipoR1, and AdipoR2 mirrored the protein expression (data not shown).

19.4 Discussion

The major finding from this preliminary study is that APN and AdipoR1 and AdipoR2 protein expression in the lung undergo significant changes during the first 3 weeks of development in the newborn rat. Specifically, we found that total lung (1) APN protein expression does not significantly increase until PD4, (2) AdipoR1 receptors remain low throughout the phase of alveolarization (PD4-PD21), while AdipoR2 expression is consistently low from PD7-PD21. At birth, the newborn rat is in the saccular stage of lung development, alveolarization starts at PD4, peaks at PD10, and is complete by PD21. Infants who are born preterm at 23–26 weeks of gestation are in the saccular stage of lung development while alveolarization continues throughout the first week of life.

Infants born prematurely sustain interruptions in lung development and alveolarization and are more susceptible to developing chronic lung disease, such as BPD (Jobe and Ikegami 2000). This increased susceptibility can be attributed to the immaturity of lung development at birth and an imbalance between anti-inflammatory and pro-inflammatory mediators, favoring the latter (reviewed in Speer 2003). As a result of this imbalance, the inflammatory responses that occur in the lung can overwhelm the system and create acute lung injury presenting as lung edema. This initial damage then leads to further injury affecting alveolar development and abnormal development of the pulmonary vasculature leading to BPD (reviewed in Speer 2003). Infants who are SGA from intrauterine growth restriction are at particular risk for developing BPD and have the lowest levels of plasma APN at birth (Hansen-Pupp et al. 2015). APN signaling is complex but appears to mediate lung protection in multiple ways: decreases endothelial leak, decreases inflammation, and decreases oxidative stress. Although the mechanism has not been elucidated, APN signaling also does appear to be involved in normal lung development since APN knockout mice have abnormal alveolarization (Bianco et al. 2013). Zana-Taieb et al. (2015), using postnatal genome-wide analysis, reported that in a newborn rat model of growth restriction alveolarization is impaired, and the peroxisome proliferator-activated receptor (PPAR) pathway is affected. PPARs are ligand-activated nuclear transcription factors that regulate the expression of genes encoding APN, AdipoR1, and AdipoR2 (Zana-Taieb et al. 2015). Moreover, giving PPAR agonists increased the expression of APN, and nebulization of the agonist blocked hyperoxic lung injury in a newborn rat model of BPD (Morales et al. 2014).

We found that total lung APN levels are lowest immediately before birth and during the first week of life in our animal model. APN modulates innate immunity, decreasing inflammation by binding to receptors on macrophages (M1 and M2) and neutrophils. APN suppresses M1 activation which produces pro-inflammatory cytokines, and promotes M2 macrophage proliferation which is involved in tissue repair and produces anti-inflammatory cytokines (Luo and Liu 2016). APN also reduces the production of reactive oxygen species (ROS) from activated neutrophils. Neutrophil infiltration from the blood into the lung produces a significant inflammatory response and contributes to the development of acute and chronic lung injury. Low levels of APN in the lung during the saccular stage of lung development may increase the vulnerability to inflammatory injury causing abnormal alveolarization, which occurs in infants with BPD. Moreover, we detected the lowest level of APN expression at FD19, suggesting that the lungs are at risk during inflammation that may occur prior to birth.

AdipoR1s are expressed on human airway epithelial cells (Miller et al. 2009) and both AdipoR1s and AdipoR2s are expressed on alveolar epithelial cells (A549 cells) (Nigro et al. 2013). However, it appears that AdipoR1 is the primary receptor protecting alveolar epithelial cells from inflammatory injury and apoptosis. More specifically, the AdipoR1 receptor blocked TNF-α-mediated translocation of NF-Kappa β in alveolar epithelial cells and increased the production of IL-10, an anti-inflammatory cytokine (Nigro et al. 2013). We found that AdipoR1 levels were lowest throughout the stage of alveolarization in the rat lung. Less is known about the function of AdipoR2 receptors on cells and tissues other than the liver, although they are present on cells in multiple tissues including the lung. Globular APN activates AdipoR1 while the full length HMW APN binds AdipoR2. From the available literature, AdipoR1 appears to be the most functional receptor on lung cells, including airway and alveolar epithelial cells (Miller et al. 2009), pulmonary artery smooth muscle cells (Weng et al. 2011), and vascular endothelial cells (Chen et al. 2015). We speculate that AdipoR1 receptors are expressed on both airway and alveolar epithelial cells during lung development in our model. The significance of our findings that AdipoR2 are also present in the lung during early development requires additional studies to determine the localization and function of these receptors. Of note, at PD4, the start of alveolarization, AdipoR1 protein expression is lower than AdipoR2.

Interpretation of our data is limited by the fact that we measured gene and protein expression for

APN, AdipoR1, and AdipoR2 in lung homogenates and thus, we are unable to know the specific cells that are expressing these proteins and their function. Neither do we know whether APN might have anti-inflammatory or pro-inflammatory properties since APN can, in fact, have pro-inflammatory effects, by stimulating the expression of cytosolic phospholipase A_2, cyclooxygenase-2, and ROS such as it does in human alveolar type II cells (Chen et al. 2016). Nevertheless, we observed significant changes in total lung APN and APN receptor protein and mRNA expression, especially during alveolarization. Although we are the first to characterize the pattern of expression of this important adipokine and its receptors during early lung development, we know that key experiments will need to be done to determine APN and APN receptor localization and its role in lung protection in a model of lung injury.

In conclusion, the youngest and the smallest infants are at risk for significant lung injury that contributes to the expression of CIH. We know that plasma and serum APN are low in these infants during the first weeks of life. If the developmental profile of the expression of APN, AdipoR1, and AdipoR2 protein in the human infant is similar to what we described for the rat, low levels of APN protein expression during the first 3 days of life and low levels of AdipoR1 receptors throughout alveolarization could increase the risk of developing inflammatory and oxidative lung injury.

References

Bianco A, Mazzarella G, Turchiarelli V, Nigro E, Corbi G, Scudiero O, Sofia M, Daniele A (2013) Adiponectin: an attractive marker for metabolic disorders in chronic obstructive pulmonary disease (COPD). Nutrients 5(10):4115–4125

Chen C, Huang J, Li H, Zhang C, Huang X, Tong G, Xu Y (2015) MicroRNA-221 regulates endothelial nitric oxide production and inflammatory response by targeting adiponectin receptor 1. Gene 565(2):246–251

Chen H, Yang C, Chang J, Wu C, Sia K, Lin W (2016) AdipoR-increased intracellular ROS promoted cPLA2and COX-2 expression via activation of PKC and p300 in adiponectin-stimulated human alveolar type II cells. Am J Physiol Lung Cell Mol Physiol 311(2):L255–L269

Di Fiore JM, Martin RJ, Gauda EB (2013) Apnea of prematurity – perfect storm. Respir Physiol Neurobiol 189(2):213–222

Gauda EB, Master Z (2017) Contribution of relative leptin and adiponectin deficiencies in premature infants to chronic intermittent hypoxia: exploring a new hypothesis. Respir Physio Neurobiol (Epub ahead of print). https://doi.org/10.1016/j.resp.2017.12.003

Hansen-Pupp I, Hellgren G, Hård AL, Smith L, Hellström A, Löfqvist C (2015) Early surge in circulatory adiponectin is associated with improved growth at near term in very preterm infants. J Clin Endocrinol Metab 100(6):2380–2387

Jobe AH, Ikegami M (2000) Lung development and function in preterm infants in the surfactant treatment era. Annu Rev Physiol 62(1):825–846

Luo Y, Liu M (2016) Adiponectin: a versatile player of innate immunity. J Mol Cell Biol 8(2):120–128

Miller M, Cho JY, Pham A, Ramsdell J, Broide DH (2009) Adiponectin and functional adiponectin receptor 1 are expressed by airway epithelial cells in chronic obstructive pulmonary disease. J Immunol 182(1):684–691

Morales E, Sakurai R, Husain S, Paek D, Gong M, Ibe B, Li Y, Husain M, Torday JS, Rehan VK (2014) Nebulized PPARγ agonists: a novel approach to augment neonatal lung maturation and injury repair in rats. Pediatr Res 75(5):631–640

Nigro E, Scudiero O, Sarnataro D, Mazzarella G, Sofia M, Bianco A, Daniele A (2013) Adiponectin affects lung epithelial A549 cell viability counteracting TNFa and IL-1ß toxicity through AdipoR1. Int J Biochem Cell Biol 45(6):1145–1153

Sliman SM, Patel RB, Cruff JP, Kotha SR, Newland CA, Schrader CA, Sherwani SI, Gurney TO, Magalang UJ, Parinandi NL (2013) Adiponectin protects against hyperoxic lung injury and vascular leak. Cell Biochem Biophys 67(2):399–414

Speer CP (2003) Inflammation and bronchopulmonary dysplasia. Semin Neonatol 8(1):29–38

Weng M, Raher MJ, Leyton P, Combs TP, Scherer PE, Bloch KD, Medoff BD (2011) Adiponectin decreases pulmonary arterial remodeling in murine models of pulmonary hypertension. Am J Respir Cell Mol Biol 45(2):340–347

Xu L, Bao H, Si Y, Han L, Zhang R, Cai M, Shen Y (2013) Effects of adiponectin on acute lung injury in cecal ligation and puncture–induced sepsis rats. J Surg Res 183(2):752–759

Zana-Taieb E, Pham H, Franco-Montoya ML, Jacques S, Letourneur F, Baud O, Jarreau PH, Vaiman D (2015) Impaired alveolarization and intra-uterine growth restriction in rats: a postnatal genome-wide analysis. J Pathol 235(3):420–430

Myo-inositol Effects on the Developing Respiratory Neural Control System

20

Peter M. MacFarlane and Juliann M. Di Fiore

Abstract

Myo-inositol is a highly abundant stereoisomer of the inositol family of sugar alcohols and forms the structural basis for a variety of polyphosphate derivatives including second messengers and membrane phospholipids. These derivatives regulate numerous cell processes including gene transcription, membrane excitability, vesicular trafficking, intracellular calcium signaling, and neuronal growth and development. Myo-inositol can be formed endogenously from the breakdown of glucose, is found in a variety of foods including breastmilk and is commercially available as a nutritional supplement. Abnormal myo-inositol metabolism has been shown to underlie the pathophysiology of a variety of clinical conditions including Down Syndrome, traumatic brain injury, bronchopulmonary dysplasia (BPD), and respiratory distress syndrome (RDS). Several animal studies have shown that myo-inositol may play a critical role in development of both the central and peripheral respiratory neural control system; a notable example is the neonatal apnea and respiratory insufficiency that manifests in a mouse model of myo-inositol depletion, an effect that is also postnatally lethal. This review focuses on myo-inositol (and some of its derivatives) and how it may play a role in respiratory neural control; we also discuss clinical evidence demonstrating a link between serum myo-inositol levels and the incidence of intermittent hypoxemia (IH) events (a surrogate measure of apnea of prematurity (AOP)) in preterm infants. Further, there are both animal and human infant studies that have demonstrated respiratory benefits following supplementation with myo-inositol, which highlights the prospects that nutritional requirements are important for appropriate development and maturation of the respiratory system.

Keywords

Myo-inositol · Respiratory neural development · Control of breathing

20.1 Introduction

Myo-inositol is a naturally occurring sugar derivative found in most foods including fruits, beans, and breast milk. Dietary/nutritional habits, therefore, can be an important determinant of biological myo-inositol expression. Endogenous formation of myo-inositol, however, can come from the breakdown of glucose and the conversion

P. M. MacFarlane (✉) J. M. Di Fiore
Case Western Reserve University, Rainbow Babies & Children's Hospital, Cleveland, OH, USA
e-mail: pmm71@case.edu

E. B. Gauda et al. (eds.), *Arterial Chemoreceptors*, Advances in Experimental Medicine and Biology 1071, https://doi.org/10.1007/978-3-319-91137-3_20

of D-glucose-6-phosphate, as well as through recycling of cellular phosphatidylinositol. It is the most abundant form of eight other stereoisomers of inositol which play major roles in a variety of metabolic pathways. Importantly, its phosphorylated derivatives comprise a variety of phosphoinositide stereoisomers, including inositol-1-4-5-trisphosphate (IP_3) and diaglycerol (DAG), which act as second messengers regulating Ca^{2+} signaling and PKC activity, respectively. Further variations in phosphorylation of the hydroxyl group(s) yield several different myo-inositol containing phosphatidylinositol phosphate (PIP) derivatives (Halub 1986).

The multitude of various derivatives play important roles in regulating membrane vesicle fission and fusion, lipid raft formation, structure and function of the cytoskeleton, chemotaxis, osmoregulation, and glucose homeostasis (Ogimoto et al. 2000; Collins et al. 2002; Harlan et al. 1994; De Camilli et al. 1996; Wurmser et al. 1999). Biologically there is a high abundance of myo-inositol in a variety of different tissues throughout the body. Millimolar concentrations of myo-inositol have been detected within brain neurons (Godfrey et al. 1982) and the sixfold higher brain concentration compared to whole body levels at birth highlights a heavy dependence on in utero brain development (Berry et al. 2003). It is not surprising, therefore, that myo-inositol may play a critical role in development of the respiratory control system. There are several types of membrane-bound transporters that regulate intracellular levels of myo-inositol of which the sodium(Na^+)/myo-inositol cotransporter 1 (SMIT1) and 2 (SMIT2) are the most abundant. SMIT1 is expressed in a variety of tissue including kidney, skeletal muscle, pancreas, heart, lung, placenta and brain (Berry et al. 1995) whereas SMIT2 is relatively weakly expressed in the brain (Roll et al. 2002).

Abnormalities in myo-inositol metabolism have been implicated in the pathophysiology of a variety of clinical settings including diabetic neuropathy, Down Syndrome, gestational diabetes, polycystic ovary syndrome, bronchopulmonary dysplasia, and acute respiratory distress syndrome. In the present review, we discuss the limited available literature describing a link between myo-inositol (and its derivatives) and respiratory neural control.

20.2 Myo-inositol and the Respiratory System

There are several examples demonstrating either a direct or indirect role for myo-inositol in regulating development and function of the respiratory neural control system. One example is the SMIT1 knockout mice which are almost completely depleted of myo-inositol (Berry et al. 2003; Chau et al. 2005; Buccafusca et al. 2008); the neonate expires soon after birth arguably from respiratory failure – although there are ambiguities over the cause of death and whether the respiratory insufficiency is due to peripheral or central abnormalities in respiratory control. Myo-inositol could also exhibit an indirect effect on breathing via its derivatives, which have been shown to influence excitability of central (brainstem) and peripheral (carotid body chemoreceptor) respiratory-related neurons. The following section briefly describes some of these findings, and concludes with a clinical example in which there is an association between a postnatal decline in endogenous myo-inositol levels in preterm infants, which parallels an increase in the incidence of intermittent hypoxemia events.

20.2.1 SMIT 1 KO Mice and Myo-inositol

One intriguing example demonstrating how myo-inositol affects the respiratory neural control system is through genetic deletion of one of its co-transporters, SMIT1 (also known as Slc5a3). In this mouse model, SMIT1 null mice (SMIT1$^{-/-}$) exhibit >90% reduction in brain, and to a similar extent peripheral, myo-inositol levels (Berry et al. 2003; Chau et al. 2005; Buccafusca et al. 2008). The embryo survives gestation although the newborn appears to exhibit major respiratory insufficiency, expressed as irregular breathing, hypoventilation and apnea. The neonates die several days after birth which has largely been attributed to respiratory failure, although there are conflicting data about the precise nature of the respiratory effects of myo-inositol and the exact cause of death. Regardless, the following section briefly assesses some of the phenotypic characteristics of the SMIT1 mouse to highlight the profound (and in some cases controversial) effects myo-inositol deficiency has on the respiratory neural control system.

20.2.1.1 Development of the Peripheral Nervous System

There are several versions of SMIT1 KO mice, which are characteristically almost depleted of myo-inositol. There are however, discrepancies over whether it is accompanied by functional and/or morphological abnormalities of the peripheral nervous system and whether such abnormalities contribute to the lethality. Chau et al. (2005) demonstrated gross morphological deficiencies in vagal, phrenic and intercostal nerve development of the fetus compared to WT littermates. The sciatic nerve, which was also poorly developed, had a marked reduction in nerve conduction velocity, which correlates functional abnormalities with morphological ones. Clearly, poorly developed phrenic and intercostal nerves would compromise motor control of breathing, whereas the vagal abnormalities could diminish pulmonary afferent inputs. The possibility that carotid body chemoafferents could also be impaired is of interest from the viewpoint that neonatal mortality has been observed following transection of the carotid sinus nerves (Serra et al. 2001). However, functional abnormalities in the SMIT1 mice have not been investigated (Chau et al. 2005). Further, abnormalities in peripheral nerve morphological were not observed in a different line of SMIT1 KO mice, which questions whether the lethality is the result of abnormal peripheral nerve development (Chau et al. 2005). There is perhaps more convincing evidence of abnormalities in brainstem respiratory control regions.

20.2.1.2 Development of Brainstem Respiratory Control Regions

There are compelling data that respiratory insufficiency (apnea and hypoventilation) of neonatal SMIT1 mice originates from abnormalities in central (e.g., brainstem) respiratory neural control regions. Diaphragm EMG recordings from the brainstem-spinal cord preparation of E18.5 mice revealed an irregular respiratory pattern compared to WT mice, which consisted of prolonged apneas and augmented breaths (Berry et al. 2003; Buccafusca et al. 2008). Further, electrophysiological recordings from neurons of the preBötzinger complex (preBötC) using the brainstem slice preparation (Berry et al. 2003) revealed abnormal inspiratory activity (Fig. 20.1). Collectively, these data suggest that the respiratory abnormalities of the neonate are likely of central origin. Interestingly, however, gross analysis of neurokinin-1 receptor positive neurons (NK1R), which label pre-BötC neurons (and adjacent *nucleus ambiguus*) revealed no apparent morphological abnormalities (Buccafusca et al. 2008). It was concluded, therefore, that the postnatal lethality of myo-inositol deficiency likely results from respiratory failure due to functional abnormalities in central mechanisms of respiratory control (Buccafusca et al. 2008). These data raise the question of whether myo-inositol supplementation could rescue the respiratory insufficiency and prevent the associated lethality.

20.2.1.3 Myo-inositol Supplementation

Given the profound respiratory disturbances of SMIT1 KO mice, one impending question is whether supplementation with myo-inositol confers protection against its genetic deficiency. A notable benefit of myo-inositol supplementation early in utero (beginning earlier than E0.9; Chau et al. 2005) was improved survival (Buccafusca et al. 2008). However, later supplementation (E15.5) appeared to be ineffective, suggesting the window of opportunity for whatever mechanisms of development myo-inositol might be necessary for has passed. Indeed, gestation appears to be a critical period for peripheral nerve development

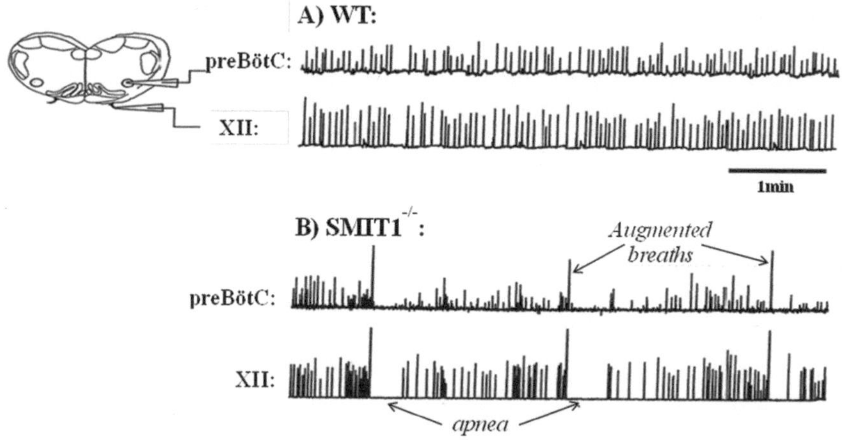

Fig. 20.1 Recordings of integrated and rectified inspiratory neuronal discharge in brainstem slice preparations isolated from E18.5 wildtype (**a**) and SMIT1−/− (**b**) fetuses. Representative recordings of the preBötzinger Complex (preBötC; the putative respiratory rhythm generating center) represent neuronal populations; the hypoglossal (XII) recordings were obtained directly from the nerve rootlet. Note the irregular respiratory rhythm and higher prevalence of apnea and also augmented breaths in the SMIT1−/− mouse compared to wildtype (WT). (Modified with permission from Berry et al. 2003)

with, for example, the phrenic nerve beginning to innervate the diaphragm around E13–14 (Greer et al. 1999). Although the mortality could be prevented by maternal supplementation with myo-inositol during pregnancy, reversal of apnea or the abnormal respiratory rhythm has not been assessed either *in vivo* or using the *in vitro* preparations outlined above. An understanding of the biochemical mechanisms underlying such benefits is also difficult to determine because maternal supplementation only modestly increased myo-inositol levels in the offspring. Further, neither depletion or supplementation with myo-inositol translated into changes in expression of any of its derivatives (IP5 or IP6) or mRNA levels for several components involved in myo-inositol metabolism (Buccafusca et al. 2008). This is surprising since there are several studies that have investigated the direct effects of various derivatives of myo-inositol on both central and peripheral mechanisms of respiratory neural control.

20.3 Derivatives of Myo-inositol

Although the above studies showed very little effect of myo-inositol depletion (or supplementation) on the expression of its derivatives in either the central or peripheral nervous system, several studies have demonstrated a direct effect of derivatives on various sites of respiratory control.

20.3.1 Respiratory Rhythm Generation

Acute application of bath-applied myo-inositol (1 mM) to the E18.5 mouse brainstem-spinal cord preparation had no immediate effect on rhythmic respiratory activity in either WT or SMIT1 KO mice (Buccafusca et al. 2008), which suggests it does not play a direct role in respiratory rhythm generation or pattern of respiratory bursting. One study, however, demonstrated that various derivatives of myo-inositol influence bursting of preBötC neurons. Using the *in vitro* brainstem slice preparation from neonatal mice, micro-injection of a PIP2 analogue directly into a preBötC neuron increased the amplitude and frequency of inspiratory drive potentials whereas depletion of PIP2 had the opposite effect (Crowder et al. 2007). Inhibiting IP3 receptors also decreased inspiratory drive potential, which suggests that these derivatives may play an important role in maintaining the robustness of respiratory rhythm generation (Crowder et al. 2007).

20.3.2 Carotid Body Signaling

Presumably the SMIT transporters are present in the carotid body since inositol phosphates and phosphoinositides have been labeled using 3H-myo-inositol (Rigual et al. 1999). Although there are no data that we are aware of that has demonstrated direct effects of myo-inositol on carotid body function, there are indirect evidence for a role of its derivatives. This may be obvious since carotid body sensitivity requires Ca^{2+} signaling. Hypoxia exposure, for example increased phospholipase C which catalyzes IP3 and DAG formation (Pokorsky and Strosznajder 1993; Berridge 1987). Rigual et al. (1999), however, showed decreased expression of various inositol phosphates (IP1, IP2, IP3) and phosphoinositides (PIP2, PIP4) in the rabbit carotid body during hypoxia. How these changes affect hypoxia sensitivity or afferent signaling is unclear. However, endothelin increased carotid body IP3 levels, which was associated with a potentiated carotid sinus nerve response to acute hypoxia (He et al. 1996; McQueen et al. 1995; Chen et al. 2000). Thus, the role of myo-inositol and its derivatives on carotid body function may depend on the initiating stimulus, which is important in the context that there is emerging evidence that the carotid body serves several functions beyond hypoxia sensing, including immuno- and glucose-sensing (Kumar and Prabhakar 2012).

20.3.3 Nucleus Tractus Solitarius (nTS)

The nTS is the site of integration of afferent information from the vagus nerve and is an important brainstem respiratory neural control center. The caudal region comprises neurons that receive and relay afferent inputs from the carotid body and, therefore, plays an important role in the reflex response to hypoxia. Several studies have explored the direct and immediate effects of myo-inositol derivatives on nTS neuron excitability and associated physiological responses including heart rate and respiratory activity. For example, microinjections of IP6 into the caudal nTS of spontaneously breathing anesthetized rats dose-dependently decreased both mean arterial blood pressure and respiratory rate, and caused transient apnea (Baracco et al. 1989). Both IP5 and IP6 also caused bradycardia although IP6 appears to exhibit the most potent properties (Vallejo et al. 1987; Hanley et al. 1988). Further, micro-injections of vasopressin into the nTS of rats increased myo-inositol-derived phosphoinositide turnover (Moratalla et al. 1988). These data emphasize not just the important role that myo-inositol derivatives may play in regulating central mechanisms of respiratory control, but their roles extend into modulating other physiological mechanisms underlying important cardiovascular responses to adverse challenges.

20.4 Associations Between Endogenous Myo-inositol and Intermittent Hypoxia in Preterm Infants

Major consequences of preterm birth (<38 weeks gestation) are the respiratory complications associated with the infant's immature, poorly developed respiratory system. Respiratory related concerns such as periodic breathing, apnea, and accompanying falls in oxygenation/intermittent hypoxemia (IH), are universal in extremely preterm infants (Di Fiore et al. 2010). IH, a surrogate measure of apnea, is of primary concern as it may exacerbate other morbidities (Poets et al. 2015); however, apnea and IH continue to be major problems despite the use of various modes of respiratory interventions including supplemental O_2, positive pressure support and caffeine (Di Fiore et al. 2010; Poets et al. 2015). Pulse oximeter monitoring of O_2 saturation has revealed a dynamic profile in the postnatal incidence of IH events. For unknown reasons the number of IH events is "low" in the first week of postnatal life in preterm infants, then increases sevenfold over the following 2–3 weeks before reaching a plateau (~4 weeks of age) and gradually declining thereafter (Di Fiore et al. 2010). These postnatal changes in IH events define several developmental "windows" or "phases" during which the vul-

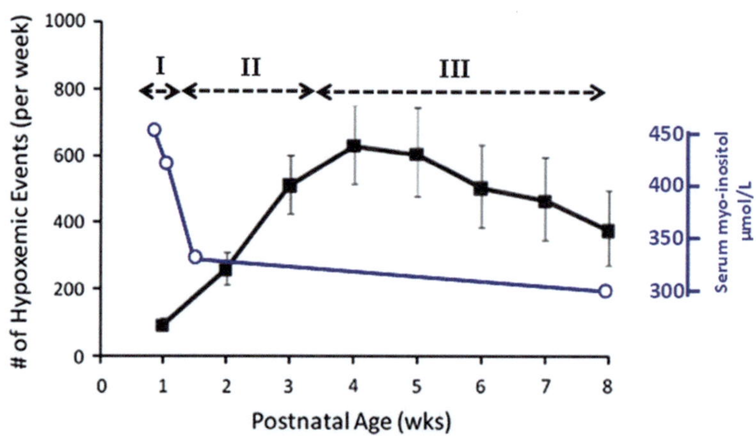

Fig. 20.2 Association between postnatal changes in IH events in preterm infants and serum myo-inositol levels. Note the incidence of IH (square symbols) events in the first week of life is low (phase I), but increases markedly over the following weeks (phase II), reaching a plateau at ~4 weeks of age and declining thereafter (phase III). These specific changes are highlighted by the 3 different phases, which could represent different windows of respiratory neural development (infants were between 24 and 28 gestation; data were obtained via pulse oximetry; values obtained from Di Fiore et al. 2010). Further, note the inverse relationship between the IH events and serum (open circles) myo-inositol levels of preterm infants (24–32 weeks gestational age) over the same 2 month postnatal time period (values obtained from Hallman et al. 1992)

nerability of respiratory neural control mechanisms underlying breathing stability may vary (Fig. 20.2). Further, throughout this 4 week time period of changes in IH events, there is an inverse relationship in serum myo-inositol levels. Specifically, serum myo-inositol are high in the first week of postnatal life followed by a rapid decrease of about 65% (Hallman et al. 1992) in the first postnatal weeks when the IH events increase (Fig. 20.2). Whether changing myo-inositol expression is a cause or consequence of the IH events is unknown, but this association, together with the animal literature described above, raises the possibility that there may be therapeutic benefit from myo-inositol supplementation as a mechanism to improve breathing stability and prevent apnea and IH.

Data relating myo-inositol to respiratory stability in preterm infants is limited but worthy of exploration. For example, myo-inositol has been shown to improve pulmonary function in premature infants with respiratory distress syndrome (Hallman et al. 1992). Additional factors including "low" levels of continuous baseline O_2 saturation (Di Fiore et al. 2012) and infection may exacerbate apnea and IH (Hofstetter et al. 2008; Stock et al. 2010). Interestingly, animal studies have shown that neonatal hypoxia exposure (Raman et al. 2005) and respiratory viral infection (Conrad et al. 2015) decreased brain myo-inositol suggesting disruption of brain energy metabolism. Taken together, these data suggest that myo-inositol levels may play a role in respiratory stability in preterm infants, but the relationship may be confounded by the clinical status and postnatal age.

20.5 Summary

Although the observations that decreased myo-inositol may or may not result in inappropriate peripheral (or central) morphological abnormalities in respiratory neural control, even modest changes in myo-inositol can translate into major functional complications. This may be particularly important to the pathophysiological mechanisms underlying various respiratory morbidities including apnea of prematurity when myo-inositol levels decrease dramatically during the first postnatal weeks. The possibility that such changes in myo-inositol can contribute to the etiology of respiratory morbidities such as apnea of prematurity either directly or indirectly is an attractive hypothesis. The high prevalence of apnea and IH events in this vulnerable population offers unique opportunities to test for potential beneficial effects of myo-inositol supplementation.

Acknowledgements P.M.M. and J.M.D. are supported by the Gerber Foundation. Reference# 1082-4005.

References

Barraco RA, Phillis JW, Simpson LL (1989) Cardiorespiratory effects of inositol hexakisphosphate following microinjections into the nucleus tractus solitarii. Eur J Pharmacol 173(1):75–84

Berridge MJ (1987) Inositol trisphosphate and diacylgycerol: two interacting second messengers. Ann Rev Biochem 56:159–193

Berry GT, Mallee JJ, Kwon HM, Rim JS, Mulla WR, Muenke M, Spinner NB (1995) The human osmoregulatory Na+/myo-inositol cotransporter gene (SLC5A3): molecular cloning and localization to chromosome 21. Genomics 25(2):507–513

Berry GT, Wu S, Buccafusca R, Ren J, Gonzales LW, Ballard PL, Golden JA, Stevens MJ, Greer JJ (2003) Loss of murine Na+/myo-inositol cotransporter leads to brain myo-inositol depletion and central apnea. J Biol Chem 278(20):18297–18302

Buccafusca R, Venditti CP, Kenyon LC, Johanson RA, Van Bockstaele E, Ren J, Pagliardini S, Minarcik J, Golden JA, Coady MJ, Greer JJ, Berry GT (2008) Characterization of the null murine sodium/myo-inositol cotransporter 1 (Smit1 or Slc5a3) phenotype: myo-inositol rescue is independent of expression of its cognate mitochondrial ribosomal protein subunit 6 (Mrps6) gene and of phosphatidylinositol levels in neonatal brain. Mol Genet Metab 95(1–2):81–95

Chau JF, Lee MK, Law JW, Chung SK, Chung SS (2005) Sodium/myo-inositol cotransporter-1 is essential for the development and function of the peripheral nerves. FASEB J 19(13):1887–1889

Chen J, He L, Dinger B, Fidone S (2000) Cellular mechanisms involved in rabbit carotid body excitation elicited by endothelin peptides. Respir Physiol 121(1):13–23

Collins BM, McCoy AJ, Kent HM, Evans PR, Owen DJ (2002) Molecular architecture and functional model of the endocytic AP2 complex. Cell 109(4):523–535

Conrad MS, Sutton BP, Larsen R, Van Alstine WG, Johnson RW (2015) Early postnatal respiratory viral infection induces structural and neurochemical changes in the neonatal piglet brain. Brain Behav Immun 48:326–335

Crowder EA, Saha MS, Pace RW, Zhang H, Prestwich GD, Del Negro CA (2007) Phosphatidylinositol 4,5-bisphosphate regulates inspiratory burst activity in the neonatal mouse preBötzinger complex. J Physiol 582(Pt 3):1047–1058

De Camilli P, Emr SD, McPherson PS, Novick P (1996) Phosphoinositides as regulators in membrane traffic. Science 271(5255):1533–1539

Di Fiore JM, Bloom JN, Orge F, Schutt A, Schluchter M, Cheruvu VK, Walsh M, Finer N, Martin RJ (2010) A higher incidence of intermittent hypoxemic episodes is associated with severe retinopathy of prematurity. J Pediatr 157(1):69–73

Di Fiore JM, Walsh M, Wrage L, Rich W, Finer N, Carlo WA, Martin RJ (2012) Support study group of Eunice Kennedy Shriver national institute of child health and human development neonatal research network. Low oxygen saturation target range is associated with increased incidence of intermittent hypoxemia. J Pediatr 161(6):1047–1052

Godfrey DA, Hallcher LM, Laird MH, Matschinsky FM, Sherman WR (1982) Distribution of myo-inositol in the cat cochlear nucleus. J Neurochem 38(4):939–947

Greer JJ, Allan DW, Martin-Carabello M, Lemke RP (1999) An overview of phrenic nerve and diaphragm muscle development in the perinatal rat. J Appl Physiol 86:779–786

Hallman M, Bry K, Hoppu K, Lappi M, Pohjavuori M (1992) Inositol supplementation in premature infants with respiratory distress syndrome. N Engl J Med 326(19):1233–1239

Hanley MR, Jackson TR, Vallejo M, Patterson SI, Thastrup O, Lightman S, Rogers J, Henderson G, Pini A (1988) Neural function: metabolism and actions of inositol metabolites in mammalian brain. Philos Trans R Soc Lond B Biol Sci 320(1199):381–398

Harlan JE, Hajduk PJ, Yoon HS, Fesik SW (1994) Pleckstrin homology domains bind to phosphatidylinositol-4,5-bisphosphate. Nature 371(6493):168–170

He L, Chen J, Dinger B, Fidone S (1996) Endothelin modulates chemoreceptor cell function in mammalian carotid body. Adv Exp Med Biol 410:305–311

Hofstetter AO, Legnevall L, Herlenius E, Katz-Salamon M (2008) Cardiorespiratory development in extremely preterm infants: vulnerability to infection and persistence of events beyond term-equivalent age. Acta Paediatr 97:285–292

Holub BJ (1986) Metabolism and function of myo-inositol and inositol phospholipids. Ann Rev Nutr 6:563–597

Kumar P, Prabhakar NR (2012) Peripheral chemoreceptors: function and plasticity of the carotid body. Compr Physiol 2(1):141–219

McQueen DS, Dashwood MR, Cobb VJ, Bond SM, Marr CG, Spyer KM (1995) Endothelins and rat carotid body: autoradiographic and functional pharmacological studies. J Auton Nerv Syst 53:115–125

Moratalla R, Vallejo M, Lightman SL (1988) Vasopressin stimulates inositol phospholipid metabolism in rat medulla oblongata in vivo. Brain Res 450(1–2):398–402

Ogimoto G, Yudowski GA, Barker CJ, Köhler M, Katz AI, Féraille E, Pedemonte CH, Berggren PO, Bertorello

AM (2000) G protein-coupled receptors regulate Na+,K+-ATPase activity and endocytosis by modulating the recruitment of adaptor protein 2 and clathrin. Proc Natl Acad Sci 7(7):3242–3247

Poets CF, Roberts RS, Schmidt B, Whyte RK, Asztalos EV, Bader D, Bairam A, Moddemann D, Peliowski A, Rabi Y, Solimano A, Nelson H (2015) Canadian oxygen trial investigators. Association between intermittent hypoxemia or bradycardia and late death or disability in extremely preterm infants. JAMA 314(6):595–603

Pokorsky M, Strosznajder R (1993) PO2-dependence of phospholipase C in the cat carotid body. Adv Exp Med Biol 337:191–195

Raman L, Tkac I, Ennis K, Georgieff MK, Gruetter R, Rao R (2005) In vivo effect of chronic hypoxia on the neurochemical profile of the developing rat hippocampus. Res Rep 156:202–209

Rigual R, Cachero MT, Rocher A, González C (1999) Hypoxia inhibits the synthesis of phosphoinositides in the rabbit carotid body. Pflugers Arch 437(6):839–845

Roll P, Massacrier A, Pereira S, Robaglia-Schlupp A, Cau P, Szepetowski P (2002) New human sodium/glucose cotransporter gene (KST1): identification, characterization, and mutation analysis in ICCA (infantile convulsions and choreoathetosis) and BFIC (benign familial infantile convulsions) families. Gene 285:141–148

Serra A1, Brozoski D, Hedin N, Franciosi R, Forster HV (2001) Mortality after carotid body denervation in rats. J Appl Physiol (1985) 91(3):1298–1306

Stock C, Teyssier G, Pichot V, Goffaux P, Barthelemy JC, Patural H (2010) Autonomic dysfunction with early respiratory syncytial virus-related infection. Auton Neurosci 156:90–95

Vallejo M, Jackson T, Lightman S, Hanley MR (1987) Occurrence and extracellular actions of inositol pentakis- and hexakisphosphate in mammalian brain. Nature 330(6149):656–658

Wurmser AE, Gary JD, Emr SD (1999) Phosphoinositide 3-kinases and their FYVE domain-containing effectors as regulators of vacuolar/lysosomal membrane trafficking pathways. J Biol Chem 274(14):9129–9132

Adrenal Medulla Chemo Sensitivity Does Not Compensate the Lack of Hypoxia Driven Carotid Body Chemo Reflex in Guinea Pigs

21

Elena Olea, Elvira Gonzalez-Obeso, Teresa Agapito, Ana Obeso, Ricardo Rigual, Asuncion Rocher, and Angela Gomez-Niño

Abstract

Guinea pigs (GP), originally from the Andes, have absence of hypoxia-driven carotid body (CB) reflex. Neonatal mammals have an immature CB chemo reflex and respond to hypoxia with metabolic changes arising from direct effects of hypoxia on adrenal medulla (AM). Our working hypothesis is that adult GP would mimic neonatal mammals. Plasma epinephrine (E) has an AM origin, while norepinephrine (NE) is mainly originated in sympathetic endings, implying that specific GP changes in plasma E/NE ratio, and in blood glucose and lactate levels during hypoxia would be observed. Experiments were performed on young adult GP and rats. Hypoxic ventilation (10% O_2) increased E and NE plasma levels similarly in both species but PaO_2 was lower in GP than in rats. Plasma E/NE ratio in GP was higher (≈1.0) than in rats (≈0.5). The hypoxia-evoked increases in blood glucose and lactate were smaller in GP than in the rat. The AM of both species contain comparable E content, but NE was four times lower in GP than in rats. GP superior cervical ganglion also had lower NE content than rats and an unusual high level of dopamine, a negative modulator of sympathetic transmission. Isolated AM from GP released half of E and one tenth of NE than the rat AM, and hypoxia did not alter the time course of CA outflow. These data indicate the absence of direct effects of hypoxia on AM in the GP, and a lower noradrenergic tone in this species. Pathways for hypoxic sympatho-adrenal system activation in GP are discussed.

Keywords

Sympathetic activity · Hypoxia · Blood glucose · Blood lactate · Catecholamine

E. Olea
Departamento de Enfermeria, IBGM-CSIC, Universidad de Valladolid, Valladolid, Spain

CIBER de Enfermedades Respiratorias (CIBERES), Instituto de Salud Carlos III, Madrid, Spain

E. Gonzalez-Obeso
Servicio de Anatomia Patologica, Hospital Clinico Universitario, Valladolid, Spain

CIBER de Enfermedades Respiratorias (CIBERES), Instituto de Salud Carlos III, Madrid, Spain

T. Agapito · A. Obeso · R. Rigual · A. Rocher (✉)
Departamento de Bioquimica, Biologia Molecular y Fisiologia-IBGM, Universidad de Valladolid-CSIC, Valladolid, Spain

CIBER de Enfermedades Respiratorias (CIBERES), Instituto de Salud Carlos III, Madrid, Spain
e-mail: rocher@ibgm.uva.es

A. Gomez-Niño
Departamento de Biologia Celular, Histologia y Farmacologia/IBGM, Universidad de Valladolid-CSIC, Valladolid, Spain

CIBER de Enfermedades Respiratorias (CIBERES), Instituto de Salud Carlos III, Madrid, Spain

E. B. Gauda et al. (eds.), *Arterial Chemoreceptors*, Advances in Experimental Medicine and Biology 1071, https://doi.org/10.1007/978-3-319-91137-3_21

21.1 Introduction

The carotid body (CB) chemoreceptors are arterial oxygen sensors, suited for monitoring changes in arterial blood PO_2 and capable of reflexly adjust ventilation to correct the drop of PO_2. It is currently accepted that hypoxia directly or indirectly through mitochondrial mechanisms, inhibits K^+-channels in the chemoreceptor cell membrane, leading to depolarization and release of neurotransmitters that activate the sensory nerve terminals of the carotid sinus nerve (Gonzalez et al. 1994). Afferent information from the carotid body is transmitted to the nucleus of the tractus solitaries (NTS) and rostroventrolateral medulla (RVLM) to increase breathing and sympathetic activation, respectively, during hypoxia (Marshall 1994; Guyenet 2000).

Although the CB chemoreceptor reflex is not completely functional at birth, direct activation of the chromaffin cells of the adrenal medulla (AM) by hypoxia in the fetus and newborn animals (Seidl and Slotkin 1985) leads to catecholamine (CA) release, which aside from cardiovascular and respiratory adjustments promote glycogenolysis and lipolysis. The resulting increase in glycemia and free fatty acids provides metabolic fuels to the newborn (Padbury et al. 1987). The direct CA secretory response to hypoxia from AM in neonatal rats is greatly attenuated in 14 days old animals, when maturation of splanchnic innervation occurs (Seidl and Slotkin 1986; Rico et al. 2005). Postnatal maturation of the CB and AM causes the hypoxic activation of the AM and, sympathetic activation, to become a reflex response (Marshall 1994). Only chromaffin cells isolated from young adult guinea pig (GP) apparently retain some oxygen sensing properties (Inoue et al. 1998).

Guinea pigs have a poor or no hyperventilatory response to hypoxia, when compared to other rodents (Schwenke et al. 2007; Gonzalez-Obeso et al. 2017). We have tested the hypothesis that adult GP could mimic neonatal mammals, using adrenal medullary derived CA to drive metabolic responses to defend from hypoxia. Animals were subjected to episodes of acute hypoxia, and blood PO_2, glucose and lactate and plasma CA were monitored; we also measured the direct effect of hypoxia on the release of CA in isolated *in vitro* AM. To strengthen the potential specificity and significance of our observations in GP, we have run parallel experiments on rats. Our findings fail to confirm the hypothesis; general responses in the intact animal and *in vitro* AM were identical in both species, including the lack of AM "direct" chemo sensitivity to hypoxia.

21.2 Methods

Animals Male Hartley guinea pigs (2–4 months old) and male adult Wistar rats (2–4 months old) with free access to food and water, were maintained under controlled conditions of temperature and humidity and a stationary 12-h light–dark cycle. Protocols were approved by the University of Valladolid Institutional Committee for Animal Care and Use following international laws and policies (European Union Directive for Protection of Vertebrates Used for Experimental and Other Scientific Ends (2010/63/EU). Animals were euthanized by a cardiac overdose of sodium pentobarbital.

Blood Gas Plasma CA, Glucose and Lactate Measurements Animals were anesthetized with intraperitoneal administration of Ketamine (100 mg/Kg) and diazepam (2 mg/Kg) tracheostomized and mechanically ventilated with room air (CL Palmer; 60 cycles/min, and a positive expiratory pressure of 2 cm H_2O). To examine the effect of hypoxia, the inlet of the respirator was connected to a hypoxic gas mixture (10% O_2, 90% N_2). Blood PO_2, PCO_2 and pH were measured in blood samples taken during normoxic conditions (air ventilated) or immediately after 15 and 30 min of hypoxic challenge with an Automatic Blood Gas System (ABL5, Radiometer Copenhagen). To measure CA content, 2 mL aliquot of blood samples were withdrawn in the same conditions. Citrated-blood was centrifuged at 1000 g for 5 min at room temperature. Supernatant was transferred to tubes containing 60 mg/ml sodium metabisulfite and frozen at −80 °C until the HPLC analysis of CA was performed. Detailed procedures for measurement of CA by HPLC have been described in detail elsewhere (Olea et al. 2014). Glucose and lactate in blood were measured with glucose and lactate meters (Ascensia Breeze 2, Bayer; Lactate Pro, Arkray).

CA Secretion and Endogenous Content from AM After anesthesia, adrenal glands were removed and placed in a Lucite chamber filled with ice-cold Tyrode solution (in mM: 140 NaCl, 5 KCl, 2 $CaCl_2$, 1.1 $MgCl_2$, and 5 glucose; pH 7.4). Adrenal glands were de-capsulated under a dissecting microscope; AM were transferred to a glass tube with Tyrode-HCO_3^- solution equilibrated with 20% O_2-5% CO_2-remainder N_2 at 37 °C and maintained in a shaker bath during 30 min, to recover from surgical stress. Later, each AM was transferred to vials containing 500 μl of Tyrode-HCO_3^- solution equilibrated either with 20% O_2–5% CO_2-N_2 (normoxic solution) or with 5% O_2-5% CO_2-N_2 (hypoxic solution). Superfusion media were collected every 10 min in eppendorf tubes containing 200 μl of 0.4 N perchloric acid (PCA), 0.1 mM EDTA and frozen at −80 °C until the HPLC-ED analysis of CA was performed. For analysis of endogenous CA content of AM and superior cervical ganglia (SCG), supernatants of glass-to-glass (0.1 N PCA, 0.1 mM EDTA) homogenized tissue were directly injected in the HPLC-ED system and chromatograms were analyzed with Peak Sample Data Chromatography System software (Buck Scientific, East Norwalk, CT).

Statistical Analysis Data are presented as mean ± SEM. The significance of differences between means was calculated by one and two-way analysis of variance (ANOVA) for repeated measures (when applicable) with Dunnett and Bonferroni multiple comparison tests, respectively. Significance was taken as $p < 0.05$.

21.3 Results

Arterial Blood PO_2 in Normoxic and Hypoxic Conditions Figure 21.1a shows levels of PaO_2 in GP and rats under normoxic and hypoxic conditions. Animals were mechanically ventilated with room air for 10 min, followed by 30 min of hypoxic ventilation (10% O_2, remainder N_2). Blood samples were withdrawn just before starting the hypoxic ventilation, and after 15 and 30 min of hypoxia, to measure blood PO_2, glucose and lactate and plasma CA. Arterial PO_2 values under normoxia were lower in GP than in rats (PO_2 66 ± 3 mmHg in GP and 80 ± 3 mmHg in rat). After 15 and 30 min of hypoxic ventilation, PaO_2 decreased to 20 ± 1 mm Hg and 42 ±3 mm Hg in GP and rats, respectively, GP showing a 71% and rats a 48% decrease in PaO_2 (***$p < 0.001$ *vs.* air in both cases). However, $PaCO_2$ values were comparable between both species.

Blood Glucose and Lactate Levels Measurements of glucose and lactate in blood were made in the same conditions as those used to measure PaO_2. Glucose levels in normoxia were similar in both species (146 ± 9 in GP and 188 ± 22 mg/dl in rats; n = 5–8; Fig. 21.1b). In GP, hypoxia increased glucose levels to 282 ± 31 after 15 min and to 407 ± 71 mg/dl after 30 min. In rats, glucose levels were increased to 451 ± 44 and 588 ± 12 mg/dl at 15 and 30 min of hypoxic exposure, respectively. Changes in lactate levels paralled with those of glucose (Fig. 21.1c). Hypoxia elevated lactate from 1.4 ± 0.1 to 3.5 ± 0.5 (15 min hypoxia) and to 5.1 ± 1.1 mmol/L (30 min hypoxia; $p < 0.001$) in GP. Rat lactate level in normoxia was higher than in GP, 2.8 ± 0.5 mmol/L, and after 15 and 30 min of hypoxia were 11.5 ± 1.0 and 9.9 ±1.1 mmol/L, respectively ($p < 0.001$).

Effect of Hypoxia on Plasma CA Levels Plasma CA content was measured in the same blood samples used for determining glucose and lactate.

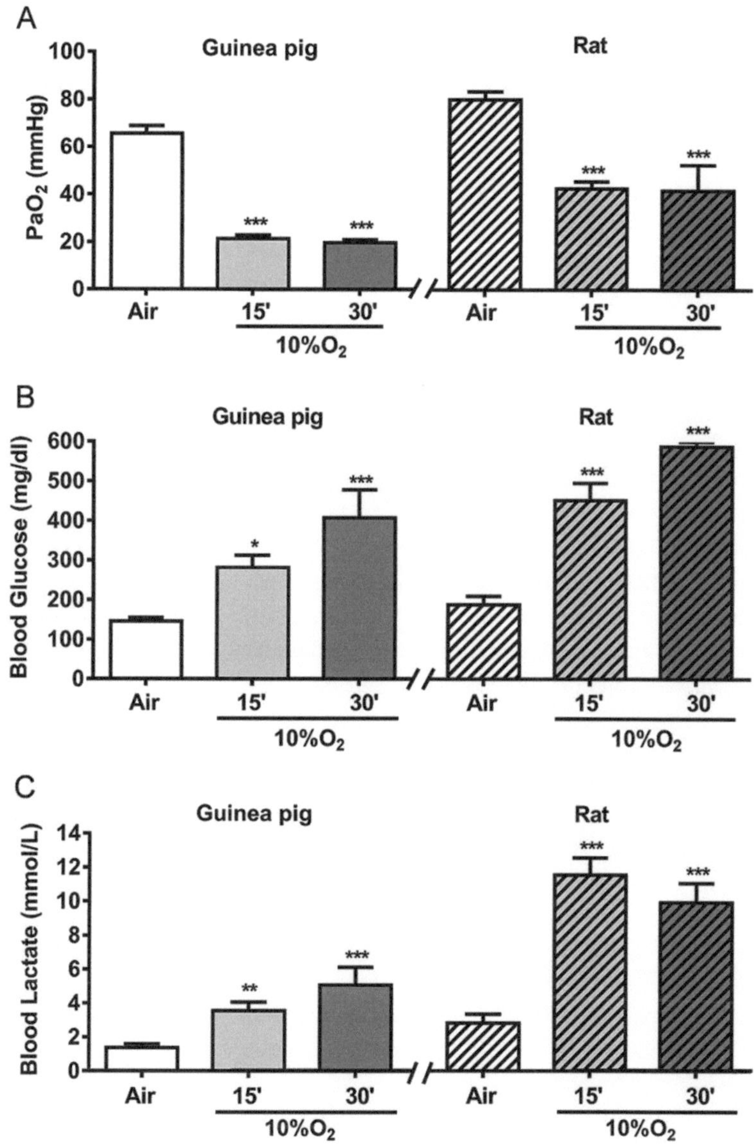

Fig. 21.1 (**a**). PaO_2 from guinea pigs and rats mechanically ventilated, breathing in normoxic and hypoxic (10% O_2) conditions (10 min air, 15 min 10% O_2 and additional 15 min 10% O_2). (**b**). Glucose and (**c**), lactate blood levels in guinea pigs and rats in normoxic and hypoxic conditions. Data are mean ± SEM; n = 9, except for the last period of hypoxia in rats (n = 3; animals died before 30 min of hypoxic exposure). *$p < 0.05$; **$p < 0.01$; *$p < 0.001$ *vs* air breathing

Data presented in Fig. 21.2a show responses of NE and E were comparable between both species. Levels of NE in GP ventilated with air were 3.1 ± 0.3 pmol/ml of plasma and they increased to 4.7 ± 0.6 and 6.9 ± 1.6 after 15 and 30 min of hypoxia. Levels of E followed a similar pattern from 3.1 ± 0.5 to 5.6 ± 0.3 and to 9.1 ± 0.9 pmol/ml of plasma. This pattern of plasma CA changes yielded E/NE ratios of 1.01, 1.20, and 1.32 for normoxia and hypoxia (15 and 30 min) respectively. Comparable values of both CA in GP have been reported by other authors (Macquin-Mavier et al. 1988). Plasma DA levels were around 5 times lower and not significantly modified by hypoxia (data not shown). In normoxic conditions, rat plasma NE levels were higher (4.7 ± 1.5 pmol/ml) and E levels were lower (2.4 ± 0.9) than those found in GP. After 15 and 30 min of hypoxia, NE levels were about double while E levels increased by 2 and 5 times, respectively. The E/NE ratios obtained during normoxia and hypoxia (15 and 30 min) were, 0.52, 0.57, and 1.30, respectively. The data obtained in the last hypoxic period from rats

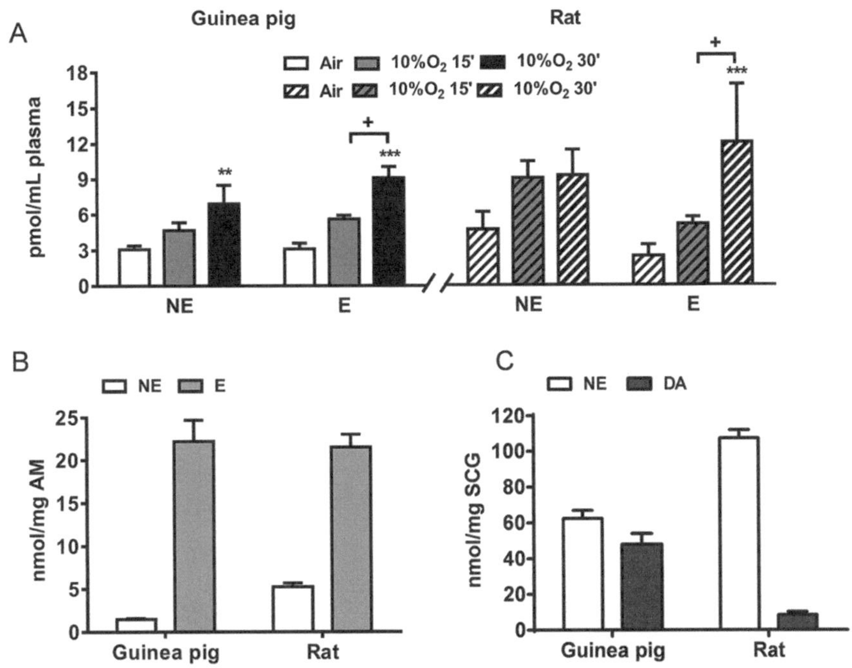

Fig. 21.2 (**a**). Plasma CA levels from guinea pigs and rats. CA were measured when animals were mechanically ventilated with room air and at the end of 15 and 30 min of the hypoxic period. Data are mean ± SEM (n = 9) except for the last period of hypoxia in rats (n = 3). (**b**). CA levels in AM from guinea pigs and rats. (**c**). CA levels in SCG from guinea pigs and rats. Data are mean ± SEM (n = 14 in GP; n = 10 in rats). $^{**}p < 0.01$; $^{***}p < 0.001$ *vs* air; $^{+}p < 0.05$ hypoxia 15 min *vs* 30 min

should be interpreted cautiously because only 3 out the 7 animals completed the 30 min of hypoxia. Levels of plasma DA (0.4 ± 0.1 pmol/ml), were minimally modified in the initial 15 min of hypoxia and augmented disproportionally after 30 min (data not shown).

Adrenal Medulla and Superior Cervical Ganglion Catecholamine Content Figure 21.2b shows CA content in AM from GP and rat. NE content in GP adrenal medulla was 1.5 ± 0.1 nmol/mg of tissue (n = 14) and rats had levels almost 4 times higher, 5.2 ± 0.4 nmol NE / mg tissue (n = 10). However, E content was similar in both species, 22.2 ± 2.5 and 21.5 ± 1.5 nmol/mg of tissue in GP and rat, respectively. Comparable levels of E and NE content in GP (Macquin-Mavier et al. 1988) and rat (Kumar et al. 2006) have been reported. The ratio E/NE was 15 in GP and 4.2 in rats, suggesting a potentially more intense role of GP AM in metabolism control, whereas in rats, AM would contribute more to the vascular control (Goldstein et al. 2003).

Figure 21.2c shows the CA content of the sympathetic ganglia. Levels of NE in the SCG of GP were about half of those of rats (62.4 ± 4.5 *vs.* 107.3 ± 4.6 pmole/mg tissue), while DA levels were 6 times higher in GP (48.8 ± 6.1 *vs.* 8.3 ± 1.9 pmole/mg tissue). DA/NE ratio was much higher in GP than in rats (0.78 vs 0.08), suggesting a strong dopaminergic modulation of sympathetic activity mediated by the dopaminergic interneurons, small intensely fluorescent cells (SIF cells; Eranko 1978).

Release of CA from the AM To further test our hypothesis of the possible direct hypoxic sensitivity of GP adrenomedullary chromaffin cells, we studied the effects of hypoxia in the isolated AM time course of CA outflow from

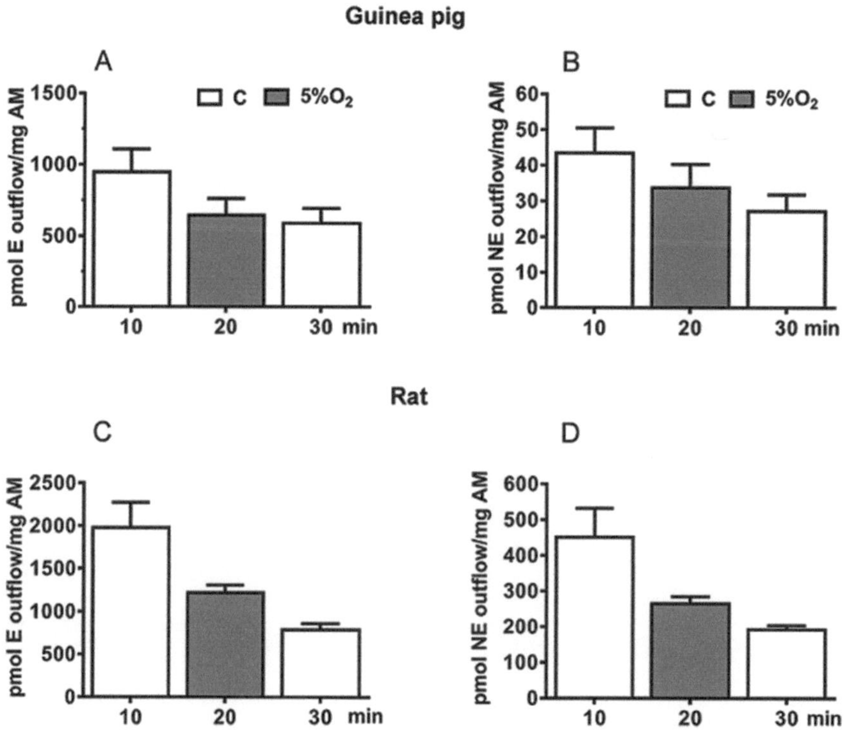

Fig. 21.3 Time course of CA outflow. Effect of hypoxia (5% O_2) on the outflow of E (Fig. 21.3a) and NE (Fig. 21.3b) from isolated guinea pig AM. Effect of hypoxia (5% O_2) on the outflow of E (Fig. 21.3c) and NE (Fig. 21.3d) from isolated rat AM. Data are mean ± SEM (n = 10 in GP; n = 8 in rats)

both species. Results presented in Fig. 21.3 show that hypoxia does not elicit increased CA efflux (i.e., does not alter the ongoing time course of the outflow) from AM in either of the species. In normoxic conditions (C), the amount of E in the incubation solution was 946 ± 160 pmol/mg of tissue in GP, and NE was much lower, 43.5 ± 7 pmol/mg of tissue. The E/NE ratio was 15 in the AM and in the incubation solutions it was around 22, suggesting that E turnover is faster than NE.

Under the same conditions the amount of E and NE in the effluent of rat AM were 1981 ± 291 pmol/mg tissue and 451 ± 81 pmol/mg tissue. The E/NE ratio was 4.2 in AM tissue and 4.4 in the incubating solution, with a comparable turnover for both CA. Data showed that in vitro hypoxia does not increase CA release from AM neither in GP nor in rats. The release of DA was much smaller in both species, nearly paralleling tissue content (data not shown).

21.4 Discussion

Main findings in the present study include: (1) The AM of GP and rats have comparable E levels, NE being four times higher in rats; E/NE ratios were 15 in GP and 4 in rat. In the SCG of GP, NE levels were half of those in rats. These data would suggest a lower noradrenergic tone in GP. The SCG of the GP has an unusual high level of DA, a negative modulator of sympathetic transmission. (2) GP has unexpectedly low PaO_2. (3) Hypoxia seems to be less effective increasing glucose and lactate levels in GP than in rats. (4) Plasma levels of E and NE are nearly identical in GP (E/NE $\approx$ 1), while in rat E was lower and NE was higher (E/NE $\approx$ 0.5). (5) In vitro, E and NE outflows from AM of the GP are not proportional to tissue content: NE outflow is slower than expected, suggesting a slower turnover rate of NE in GP. In rats, the outflow of E is nearly twice as high as in GP, and both CA appear in the effluents in proportion identical to tissue content. In neither of the species hypoxia altered the time course of the outflow.

The sympathetic nervous system (SNS) and AM are considered to function as a unit, and indeed they do so in adjusting the bodily functions in the fight and flight response. NE is the SNS neurotransmitter responsible for the maintenance of resting tone for cardiovascular function, as well as for the reflex responses that endanger cardiovascular normal functioning. E, the principal AM hormone, is mainly responsible for adjusting and coordinating metabolic responses in any situation that menaces the homeostasis of the organism (Goldstein et al. 1983). Systemic hypoxia represents a stressful condition endangering cell survival, and probably the most efficient response of mammals is hyperventilation triggered by CB chemoreceptors, a response that GP lacks (Schwenke et al. 2007; Gonzalez-Obeso et al. 2017). Therefore, we envisioned the possibility that non-ventilatory, mostly metabolic responses would be more developed in GP as a result of the intrinsic sensitivity of AM to hypoxia, mimicking the situation of neonatal animals. However, findings have not uncovered striking differences in the sympatho-adrenal mediated responses between GP and rats, the later with well-developed CB chemoreflex. On one hand, hypoxia causes comparable trends in plasma CA, glucose and lactate in GP and rats; on the other hand, in vitro experiments indicate that hypoxia does not activate CA release from the *in vitro* AM of either species. The lack of response of adult rat AM to hypoxia was expected from previously published findings in the intact animal and in isolated adrenomedullary cells (Seidler and Slotkin 1985, 1986), but in adult GP it would appear that chromaffin cells could directly be sensitive to hypoxia (Inoue et al. 1998). However, these authors called mild hypoxia to a superfusion with N_2-equilibrated solution that produced a small or no secretory response in the GP chromaffin cells; secretory response was only consistent when they added sodium dithionite to the N_2-equilibrated medium (i.e, pure anoxia and a strongly reducing milieu). Similarly, the level of chemical (histotoxic) hypoxia used by Inoue et al. (1998) was really strong as evidenced by a dose-response curve to cyanide shifted to the right by nearly two orders of magnitude in comparison to the curve in the CB (Obeso et al. 1989).

The near equivalence of the sympathetic-AM response to hypoxia in adult rats and GP forces to question how the sympatho-adrenal activation is achieved in GP. In the rat most of the response would be reflexly triggered by the CB (Marshall 1994), but in GP, owing to the lack of functional CB, there must be additional hypoxia-sensitive structures capable of triggering the sympatho-adrenal activation. In this regard, several studies have identified regions of the caudal hypothalamus and rostral ventrolateral medulla (RVLM) in some species that are directly excited by hypoxia which, when activated, increase sympathetic discharges and cause increased blood pressure and heart rate (Guyenet 2000). A recent study has shown that in brainstem slices comprising RVLM, hypoxia releases lactate and ATP which augment recorded neuronal activity; the study also shows that in vivo application of lactate and ATP in the ventral surface of the medulla oblongata increases sympathetic nerve activity and arterial blood pressure (Marina et al. 2015). These considerations lead to suggest that while in the rat the sympatho-adrenal activation elicited by hypoxia would mostly be a reflex in origin, it would be directly mediated by hypothalamic and RVLM sensitivity to hypoxia in GP. In this context, it might be suggested that GP could represent a simplified model to study the brainstem, and, in general, the central nervous system, sensitivity to hypoxia and the cardiovascular-endocrine responses elicited by low PaO_2.

Acknowledgements This work was supported by Grants BFU2015-63706R (MINECO, FEDER-UE) and CIBER CB06/06/0050 from ISCiii (Spain).

References

Eränkö O (1978) Small intensely fluorescent (SIF) cells and nervous transmission in sympathetic ganglia. Annu Rev Pharmacol Toxicol 18:417–430

Goldstein DS, McCarty R, Polinsky RJ, Kopin IJ (1983) Relationship between plasma norepinephrine and sympathetic neural activity. Hypertension 5(4):552–559

Goldstein DS, Eisenhofer G, Kopin IJ (2003) Sources and significance of plasma levels of catechols and their metabolites in humans. J Pharmacol Exp Ther 305(3):800–811

Gonzalez C, Almaraz L, Obeso A, Rigual R (1994) Carotid body chemoreceptors: from natural stimuli to sensory discharges. Physiol Rev 74:829–898

Gonzalez-Obeso E, Docio I, Olea E, Cogolludo A, Obeso A, Rocher A, Gomez-Niño A (2017) Guinea pig oxygen-sensing and carotid body functional properties. Front Physiol 8:285

Guyenet PG (2000) Neural structures that mediate sympathoexcitation during hypoxia. Respir Physiol 121(2–3):147–162

Inoue M, Fujishiro N, Imanaga I (1998) Hypoxia and cyanide induce depolarization and catecholamine release in dispersed guinea-pig chromaffin cells. J Physiol 507(3):807–818

Kumar GK, Rai V, Sharma SD, Ramakrishnan DP, Peng YJ, Souvannakitti D, Prabhakar NR (2006) Chronic intermittent hypoxia induces hypoxia-evoked catecholamine efflux in adult rat adrenal medulla via oxidative stress. J Physiol 575(Pt 1):229–239

Macquin-Mavier I, Clerici C, Franco-Montoya ML, Harf A (1988) Mechanism of histamine-induced epinephrine release in guinea pig. J Pharmacol Exp Ther 247(2):706–709

Marina N, Ang R, Machhada A, Kasymov V, Karagiannis A, Hosford PS, Mosienko V, Teschemacher AG, Vihko P, Paton JF, Kasparov S, Gourine AV (2015) Brainstem hypoxia contributes to the development of hypertension in the spontaneously hypertensive rat. Hypertension 65(4):775–783

Marshall JM (1994) Peripheral chemoreceptors and cardiovascular regulation. Physiol Rev 74(3):543–594

Obeso A, Almaraz L, Gonzalez C (1989) Effects of cyanide and uncouplers on chemoreceptor activity and ATP content of the cat carotid body. Brain Res 481(2):250–257

Olea E, Agapito MT, Gallego-Martin T, Rocher A, Gomez-Niño A, Obeso A, Gonzalez C, Yubero S (2014) Intermittent hypoxia and diet-induced obesity: effects on oxidative status, sympathetic tone, plasma glucose and insulin levels, and arterial pressure. J Appl Physiol 117(7):706–719

Padbury J, Agata Y, Ludlow J, Ikegami M, Baylen B, Humme J (1987) Effect of fetal adrenalectomy on catecholamine release and physiologic adaptation at birth in sheep. J Clin Invest 80(4):1096–1103

Rico AJ, Prieto-Lloret J, Gonzalez C, Rigual R (1987) Hypoxia and acidosis increase the secretion of catecholamines in the neonatal rat adrenal medulla: an in vitro study. Am J Physiol Cell Physiol 289:C1417–C1425. 2005

Schwenke DO, Bolter CP, Cragg PA (2007) Are the carotid bodies of the guinea-pig functional? Comp Biochem Physiol A Mol Integr Physiol 146(2):180–188

Seidler FJ, Slotkin TA (1985) Adrenomedullary function in the neonatal rat: responses to acute hypoxia. J Physiol 358:1–16

Seidler FJ, Slotkin TA (1986) Ontogeny of adrenomedullary responses to hypoxia and hypoglycemia: role of splanchnic innervation. Brain Res Bull 16:11–14

A Tribute to Machiko Shirahata, M.D., D.M.Sc. (1950–2016)

James S. K. Sham and Estelle B. Gauda

Machiko Shirahata, M.D., D.M.Sc., an internationally recognized leader in the study of carotid body chemoreception, and a beloved member and friend of ISAC died of lung cancer on April 25, 2016 in Baltimore. She was known to many as a dedicated and innovative scientist, a kind and generous colleague, a devoted and beloved mentor, and a very dear friend.

Machiko was born April 28, 1950 in Saitama, Japan. Her name in Japanese (眞知子) literally means a person of true understanding and real knowledge, a perfect description of her as a scientist and a teacher. Machiko received her M.D. in 1976 and residency training in anesthesiology from Chiba University School of Medicine. She developed strong interest in research early in her career, and began research training in cardiac metabolism and sympathetic regulation of cardiovascular function in the National Cardiovascular Research Institute, Osaka, from 1979 to 1982. She returned to Chiba, served as the chief anesthesiologist at the National Chiba Hospital from 1982 to 1985. In the same period of time, she completed her thesis research on the effects of hypercapnia on renal nerve activity and received the degree of Doctor of Medical Sciences (Ph.D. equivalent) from Chiba University in physiology and anesthesiology. In 1985, she came to the United States for postdoctoral research training at the Department of Physiology of University of Pennsylvania with Professor Sukhamay Lahiri, a world leader in the study of oxygen sensing and carotid body. In 1988, she joined the faculty of the Department of Anesthesiology in the Johns Hopkins University School of Medicine. She subsequently joined the faculty of the Department of Environmental Health Sciences in the Johns Hopkins School of Public health in 1990, advanced to full professor in 2005 and spent her entire career there until she passed.

During her career, Machiko published over 100 research articles and book chapters. She made seminal contributions and led the way in defining many critical aspects of the chemosensing properties of the carotid body. In the

J. S. K. Sham (✉)
Division of Pulmonary and Critical Care Medicine
Johns Hopkins University School of Medicine
Baltimore, MD, USA
e-mail: drprade@sprivail.org

E. B. Gauda
Department of Pediatrics, The Hospital for Sick Children
University of Toronto
Toronto, ON, Canada
e-mail: estelle.gauda@sickkids.ca

E. B. Gauda et al. (eds.), *Arterial Chemoreceptors*, Advances in Experimental Medicine and Biology 1071, https://doi.org/10.1007/978-3-319-91137-3

90's, Machiko and Professor Robert Fitzgerald revisited the previously rejected cholinergic hypothesis of carotid body chemoreception (Fitzgerald and Shirahata 1994). Through a series of elegant studies, she identified the expression of various nicotinic receptor subunits and muscarinic receptor subtypes, and demonstrated the acetylcholine-induced modulation of ion currents, intracellular Ca^{2+} and membrane excitability in the glomus cells and afferent neurons of petrosal ganglion. Substantiating with the observations of hypoxia-induced acetylcholine release from carotid bodies, they successfully established an important role of acetylcholine in hypoxic neurotransmission in carotid body (Shirahata et al. 2007).

In the early 2000's, Machiko started a new research direction for studying the genetic regulation of carotid body morphology, physiology and development (Yamaguchi et al. 2003). Mouse was the model of choice because of the availability of different genetic inbred stains and knockout mice. But the experimentation was inherently difficult and even formidable due to the extremely small size of the mouse carotid body. To overcome this limitation, Machiko and her associates systematically developed a set of innovative techniques for Ca^{2+} measurement of cultured glomus cells; patch-clamp recording of ion currents in glomus cells of an intact carotid body; and recording carotid sinus nerve activity *in vitro* and in anesthetized mice *in vivo*. By applying these techniques on different strains of inbred mice in conjunction with the morphological analysis and the microarray analysis to determine the differential gene expression in carotid body, Machiko's team provided the most convincing evidence for the important genetic influence on the carotid body functions and development (Balbir et al. 2007; Kostuk et al. 2012).

In the personal side, Machiko was a very kind and generous person. She always selflessly shared her ideas, resources and expertise on the carotid body with her colleagues; and eagerly guided young faculty members in their career. These interactions often turned into long lasting collaborations, and the ideas evolved into many still on-going projects related to the novel roles of carotid body in a variety of pathophysiological conditions and human diseases, including sleep apnea and intermittent hypoxia; inflammation and carotid body development in neonates; epigenetic modification of carotid body functions in neonates; and obesity-induced hypertension. Her devotion to her work and colleagues was truly admirable. Even during the advanced stage of her illness, she continued to participate in the studies and successfully wrote grants for the collaborative projects. Her passion for science and her faithfulness in friendship were unsurpassed. Machiko was a beloved mentor for numerous doctoral and postdoctoral trainees. Her devotion for mentoring trainees in science and her tireless support of her trainees was known to all. It was a common feeling for all her trainees that they were not only joined Machiko's laboratory but reborn into her family. As a former chief anesthesiologist, Machiko maintained close interaction with her colleagues in Japan, and provided research training for many young Japanese anesthesiologists who later developed very successful careers in Japan.

It is important to note that Machiko's passion for science is a reflection for her passion for life in general. She was a loving mother and was the number one fan of her son, Akira Fitzgerald, who is a professional soccer player. She would never miss any Akira's game by either traveling long distance to watch the game in person or inviting friends to come over to enjoy on television. Machiko loved to exercise and was a very good horseback-rider. She enjoyed singing and playing the piano and was a huge supporter of the Baltimore School of Music. Most importantly, she was a dear friend for those who had the good fortune to know her. There were so many wonderful memories of her working with colleagues and trainees in the laboratory and the department, giving stimulating lectures and sage discussions in scientific meetings (especially in the every ISAC meetings she attended since 1989), and cordially sharing in many parties and home dinners. Machiko had influenced the people around her in many positive ways both scientifically and personally. She will be missed by all very dearly.

References

Balbir A, Lee H, Okumura M, Biswal S, Fitzgerald RS, Shirahata M (2007) A search for genes that may confer divergent morphology and function in the carotid body between two strains of mice. Am J Physiol Lung Cell Mol Physiol 292(3): L704–715

Fitzgerald RS, Shirahata M (1994) Acetylcholine and carotid body excitation during hypoxia in the cat. J Appl Physiol (1985) 76(4): 1566–1574

Kostuk EW, Balbir A, Fujii K, Fujioka A, Pichard LE, Shirahata M (2012) Divergent postnatal development of the carotid body in DBA/2J and A/J strains of mice. J Appl Physiol (1985) 112(3): 490–500

Shirahata M, Balbir A, Otsubo T, Fitzgerald RS (2007) Role of acetylcholine in neurotransmission of the carotid body. Respir Physiol Neurobiol 157(1): 93–105

Yamaguchi S, Balbir A, Schofield B, Coram J, Tankersley CG, Fitzgerald RS, O'Donnell CP, Shirahata M (2003) Structural and functional differences of the carotid body between DBA/2J and A/J strains of mice. J Appl Physiol (1985) 94(4): 1536–1542

Machiko Shirahata Awardees

Bernardete F. Melo
Erin Leonard
Hortensia Torres-Torrelo
Ian Wenker
Inmaculada Docio
Maria Paz Oyarce
Nicholas Jendzjowsky
Robert Lewis
Ryan Rakoczy
Tara Janes

A Tribute Constancio González Martinez, M.D., Ph.D. (1949–2015)

Nanduri R. Prabhakar and Estelle B. Gauda

In 2015, ISAC lost an amazing person and scientist, Constancio González Martinez, M.D. Constancio served as ISAC President from 2005 to 2008, was a superb scientist and dear friend to many, whose seminal work on the carotid body chemotransduction, neurotransmission and oxygen sensing, informed so many of us. His remarkable scientific career served as a model to inspire many to pursue work in the area of arterial chemoreception.

Constancio, was born in 1949 in the village of Renedo de Valderaduey in the Spanish province of León, a place that he had many fond memories of and a place he spoke of with incredible pride. He received his medical degree in 1974 from the University of Valladolid in Spain. He published his thesis, entitled Neurotransmission in the Carotid Body (under Professor Carlos Belmonte) in 1977 and pursued postdoctoral training with Professor Carlos Eyzaguirre at the University of Utah from 1976 to 1980. Subsequently, Constancio returned to Valladolid where he advanced to the rank of Professor of Physiology, becoming the Director of the Department of Biochemistry and Molecular Biology and Physiology of the Medicine Faculty in 1995, a position he held until his death.

Constancio received numerous accolades and awards. To mention a few, he was the President of the Spanish Society of Physiological Sciences and Editor of Respiratory Physiology of Neurobiology. In 2011, he received the prestigious Castilla y León Award for Scientific and Technical Research, for his excellence in research, teaching and his contributions to arterial chemoreception that laid the foundation for understanding other pathologies. He has co-authored over 200 articles in international journals, books and monographs, publishing landmark articles on arterial chemoreceptors and state-of-the-art reviews for the Handbook of Physiology, Physiological Reviews and the Neurosciences Encyclopedia.

In 1977, he along with Professor Sal J. Fidone published a seminal paper demonstrating that hypoxia stimulated the release of dopamine from the carotid body (González C & Fidone S. Neurosci. Lett. 6: 95–99, 1977), a finding which we all use today as an index of cellular response of the carotid body. His discoveries also contributed to the "membrane hypothesis" of hypoxic sensory transduction demonstrating O2 sensitivity of K+ channels in glomus cells.

His leadership and vision substantially contributed to the creation of the Institute of Molecular Biology and Genetics (IBGM), a joint research center of the University of Valladolid and the Consejo Superior de Investigaciones Científicas (CSIC).

N. R. Prabhakar
Institute for Integrative Physiology and Center for Systems Biology of Oxygen Sensing, Biological Sciences Division, University of Chicago, Chicago, IL, USA

E. B. Gauda (✉)
Department of Pediatrics, The Hospital for Sick Children, University of Toronto, Toronto, ON, Canada
e-mail: estelle.gauda@sickkids.ca

E. B. Gauda et al. (eds.), *Arterial Chemoreceptors*, Advances in Experimental Medicine and Biology 1071, https://doi.org/10.1007/978-3-319-91137-3

Many of us will remember him as tireless with an insatiable thirst for knowledge, as a loving husband and father, as a patient teacher and mentor, as a great dancer and as a gracious host who loved his country, his profession and life (Figs. 1 and 2).

Fig. 1 Constancio González (middle) at ISAC meeting in Valladolid, Spain in 2008 with Krishna Lahiri (left) and Sukhamay Lahiri (right)

Fig. 2 Constancio González with family at the University of Valladolid, Spain in 2011. Children (left to right) Ana, Constancio (middle), Elvira (right) and wife and scientific partner Ana Obeso

Concluding Remarks

Future generations are coming and the challenge of understanding the role of chemoreceptors in disease is still present

This volume of advances in experimental biology and medicine reflects much more than just the state of the art in the field of arterial chemoreceptors. It reflects the leadership of Professor Estelle Gauda in her role as President of the International Society of Arterial Chemoreceptors (ISAC) and Chair of the Organizing Committee of the XX ISAC meeting held July 23–27, 2017 at the Charles Commons of Johns Hopkins University, Baltimore, Maryland, USA.

The organization of the meeting coincided with the change of Professor Gauda from Baltimore to the University of Toronto where she started a new position. In these circumstances, the high quality of the meeting was only possible thanks to its enormous capacity of work and her management skills of a very efficient team. They have given us a meeting with an excellent scientific program and rich in social activities that have contributed to a very pleasant environment, conducive to the creation of new collaborations and scientific innovation. On behalf of the participants, I can not fail to emphasize my thanks to Professor Gauda and her team: Rosie Silva, Setsuko Takase, Sam Singh, George Kim and Sonia Dos Santos. Thanks also to the sponsors who made the meeting possible, highlighting the generosity of Akira Fitzgerald and Bob Fitzgerald.

The meeting was a landmark in the articulation between the past and the present in both research on the carotid body and in the responses to hypoxia and in the renewal of generations of researchers. Thanks to Professor Fitzgerald for taking up the challenge to publish a historical review of the carotid body that surpassed all expectations.

A tribute to Professors Constancio González and Machiko Shirahata who early passed away was due, but always insufficient. Professor González of the Department of Biochemistry and Molecular Biology and Physiology of the Faculty of Medicine of the University of Valladolid, Spain, was past President of ISAC and organizer of the ISAC meeting in 2008. The contribution of Constancio González to understanding carotid body function in health and disease has probably been the most consistent, comprehensive and long lasting worldwide. His passion and accurate knowledge about the carotid body is well documented not only in the landmark papers but also in minor contexts like the concluding remarks of the 2006 volume of advances in experimental biology and medicine book edited by Hayashida et al. He was an odd mentor who has inspired several generations of researchers. I would like to emphasize his detrimental role in the development of the research group in NOVA Medical School in Lisbon namely, mentoring the Portuguese Sílvia Conde, one of the most active new group leaders in the carotid body field. The meeting in Baltimore speaks for itself: six communications originated in his Valladolid group, one De Castro Award and the continuity under the clear leadership of Professor Obeso. Professor Gonzalez certainly has reasons to be proud.

The Department of Anesthesiology and Critical Care Medicine at Johns Hopkins

E. B. Gauda et al. (eds.), *Arterial Chemoreceptors*, Advances in Experimental Medicine and Biology 1071, https://doi.org/10.1007/978-3-319-91137-3

University School of Medicine has hosted Machiko Shirahata since 1988. She reciprocated giving international visibility to her department's investigation of the role of the carotid body in a variety of human diseases (sudden infant death, obstructive sleep apnea, hypertension and diabetes). The tribute to Machiko was an example of the relevance of human qualities to the impact of science. As part of her legacy she wanted to support trainees with interest in hypoxic chemosensitivity to attend the ISAC international meeting. In her honor, her son Akira Fitzgerald, has made the Machiko Shirahata Trainee ISAC Award possible. The first Machiko Shirahata PhD Student awardees supported to attend ISAC 2017 were: Bernardete F. Melo (Portugal); Erin Leonard (Canada); Hortensia Torres-Torrelo (Spain); Ian Wenker (USA); Inmaculada Docio (Spain); Maria Paz Oyarce (Chile); Nicholas Jendzjowsky (Canada); Robert Lewis (USA); Ryan Rakoczy (USA); Tara Janes (Canada).

From the sessions I highlight two points: the concept underlying their organization and the identification of cutting edge topics in the investigation of arterial chemoreceptors and hypoxic response. The organization model included 1-h plenary lectures by keynote speakers and oral presentations by young researchers. The historical evolution of research on each topic, the results with the latest technologies and the current controversies were particularly important for the training of the younger audience and to trigger a highly participated discussion by the audience. The model resulted in fullness mainly thanks to the quality of keynote speakers.

In the first session Greg Semenza presented the state of the art in the Role of the Hypoxia-Inducible Factors in the Carotid Body and the oral communications brought to the floor cell-cell interactions and microRNA expression in the carotid body; spinal oxygen sensors as cardiovascular sensors and an index to be used in acute mountain sickness.

The second session was dedicated to oxygen central chemoreception and J. M. (Nino) Ramirez gave a talk on Oxygen Sensing in the Central Respiratory Network: Relevance to Disordered Breathing and Cardiorespiratory Control. The oral communications included CNS mechanisms of cardiorespiratory homeostasis; Pro-Inflammatory Cytokines in the Nucleus of the Solitary Tract of Hypertensive Rats Exposed to Chronic Intermittent Hypoxia and Acute Hypoxia and Rhythm Generator for Lung Breathing in Brainstem Preparations.

The plenary lecture of the third session was given by Chis N. Wyatt and reviewed the pros and cons of the different hypothesis to explain the molecular mechanisms involved in Acute Oxygen-Sensing by the Carotid Body. The oral communications identified ectonucleotidases and cysteine redox cycle as emerging targets to manipulate the responses to chronic hypoxia.

The fourth session was aimed to address new Perspectives on (Mal)adaptive Responses to Hypoxia and began with the plenary lecture by J. Nanduri on Epigenetics Regulation of Carotid Body Oxygen Sensing: Physiological Consequences from Fetus to Adult. During the oral communications, we found results on mechanisms of Carotid Body Long-Term Facilitation without CIH Preconditioning; the contribution of Relative Leptin and Adiponectin Deficiencies to Chronic Intermittent Hypoxia in Premature Infants and Spontaneous Tumorigenesis in chronic intermittent hypoxia.

Session five and six were centered on the Role of Chemoreception in disease. Rodrigo Iturriaga in his lecture: Translating Carotid Body Function in Clinical Medicine mainly addressed hypertension and Michael Joyner addressed glucose homeostasis and thermo regulation. The oral communications provided new insights about the effects of Acute and Chronic Progesterone Administration on Apnea Frequency in Newborn Male and Female Rats; about the effects of Leptin on Blood Pressure Acting in the Carotid Body and about Inflammation and Carotid Sinus Nerve Activity. Recent evidences were also provided on the effects of Carotid Sinus Nerve Resection on Decrease Weight Gain and Fat Mass in a Rodent Model of Obesity. Over activation of Carotid Sinus Nerve Activity in High-Fat Animals was shown to be mediated by A2 Adenosine Receptors and Carotid Body CBS has been proposed as a

Therapeutic Target for Treating Sleep Apnea in a Pre-Clinical Model.

Session seven was focused on development. Andrea Porzionato centered the plenary lecture in the pathological features of Peripheral and Central Chemoreceptors in Sudden Infant Death Syndrome. The oral communications provided new insights about the effects of Myo-Inositol on Respiratory Neural Control During Postnatal Development and the protective effects from SIDS of milk.

The last outstanding plenary lectures were about aging and carbon dioxide chemoreception. Camillo Di Giulio gave a comprehensive revision about "The Aging Carotid Body." Patrice Guyenet showed cutting edge science about Neural Circuits and Molecular Mechanisms of CO2 Sensing controlled by the Retrotrapezoid Nucleus (RTN). Finally, Cormac Taylor dedicated his talk to the Regulation of Inflammatory Gene Expression by Hypoxia and Hypercapnia.

From the poster session, I would highlight: new insights on BK/Kv, Task, FastK, T-type Ca++, channels in glomus cells and Connexin-based hemichannel blockers; advances in Chronic Intermittent Hypoxia mechanisms; lactate Sensing by Carotid Body Glomus Cells; Effects of Systemic Erythropoietin Injections on Carotid Body Chemosensory; the interesting interaction between Chronic Heart Failure and Chemoreflexes and non-invasive measurements of human sympathetic reactivity.

In summary, the science presented in ISAC 2017 is in line with emerging translational topics and advances in understanding co-morbidities and the high quality of graduate students and junior group leaders and ensures an optimistic future for ISAC.

We thank Estelle Gauda and all the Scientific Organizing Committee members, Nanduri Prabhakar, Harold Schultz and Chris Wyatt for this amazing opportunity to discuss science in a very stimulating and friendly environment.

I would finalize, thanking Sílvia Conde for agreeing to co-organize the next meeting – ISAC XXI and to invite you all to Lisbon, Portugal in 2020.

Emilia Monteiro
NOVA Medical School
Universidade NOVA de Lisboa
Lisbon, Portugal

Composite Pictures from Conference

ISAC XX
July 23-27,2017
Johns Hopkins University
Baltimore, Maryland, USA

E. B. Gauda et al. (eds.), *Arterial Chemoreceptors*, Advances in Experimental Medicine and Biology 1071, https://doi.org/10.1007/978-3-319-91137-3

Heymans-DeCastro-Neil Awardees and Selection Committee
ISAC XX
Joana Sacramento (Portugal), Ryan Rakoczy (USA, not shown), Elena Olea Fraile (Spain)
Selection committee: Ana Obeso, Emilia Monteiro, Aida Bairam

Gala Pictures ISAC XX, 2017

Inaugural Machiko Shirahata ISAC Trainee Travel Awardees 2017:
M. Oyarce, T. Janes, H. Torres-Torelo, I Docio, B Melo, E Leonard
Not present:, I. Wenker, R. Lewis, N. Jendzjowsky
S. Polotksky (far left) and E Gauda (far right) presented the award

Gala Pictures ISAC XX, 2017

FSC www.fsc.org
MIX
Papier aus verantwortungsvollen Quellen
Paper from responsible sources
FSC® C105338

If you have any concerns about our products, you can contact us on
ProductSafety@springernature.com

In case Publisher is established outside the EU, the EU authorized representative is:
Springer Nature Customer Service Center GmbH
Europaplatz 3, 69115 Heidelberg, Germany

Printed by Libri Plureos GmbH
in Hamburg, Germany